Engineering and Managerial Economics

Engineering and Managerial Economics

Ira H. Kleinfeld

University of New Haven

Holt, Rinehart and Winston

New York Chicago San Francisco Philadelphia
Montreal Toronto London Sydney
Tokyo Mexico City Rio de Janeiro Madrid

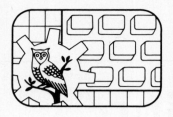

HRW
Series in
Industrial
Engineering

Leland T. Blank, *Series Editor*

To Ruth,
Barry,
and Michael

Acquisitions Editor: John Beck
Production Manager: Paul Nardi
Project Editors: Robert Greiner
                      Kathleen Nevils
Design Supervisor: Robert Kopelman
Interior Design: Caliber Design Planning, Inc.
Cover Illustration and Design: Steve Bliss, Heffernan, Hastie and
                                    Leibman, Inc.
Illustrations: Scientific Illustrators

Library of Congress Cataloging in Publication Data

Kleinfeld, Ira H.
  Engineering and managerial economics.

  Includes index.
  1. Engineering economy. I. Title.
TA177.4.K59 1986   658.1'5   85-17234
ISBN 0-03-070346-8

CBS COLLEGE PUBLISHING
Holt, Rinehart and Winston
The Dryden Press
Saunders College Publishing

Preface

Engineering economics is an important subject for the aspiring engineer, for it is not sufficient to learn the scientific and engineering principles of design work and decision-making in a world whose technologies are rapidly changing. An engineer's work must be economically sound if it is to be most useful, practical, and even feasible in an ever more competitive world. Solutions that are technically possible but uneconomic are not good engineering solutions. This has been recognized by engineering educators for some time. Therefore, engineering economics courses have evolved to fill the pedagogical need for students of engineering. Although a student at the baccalaureate level may study engineering science and design for one and one-half or two years in a typical four-year, A.B.E.T.-accredited program, he or she would ordinarily be exposed to no more than one course in engineering economics, whether it be a quarter, trimester, or semester in duration. Thus, the importance of this course should not be underestimated. Also, the first stage of the professional engineering licensing exam (EIT) contains a small percentage of questions from the engineering economics material presented in the typical course, attesting to the importance of the material to the aspiring engineer. Thus, it is incumbent that the course contain material that is sufficient to provide the student with concepts and skills that will prove useful to a practicing engineer.

Even given the importance of the subject, one may wonder, since a number of books already exist, why the author has gone to the trouble of preparing a new text. Naturally, any teacher of a subject has his or her own preferences for how the material should be presented, and I am no different. This book represents a treatment of the material that I feel comfortable with, and that includes a number of unusual features. I shall attempt to outline these special features in the paragraphs that follow.

Many of the solved examples in the text are variations of a given solved problem, sequenced to illustrate new concepts more coherently. In a similar way, end-of-chapter exercises in subsequent chapters are frequently revisits of earlier ones; thus, the student can more easily grasp the effect of the current topic. These techniques are extensively used in Chapter 3, for developing yardsticks for evaluation of alternatives, as well as in Chapter 4, for comparison of alternatives; Chapter 8, for after-tax analyses; Chapter 9, for replacement analysis; and Chapter 10, the impact of inflation. This approach mimics, in an educational vein, the idea of sensitivity analysis advocated in the text for economic analysis.

Another unusual feature is the emphasis on making replacement analyses on an after-tax basis. Although this is the correct way to perform such an analysis, most books avoid it out of expediency, since it is more cumbersome to demonstrate.

Other topics covered in this text, not generally appearing elsewhere in books of this kind, are job order costing and standard costs. These are included for practical reasons, since most engineers gathering data for an analysis will have to have some knowledge of these systems in place in their organizations. Another topic ordinarily not found in engineering economics texts is the capital-asset-pricing model for determining the cost of capital. This was included as an alternative to the weighted-cost-of-capital approach, no longer as widely used by financial analysts.

Finally, this book, more than most, incorporates the use of modern computers to aid in the analytical process. An attempt has been made to develop user-friendly and flexible software of a general-purpose sort. (Most previously shown software has been designed for specific purposes. Thus, if a slightly different problem appeared, the software was not useful for that purpose.) The time value programs in this book will handle almost all of the exercises in Chapters 2, 3, 4, 8, 9, 10, and 14. In the after-tax program written for the IBM PC, sensitivity analysis and graphic display are facilitated. (Note that the after-tax program can be used for pre-tax analyses as indicated in the documentation.) In fact, the idea has been to show what the properties of good software for engineering economics should be and to provide a satisfactory prototype for students and practitioners well into the computer age.

It is no longer sufficient to teach engineering science principles in the hope that students will develop their own software in their spare time. Good software should be a component of the course—not to make problems cook-bookish, thus camouflaging the principles expected to be learned, but instead, to permit the student or practitioner to play "what if?" more effectively. Thus, better learning will take place for the student, and better analysis will be possible for the practitioner. Such software is too complex to expect students to prepare in their spare time. It is worth pointing out that recently A.B.E.T. has shown an inclination to demand accelerated use of computers in engineering courses of all sorts, including engineering economics. This has been meant to mirror the increasing use of computers by engineers in practice.

This book has been used by the author in a one-semester (15-week) course in the engineering school at the University of New Haven. The course may be taken by students as early as the sophomore year, since it requires no more than algebra. As such, they might not yet have had a course on probability and statistics.

This was the rationale for the inclusion of Chapter 13 on risk and uncertainty. Chapter 15, on project scheduling, was included because most of the students would have had no other opportunity to be exposed to critical path techniques in their undergraduate education. My approach is to cover all the chapters in the order in which they appear in the book. Other approaches are possible, as might be necessitated by a shorter course. For example, a 10-week course might include Chapters 1, 2, 3, 4, 5, 15, 6, 7, 8, 9, and 16, in that order, as suggested by one of the reviewers. Most instructors would probably wish to include Chapter 14, Analysis of Public Projects. This can be done any time after Chapter 4.

The preparation of this book has been facilitated by the help of many people. Messrs. Bruce Bishop and Michael Keirstead were immensely helpful in developing the BASIC programs. Mr. Homayoun Mehrtash analyzed most of the exercises, providing an important check to the software. Mr. Ravi Nuthakki aided in the research. Mr. John Beck, my editor at Holt, Rinehart and Winston, had many good suggestions, which I tried to incorporate. I would also like to thank Jane M. Fraser, Purdue University; Maurice K. Kurtz, Jr., Florida Institute of Technology; Diran Apelian, Drexel University; B. Vinod, Rutgers, The State University of New Jersey; Lloyd J. Dumas, University of Texas at Dallas; and Leland T. Blank, Texas A and M University. Finally, I would like to express my appreciation to the University of New Haven, which has given me the opportunity to teach the subject of this book. As any teacher knows, it is not until one tries to convey material to others that one realizes one's own weaknesses. At that point one can remedy them. In addition, the University has provided me with computer and word-processing support, as well as summer support to do some of the writing.

Ira H. Kleinfeld
Hamden, CT, 1986

Contents

5 Computational Considerations   79

6 Break-Even Analysis   102

7 Depreciation   123

8   The Impact of Taxes on Investment Decisions   145

9   Replacement Analysis   173

13   Risk in Engineering Economic Analysis   263

14   Analysis of Public Projects   303

15   Project Scheduling and Management: Critical Path Methods   323

16 Conclusion   343

Appendixes   357

Index   445

1

Overview of Engineering Economics

Introduction

The subject matter of this book has long been recognized as an important component of the requisite skills of the practicing engineer or manager, for it is not sufficient for an engineer or technologist to base decisions regarding design or operations on technical considerations alone. Satisfactory design and sound industrial operations depend on a prudent balance between what technical and scientific principles can permit in a given situation and what economic constraints demand.

This was well expressed by Henry R. Towne, a past president of A.S.M.E. and the Yale and Towne manufacturing company, who was a colleague of Frederick W. Taylor, considered by many to be the father of industrial engineering and scientific management:

> The dollar is the final term in almost every equation which arises in the practice of engineering in any or all of its branches, except qualifiedly as to military or naval engineering, where in some cases cost may be ignored. In other words the true function of the engineer is, or should be, not only to determine how physical problems may be solved, but also how they may be solved most economically. For example, a railroad may have to be carried over a gorge or arroyo. Obviously it does not need an engineer to point out that this may be done by filling the chasm with earth, but only a bridge engineer is competent to determine whether it is cheaper to do this or to bridge it, and to design the bridge which will safely and most cheaply serve, the cost of which should be compared to that of an earth fill.

Therefore the engineer is, by nature of his vocation, an economist. His function is not only to design, but also to design as to ensure the best economical result. He who designs an unsafe structure or an inoperative machine is a bad engineer; he who designs them so that they are safe and operative, but needlessly expensive, is a poor engineer, and, it may be remarked, usually earns poor pay; he who designs good work, which can be executed at a fair cost, is a sound and usually a successful engineer; he who does the best work at the lowest cost sooner or later stands at the top of his profession, and usually has the reward which this implies.*

The concern for cost consciousness is as valid now as it was when it was addressed by Towne in 1905. However, the issue of the engineer as economist is more general than is implied by Towne's bridge example. Engineers must not only be concerned with design, component, construction or fabrication, and operating costs, but with capital costs as well. This is because investable capital funds are, by their nature, scarce. This is true irrespective of the type of organization undertaking the project: profit-seeking, nonprofit, not-for-profit, governmental, or quasi-governmental. The relative scarcity of capital imposes true economic costs on its use: both direct, such as when capital funds must be borrowed for facilitating fixed investment or for working capital; and indirect, from the fact that the use obtained from the capital funds must be justified in relation to all other potential uses. The net result is that capital costs associated with a design or proposal are a vital component of its total costs and, thus, must be included along with other factors in its economic evaluation. One of the major aims of this text is to give the reader insight as to how this objective may be accomplished.

In addition, the "cost consciousness" issue now presents itself to practicing engineers and managers in a variety of additional, specialized contexts. These include:

The producibility of designs.   Modern production, particularly large-volume or mass production, requires attention to two related and equally important issues: the suitability of the design for the performance of the intended function, and the demands that it makes on the manufacturing methods and procedures. Too often the latter issue is downgraded or ignored, thus leading to excessive costs in manufacture or fabrication.

Value engineering.   Value engineering is concerned with modification or elimination of elements that contribute to the costs of a design without being absolutely necessary for required performance, quality, safety, reliability, maintainability, standardization, or interchangeability. It usually involves an organized effort directed at analyzing the function of a design for the purpose of achieving the required performance at the lowest overall cost.

*Henry R. Towne, address to graduating engineering students at Purdue University, February, 1905, published as part of the foreword to Frederick W. Taylor, *Scientific Management*, New York, Harper & Row, 1947, pp. 6–7.

Modeling the Economic Factors Bearing on Design and Decision-Making

An important consideration prerequisite to the economic analysis of alternatives concerns the process of economic modeling. The alternatives under consideration are generally complex and can be described in great detail. Such complexity does not lend itself well to analytical treatment. In order to arrive at economic engineering designs or to make sensible operational decisions from the economic point of view, it is often necessary to simplify the vast myriad of detail which confronts the analyst. This is because the human mind often cannot cope with the large number of entities in a particular system, related to one another in a complex pattern, or because of the great quantity of data associated with a process or situation. Consequently, decision analysts often resort to representing complex alternatives in simpler terms. Rather than addressing all aspects of a process, an analyst may need to consider only its most pertinent aspects. Thus, in conceptualizing a problem, a simplifying sort of transformation commonly takes place. This is referred to as modeling, and is quite widely used by modern scientific, engineering, and managerial practitioners. The modeling process attempts to characterize or represent a real process or situation with a simpler one which can be more readily analyzed by humans (with or without the aid of machines like computers). To the extent that the models are valid, or closely representative of the processes they purport to describe, conclusions based on those models will be reliable in practice.

An illustration will serve to make the idea of economic modeling clearer. Suppose a firm is considering the purchase of a new machine which can take on a variety of tasks, requiring setup and operational tending at various levels of difficulty, depending on the task. (This has numerous implications, including those for labor costs associated with operation of the machine.) It requires maintenance, both preventive and repair. It uses power at various rates, depending on the nature of its utilization. Its output can vary in response to the type of task assigned to it, to variations in scheduling work, and to many other factors. This sort of description can go on almost ad infinitum. Although of some use in certain realms, excessive detail is counterproductive when certain decisions need to be taken because it tends to cloud the basic issues.

Hence, modeling may be called for, involving specification of certain of the machine's characteristics under given conditions. These conditions may be simplifications of reality in order to permit rational decision-making to proceed. However, it is important that the process of modeling retain sufficient fidelity to reality so that the decisions dependent on the models are valid.

Apart from validity, another pitfall to be wary of is the potential failure to distinguish between reality and its model. Too often, even experienced analysts treat their economic models as though they were reality rather than mere representations of it. This may not seem like a significant point, but the reader should bear in mind that many serious errors of oversimplification have been made that surely could have been avoided had this principle been adhered to properly.

To round out the previous illustration, Table 1–1 may serve to represent the important economic characteristics of the machine in question.

TABLE 1–1

Investment cost	$50,000.
Economic life	5 yrs.
Annual revenue associated with machine operation	$20,000.
Annual costs associated with machine operation:	
Labor	$5,627.70
Power	$1,000.
Maintenance	$500.
Taxes, insurance, misc.	$500.
Total annual costs	$7,627.70
Salvage value	$5,000.

For many decision situations such a model, or representation of the salient aspects of the problem at hand, will be sufficient to allow the engineer or analyst to proceed. Subsequent chapters of this book contain many such descriptions, for the purpose of indicating how the alternatives may be analyzed. This, too, may be thought of as part of the modeling process. The reader should bear in mind, however, that in presenting such analytical material in the book, it is necessary to start with numerical parameters such as those in the example above, which, in practice, represent a part of the modeling process. In contrast, the practicing engineer must commence "at the beginning"; he must define the problem, establish what the technical alternatives are, estimate the economic characteristics of each (which is the modeling process described above), and then evaluate each of the alternatives using models such as are described in this book. Depending on the outcome of these evaluations, he may either be ready to make a decision or recommendation, or may need to return in iterative fashion to one or more of the alternatives for redesign and reevaluation, if the results of the previous step left something to be desired.

Cash Flow Diagrams

A form of model often used in this book, and incorporated into other models, is the cash flow diagram. It is used to represent the occurrence of observed or projected monetary flows at various times. It involves two sorts of quantities: magnitudes and time of occurrence. Time is depicted linearly on a horizontal axis. The present is usually designated as time zero. Magnitudes associated with a particular time period are represented as occurring at an end point of the appropriate time interval. Suppose, for instance, that operating expenses for a certain machine are currently estimated at $4,000 per year and that they are expected to increase by $200 in each of the next four years. Although it is not possible to anticipate the exact operating characteristics and, therefore, expenses of the machine in the next year, let alone the next four, some representation or estimate of the anticipated expense may need to be made. Furthermore, it may not be necessary or feasible for analytical purposes to characterize the pattern of expense for units of time less than, say, one year. If such is the case, a simple cash flow diagram depicting annual expenses would be as in D1–1.

D1–1

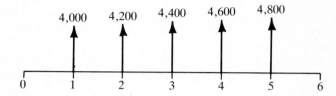

Note that convention calls for the depiction of the first year's operating expenses at the latter end of the interval for the first year, namely time 1. If a situation exists for which this procedure gives unrealistic results, alternative approaches may be taken. For example, if a lease payment is prepaid, it may be indicated at the beginning point for each period. Thus, a five-year lease in the amount of $5,000 per year may be represented as in Cash Flow Diagram D1–2.

D1–2

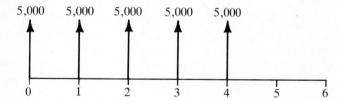

Sensitivity Analysis

An important part of the modeling process involves estimation of the values of the variables or parameters of the models employed. Clearly this introduces some risks, in that it is generally true that subsequent experience tends most often to differ, at least in some respect or degree, from the best estimates. This does not imply that the future should be ignored or that the attempt to estimate certain variables will necessarily be fruitless. Rather, it indicates that great care should be taken in estimation, and that the possible errors of judgment resulting from estimating variances should be anticipated. Sensitivity analysis helps to address this problem in the following way:

Sensitivity analysis is a technique of examining a model's characteristics, behavior, or outcomes when a particular variable in question is permitted to take on a range of likely or potential values. Typically, if a model can be evaluated under such a range of assumed values and the design, configuration, or system represented by that model is still satisfactory, then the sensitivity analysis allows greater confidence in the initial conclusions rendered by the modeling process under the best estimates. If the reverse holds—namely, that a sensitivity test of a particular variable gives differing conclusions depending on what likely values are assigned—then there is clearly less confidence about the design or system. However, this does not preclude the possibility of making decisions about the system or design. It simply means that the sources of certain risks or problem areas have been identified, and need to be dealt with further. One possible response is to turn greater attention to the estimating process, so as to determine more precisely the relative likelihood that "dangerous values" will, in fact, be encountered. Another possibility is to redesign the system so that even if such values do occur, a desirable outcome will still be assured. In any event, the analyst is still surely better off, having undertaken a sensitivity analysis, in that a deeper level of insight into his problem has been achieved.

The following hypothetical situation will serve to clarify the value and practice of sensitivity analysis.

A manufacturing company requires the service of a new 50-h.p., totally enclosed, fan-cooled electric motor for which its best estimate is that it will be run an average of 7 hours per day for 250 days per year. Past experience indicates that (1) its annual costs for property taxes and insurance can be estimated at 5 percent of investment cost, (2) it must earn 18 percent on invested capital before income-tax considerations, and (3) it must recover capital invested in such equipment within 3 years. Two motors are offered to the company. Motor A costs $1,990 and has a projected efficiency of 90.5 percent at the indicated operating load. Motor B, a high-efficiency version, costs $2,429 and has a guaranteed efficiency of 94.5 percent under the same operating load. Electric energy currently costs the company 5 cents per kilowatt-hour, and 1 h.p. = 0.746 kw.

In evaluating the relative economic desirability of the two motors it is apparent that the lower operating cost of motor B is offset, to some extent, by a higher acquisition cost. In order to weigh the relative importance of these two factors, the firm's engineering department has estimated the total annual cost of each motor. They have done so by calculating the annual equivalent capital cost for each motor, using time value concepts to be explained further on in the text. To that was added annual costs for property taxes and insurance, energy at the indicated price, taking into account expected efficiency and use, and finally arriving at total annual costs. In this case, motor A has an annual cost of $4,621, and motor B an annual cost of $4,692. The use of the best estimates thus favors motor A. In other words, at 5 cents per kilowatt-hour, the energy savings available by choosing motor B is not sufficient to justify its higher acquisition cost.

However, at least two of the relevant variables may be thought of as subject to significant variation in practice from their best estimates—namely, annual hourly use and the price of electric power. A sensitivity analysis was conducted for each of those variables. For example, if the price of electric power were to rise to 8 cents per kilowatt-hour with other factors remaining the same, the total annual cost of motor A would be $6,785, while that of motor B would be $6,765. Thus, at a higher energy cost than the original "best estimate" computation, motor B would be preferred. This is attributable to the fact that at higher energy prices the energy savings provided by motor B more than offsets its higher acquisition cost.

Computation of total annual costs under varying assumptions of electricity prices and annual hourly use give the functions illustrated by Figures 1–1 and 1–2. These constitute a form of sensitivity analysis. They show, for example, that electricity prices beyond 7 cents per kilowatt-hour (at an annual use of 1,750 hours) favor motor B. Thus, the decision concerning which motor to choose should be based not only on the current electricity price but on the likely range of prices over the three-year study period. If it is safe to assume that energy prices will not rise beyond 7 cents per kilowatt-hour within three years, then it would be safe to choose motor A.

It should be pointed out that sensitivity analysis can often be facilitated by the use of computers. Such opportunities will be illustrated later on in the text.

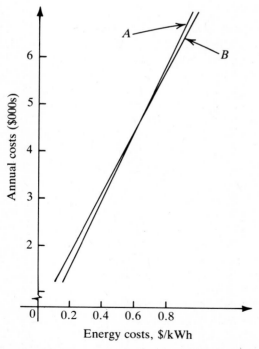

FIGURE 1–1 Annual Motor Costs vs. Electricity Prices

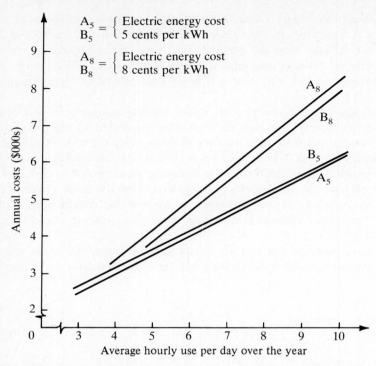

$A_5 = \begin{cases} \text{Electric energy cost} \\ \text{5 cents per kWh} \end{cases}$
B_5

$A_8 = \begin{cases} \text{Electric energy cost} \\ \text{8 cents per kWh} \end{cases}$
B_8

FIGURE 1–2   Annual Motor Costs at Various Hourly Uses

2

Time Value Concepts

Introduction

The process of modeling the economic characteristics of engineering design alternatives begins with time value models. These are described in this chapter. Subsequent chapters deal with the application of these principles to specific situations.

Basic Time Value Models

The central theme of this text is that economic resources can be put to productive use: money can be deposited in a bank to earn interest over time, a machine can be purchased to provide, over its useful life, a necessary "value added" in a stage of production, or expenditures on research and development can provide future sales or benefits for the organization.

Time value models require information concerning the initial value of the resource(s) under consideration and the time period over which to study the resources and the effects of their use. In addition, the models also require an assumption concerning the rate at which the assets could otherwise generate income or service. Furthermore, the basic models presume that that rate is constant during the period in question. Let us call that rate the *potential earnings rate* and symbolize it by the letter *i*. Then our first model may be derived as follows:

Divide a time horizon into equal periods, as suggested by the schematic time line below:

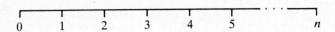

Suppose that the value of a resource is $1 at time 0. This gives Cash Flow Diagram D2–1. (Cash flow diagrams were introduced in Chapter 1.)

D2–1

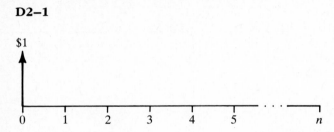

If the potential earnings rate is i per period, then the "amount earned" during the first period is $\$1 \times i$. The value at time 1 would then no longer be $1, but would be the total of the amount at the beginning of the period, or $1, and the "amount earned" or $\$1 \times i$. That total, in dollars, equals $1 + (1 \times i)$, or simply $1 + i$. That pattern may be continued in tabular form as in Table 2–1.

TABLE 2–1 Future Time Value of One Dollar

Value, Beginning of Period	Increase During Period	Value, End of Period
1	$1\,(i)$	$1 + i$
$1 + i$	$(1 + i)i$	$(1 + i) + (1 + i)i = (1 + i)^2$
$(1 + i)^2$	$i(1 + i)^2$	$(1 + i)^2 + i(1 + i)^2 = (1 + i)^3$
"	"	"
"	"	"
"	"	"
$(1 + i)^{n-1}$	$i(1 + i)^{n-1}$	$(1 + i)^{n-1} + i(1 + i)^{n-1} = (1 + i)^n$

The pattern of Table 2–1 constitutes a proof by induction, and indicates that the value of the resource by the end of period n would be $(1 + i)^n$.

To generalize the model for initial values other than $1, the reader should verify that, beginning with $\$P$, one would end the first period with $\$P(1 + i)$. In a similar fashion one would end up with a value of $\$P(1 + i)^n$ after n periods. Thus our first result can be expressed by the equation,

$$F_n = P(1 + i)^n$$

(2–1)

where F_n is the value of the resource expressed at time n and P is its value at time 0 or the base period.

This general model is often applied to find the future value of a sum of money invested at a given interest rate. (See Example 2–1 below.) Other uses of the model are found throughout the text. The exercises at the end of this chapter illustrate the alternative uses to which the model may be put.

The model is also applicable to other fields, not only for economic decision analysis. For example, it may be used as a growth model relating future amounts or values to present ones. Thus, it may be used by demographers to describe the growth of populations over time, where P is the initial or base period population, i is the rate of growth of the population, and F is the projected future population.

If $i > 0$, a plot of F_n vs. time would look like Figure 2–1.

An example of the application of the model described by Equation 2–1 to an economic situation will serve to illustrate its use in a context more relevant to the theme of this book:

EXAMPLE 2–1

A beginning sum of $2,000 (often referred to as the principal) is invested on January 1, 1983. It can be put to work at 12 percent per annum. What will it be worth in four years?

Solution Here the task is to compute F_4, or the value of the original amount four periods hence. Equation 2–1 gives

$$F_4 = 2,000(1.12)^4 = \$3,147.04$$

The reader should not only verify the result above but should compare it to a result obtained by multiplying 2,000 by 0.12 and adding 4 times the product to

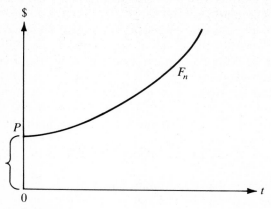

FIGURE 2–1

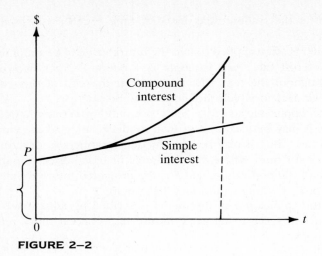

FIGURE 2–2

the principal. This latter procedure calculates simple interest, while Equation 2–1 calculates compound interest. These are compared graphically in Figure 2–2.

The Equivalence Concept

The model described by Equation 2–1 does not simply predict what a given bank account will be worth in n periods. Rather, it indicates that F_n is equivalent to P if they are separated by n periods with an earnings rate of i per period. The idea of equivalence permits the use of the models to be generalized to almost all decision situations, because most decisions involve long time spans during which outflows and inflows of resources are expected. The equivalence concept of the time value models permits rational analysis of the relative importance of each of those anticipated flows even though they occur at different times.

The concept of opportunity costs can shed light on the equivalence concept. An opportunity cost is a hypothetical value, not a record of observed historical costs. Rather, it represents what economic benefit would be foregone if an asset is put to use in some alternate way.

For example, suppose one were graciously to lend an amount of money, say $1,000, to one's favorite charity at a below-market interest rate for one year. If the interest rate received were 4 percent per annum, while the best rate that could be earned with the same risk was 14 percent per annum, then the difference of 10 percent of $1,000, or $100, would be the opportunity cost of the charitable act for the year.

The example above indicates that opportunity costs involve dimensions of both amounts and time. The equivalence concept used in time value models incorporates both types of quantity.

An understanding of opportunity costs serves to clarify the equivalence concept. For example, if money can be put to productive use, $1 of resources today is not equivalent to $1 of resources tomorrow. The lack of such equivalence is due to the productive potential of the resource. That is, $1 today should be equivalent to more than $1 tomorrow.

The following extreme illustration will serve to emphasize the importance of the equivalence concept because it spotlights the pitfalls of ignoring it, not because of the context of the application.

In a 1930s vintage movie, Groucho Marx portrayed the Dutch colonial Director General, Peter Minuit, who purchased Manhattan Island from the resident Indians (known as the Man-Hattes) in 1626 for 60 Dutch guilders. Marx's portrayal, with cigar, raised eyebrows, and all, was humorous in its implication that the Man-Hattes were foolish to part with the island for as paltry a sum as 60 guilders, which, at the time of the movie, was said to be worth $24.

Marx's genius notwithstanding, there is a flaw in the foundation of the humor for that situation. The movie, appearing in 1936, used the figures of 1626 and $24. However, doing so implies zero opportunity cost of funds or resources between those periods, which can hardly be right. In other words, $24 in 1626 is not equivalent to $24 in 1936 (irrespective of inflation and deflation). Rather, the simple model, Equation 2–1, gives an indication of the equivalent worth of the $24 the Man-Hattes received in 1936, when the movie appeared. Even if a conservative earnings rate of $i = 6$ percent per year is used, the following 1936-equivalent value is obtained:

$$F_{1936} = P_{1626} (1.06)^{310}$$

$$= \$24 (1.06)^{310}$$

$$= \$1.679 \text{ billion}$$

where F_{1936} is the equivalent value, in 1936, of the original transaction quantity, P_{1626}, and 310 periods separate the dates.

It should be noted that the 1936 equivalent value estimate of the 1626 transaction was close to the assessed value of Manhattan Island at that time! Clearly, it was the public's ignorance of time-value and equivalence concepts, as well as Groucho Marx's antics, which led to laughter.

Discounting the Future

It is just as important to be able to utilize Equation 2–1 "in reverse." Thus, dividing both sides of 2–1 by $(1 + i)^n$ yields

$$P_0 = \frac{F_n}{(1 + i)^n} \qquad\qquad (2\text{–}2)$$

where P_0 is the present value of a future amount, F_n is to be realized in n periods, with the earnings rate as i for the entire interval. Thus, it is often necessary to evaluate a future value at the present time. An example will illustrate:

EXAMPLE 2–2

Suppose an amount of $100 is desired to be received in two years, and money is worth 9 percent per year. How much should be invested now in order to have $100 available in two years, at the indicated interest rate? The model, in the form of Equation 2–2, can be used to answer the question.

$$P_0 = \frac{\$100}{(1.09)^2} = \$84.17$$

Alternatively, and more in keeping with the equivalence concept, one could have asked, what is the present worth of those funds, or, how much would one just be indifferent to receiving, at present, in place of $100 in two years, if funds are worth 9 percent? The answer would clearly be the same, namely $84.17, and represents the equivalent value now of the $100 to be received two years hence. Their equivalence is based on the earning potential of the invested funds, as explained previously.

Often it is necessary to compute the equivalent present value of a stream of future values of varying amounts. Here it is helpful for the reader to sketch a cash flow diagram, as was done previously to represent such a flow. The appropriate amount corresponding to each period can then be depicted. This is illustrated by Example 2–3.

EXAMPLE 2–3

D2–2

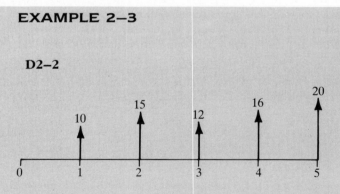

The interpretation here is that $10 will be received at the end of period 1, $15 at the end of period 2, and so on, until $20 is received at the end of period 5. The equivalent value of this flow at time zero is computed by discounting each of the above separately (at which point they are all expressed in terms of time zero value) and summing. Thus, the value at time zero can be expressed as

$$P_0 = \frac{10}{(1.10)} + \frac{15}{(1.10)^2} + \cdots + \frac{20}{(1.10)^5} = \$53.85$$

The reader should verify this computation, which is often referred to as a discounted cash flow.

Annuities

Annuities may be thought of as a special case of the time series flows depicted in Example 2–2. The difference is that the amount is always the same. In other words, an annuity is a series of equal, periodic amounts. This can be depicted in a manner similar to the previous example. Cash Flow Diagram D2–3 shows an annuity with amount A occurring over five periods.

D2–3

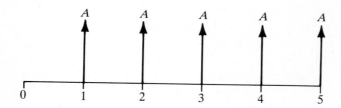

Here it is important to note that the first transaction occurs at the end of the first period. If one desires the equivalent future value of the stream at time 5, it is clear that this can be easily done by computing the future value of each A, using Equation 2–1 and then summing. However, for annuities with many payments, this is tedious and time-consuming. The alternative is to derive a simple algebraic expression for this amount, which follows below. Cash flow diagram D2–4, shown below, indicates a general annuity—in other words, one covering n periods.

D2–4

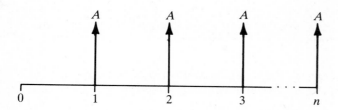

For such an annuity the lump-sum future equivalent value, which is often called a sinking fund, may be thought of as the accumulated value of n deposits into a savings account earning a rate of i per period on all balances. However, that equivalent value should be thought of more generally, since this model has many other applications beyond that of payments into a bank account. If F_n is the future equivalent value of the entire sum, it may be computed by

$$(a) \quad F_n = A + A(1 + i) + A(1 + i)^2 + \cdots + A(1 + i)^{n-1}$$

The first A in equation (a) represents the value, at time n, of the *last* transaction. Since it occurs at time n its value is unadjusted. The second entry in equation (a), $A(1 + i)$, represents the value, at time n, of the *second to the last* transaction. It is separated from time n by one period and is therefore adjusted by $(1 + i)$. The last entry corresponds to the first transaction, which is separated by $n - 1$ periods, and is therefore adjusted by $(1 + i)^{n-1}$.

Multiplying equation (a) by $(1 + i)$ gives (b).

$$(b) \quad (1 + i)F_n = A(1 + i) + A(1 + i)^2 + A(1 + i)^3 + \cdots + A(1 + i)^n$$

Note that the right sides of both equations (a) and (b) have all terms in common except two. Thus, if (a) is subtracted from (b), one obtains

$$iF_n = A(1 + i)^n - A = A[(1 + i)^n - 1]$$

Dividing both sides by i then gives the desired result

$$F_n = A\left[\frac{(1 + i)^n - 1}{i}\right] \tag{2-3}$$

EXAMPLE 2–4

As an illustration of the use of Equation 2–3, imagine the depositing of funds into a bank savings account. Suppose \$100 per year is deposited beginning with a child's first birthday, continuing each birthday thereafter until age 18. Further, suppose that the fund earns 6 percent per annum. How much will be accumulated by the 18th birthday? The answer is given by:

$$F_{18} = 100\left[\frac{(1.06)^{18} - 1}{0.06}\right] = \$3,090.57$$

Note that this is more than $18 \times \$100$ because of the time value, or earning power, of the funds. Another common use of this model is to compute the future equivalent value of a set of equal anticipated costs or revenues.

As with Equations 2–1 and 2–2, the inverse of the right side of Equation 2–3 is useful. Thus, one easily obtains from Equation 2–3,

$$A = F_n\left[\frac{i}{(1 + i)^n - 1}\right] \tag{2-4}$$

Equation 2–4 is clearly useful in determining the periodic amount necessary in order to attain a sinking fund of a given amount by period n, if the earning potential of the resources is expressed at the rate of i per period. Again, an example will clarify the point.

EXAMPLE 2–5

Suppose a company borrows $1 million at time zero by issuing bonds in that amount. The $1 million must be repaid in ten years. If it desires to deposit an amount at the end of each of the next ten years so as to accumulate the $1 million to be repaid, how much should be deposited each year if money earns 8 percent per annum? This is given by:

$$A = \$1M \left[\frac{0.08}{(1.08)^{10} - 1} \right] = \$69,029.49$$

Equation 2–4 is also useful for computing the equivalent periodic annuity value of a future capital requirement, such as replacement of an existing asset.

Capital Recovery

Equations 2–1 to 2–4 describe models which relate present and future values of individual amounts and annuities. The results of these can be combined to relate the present value of a lump sum to an annuity of equivalent value. This can be shown schematically in Cash Flow Diagram D2–5.

D2–5

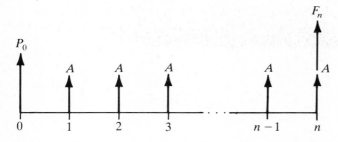

In other words, since an annuity has an equivalent future lump sum, or can be thought of as equivalent to it, and that lump sum, F_n, has an equivalent present value, P_0, then P_0 may be thought of as equivalent to the annuity. Algebraically, since

$$P_0 = \frac{F_n}{(1 + i)^n}$$

(2–2)

and F_n, from 2–3, can be substituted in 2–2,

$$F_n = A\left[\frac{(1+i)^n - 1}{i}\right] \tag{2-3}$$

one obtains

$$P_0 = A\left[\frac{(1+i)^n - 1}{i}\right]\cdot\left(\frac{1}{(1+i)^n}\right) = A\left[\frac{(1+i)^n - 1}{i(1+i)^n}\right] \tag{2-5}$$

This gives the present worth of an annuity of amount A per period for n periods. The interpretation of that present worth is the same as in Example 2–2, where the present worth was referred to as a discounted cash flow. This is extremely useful for finding the equivalent present worth of an anticipated annuity stream of costs or revenues. It can also be used to compute the balance remaining to be paid on a loan with an annuity payment schedule. (See Example 2–7 below.)

Once again, this model can be useful in inverse form. Equation 2–5 easily yields

$$A = P_0\left[\frac{i(1+i)^n}{(1+i)^n - 1}\right] \tag{2-6}$$

This is often referred to as the capital recovery factor, since it indicates what annuity income is equivalent to a capital investment of P_0 at time zero.

Equation 2–6 is very commonly used to compute the repayment annuity associated with a loan. Installment loans and conventional mortgages are common examples.

EXAMPLE 2–6

The terms of repayment of a conventional mortgage in which $75,000 is borrowed specify repayment in equal monthly installments over 25 years (300 months) at 1 percent per month. Thus the month is the relevant period corresponding to Cash Flow Diagram D2–6.

D2–6

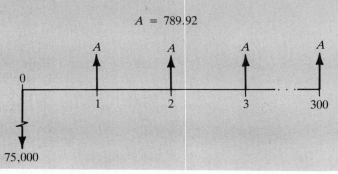

$A = 789.92$

(The convention is to refer to the interest rate on an annual basis, and to simply state it as a multiple of 12 times the monthly rate, which in this example is 12 percent. More will be said of this convention later.) The monthly amount to be repaid is given by 2–6 and is

$$A = 75,000 \left[\frac{0.01\,(1.01)^{300}}{(1.01)^{300} - 1} \right] = \$789.92$$

An interesting example of the use of Equation 2–5 is related to the previous example:

EXAMPLE 2–7

Suppose amount A, or $789.92, is paid each month for exactly one year. What balance remains to be paid at that time?

The fastest solution to this problem can be obtained by computing the equivalent value of the payments still to be paid. In other words, the balance is the present worth of the 288 monthly payments remaining.

$$P_{12} = 789.92 \left[\frac{(1.01)^{288} - 1}{0.01(1.01)^{288}} \right] = \$74,493.74$$

An alternative solution to the problem above serves to reinforce the concept of equivalence. Many people intuitively prefer to compute a balance by subtracting "what was paid" from "what was owed." Such an approach can be used in the problem above, but care must be taken to express "paid" and "owed" in terms that are equivalent. In other words, the monthly payments of $789.92 occur at different times, and all are not coincident to the principal amount of the loan. Thus, they cannot be combined directly. A correct approach would be to compute the future equivalent value, at period 12, of the 12 payments made and to subtract the value of those payments from the equivalent value of the principal, expressed at time 12. Equation 2–3 produces an equivalent value of:

$$F_{12} = \$789.92 \left[\frac{(1.01)^{12} - 1}{0.01} \right] = \$10,018.16$$

This is subtracted from the principal, also valued at time 12:

$$P_{12} = 75,000(1.01)^{12} = \$84,511.88$$

This gives $74,493.72, which agrees, within reasonable rounding error, with the preceding result.

This is not the only alternative approach to the problem using equivalence. The reader may verify other approaches that seem useful.

Notation

Equations 2–1 to 2–6 are too cumbersome to use repeatedly in a text. There is a standard shorthand notation to denote the exponential factors found on the right side of those equations. The notation to be used in this text can be summarized by rewriting those equations, using F to stand for future value, P for present value, and A for annuity amount. The subscripts indicate time periods.

$$F_n = P_0(F/P\ i\%, n) \tag{2-1}$$

$$P_0 = F_n\ (P/F\ i\%, n) \tag{2-2}$$

$$F_n = A\ (F/A\ i\%, n) \tag{2-3}$$

$$A = F_n\ (A/F\ i\%, n) \tag{2-4}$$

$$P_0 = A\ (P/A\ i\%, n) \tag{2-5}$$

$$A = P_0\ (A/P\ i\%, n) \tag{2-6}$$

The symbol i is the earnings, discount, or interest rate per period, and n is the number of periods. Thus, for example, $(F/P\ i\%, n)$ may be read as the factor yielding future value, given present value and an interest rate of i for n periods.

$$(F/P\ i\%, n) = (1 + i)^n$$

Nominal and Effective Interest Rates

Earnings or discount rates (which in many applications are interest rates) are conventionally expressed on an annual basis. Thus, in Example 2–6, convention labels the interest rate of that transaction as 12 percent per annum. Such an annual stated rate is referred to as the *nominal* rate. When the period of compounding is in equal fractions of a year, say quarterly or monthly, then the process of more frequent compounding results in an *effective* annual rate which is greater than the nominal. This is illustrated as follows:

Interest is at 12 percent per year, compounded quarterly. The convention referred to indicates that the rate applicable per period (here a quarter) is the annual rate divided by the number of periods or quarters per year. For this example, the quarterly rate would be 12%/4, or 3 percent per quarter. However, the effect of this would be an annual equivalent of

$$(1.03)^4 - 1 = 0.12551 \text{ or } 12.551 \text{ percent}$$

which is clearly greater than the nominal rate.

The result of increasing the number of compounding periods per year (or, conversely, decreasing the length of the compounding period) is easily seen. For the situation above, changing to a monthly period, as in Example 2–6, yields a period rate of 1 percent. The annual effective rate then becomes

$(1.01)^{12} - 1 = 0.12683$ or 12.683 percent

which is still higher than the quarterly rate.

Continuous Compounding—Discrete Amounts

When the width of the compounding period goes to zero (or when the number of compounding periods per year increases without bound), the result is referred to as continuous compounding. Since the effective rate increases as the period becomes smaller (or as the number of compounding periods per year becomes larger) one might infer that the effective rate can be increased without bound. This intuitive notion is incorrect, however. Figure 2–3 shows the relation between the effective interest rate and an increase in the number of compounding periods. It is evident that the effective rate approaches a limit that is depicted by the asymptote. Mathematically, that limit is derived as follows:

Suppose the nominal interest rate is i per period as before. Amounts occur at the end points of the period referred to by the nominal rate, as depicted in Cash Flow Diagram D2–7.

D2–7

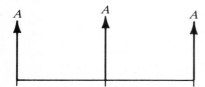

For the moment, divide each period into L discrete subperiods for compounding. Using the above formulation, which describes effective rates as a function of the nominal rate and the frequency of compounding, one has

$$c = \left(1 + \frac{i}{L}\right)^{L} - 1$$

where c is the effective equivalent rate for the period. This can alternatively be written as

$$c = \left[\left(1 + \frac{i}{L}\right)^{L/i}\right]^{i} - 1$$

Continuous compounding may be defined as the effect on c, above, if L grows without bound. Now one can utilize the definition of the base of natural logarithms in conjunction with the above.

$$e = \lim_{k \to \infty} \left(1 + \frac{1}{k}\right)^{k}$$

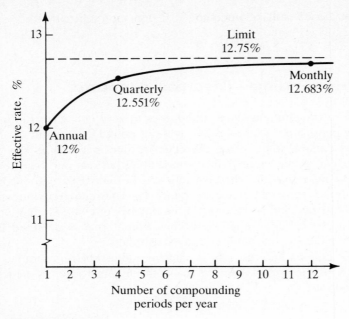

FIGURE 2–3

Thus $c = e^i - 1$, which is the limit approached by the curve of the effective rate vs. increasing compounding periods. To return to the example of Figure 2–3, if $i = 12$ percent per year nominally, the effective annual rate, if compounding is continuous, is $e^{0.12} - 1 = 1.1275 - 1 = 0.1275$ or 12.75 percent. The value 0.1275 is the value asymptotically approached by the curve.

Since $e^i - 1$ is the effective rate for continuous compounding, that rate can be used in the time value models described by Equations 2–1 to 2–6. Thus, for continuous compounding, F_n as a function of P_0 becomes,

$$F_n = P_0(1 + e^i - 1)^n = P_0 e^{in} \tag{2–1'}$$

Similarly, F_n as a function of A becomes,

$$F_n = A\left[\frac{(1 + e^i - 1)^n - 1}{e^i - 1}\right] = A\left(\frac{e^{in} - 1}{e^i - 1}\right) \tag{2–3'}$$

and P_0 as a function of A becomes,

$$P_0 = A\left[\frac{(1 + e^i - 1)^n - 1}{(e^i - 1)(1 + e^i - 1)^n}\right] = A\left[\frac{e^{in} - 1}{(e^i - 1)e^{in}}\right] \tag{2–5'}$$

Clearly, the inverse of Equations 2–1', 2–3', and 2–5' can easily be derived to round out the complement of formulas needed for time value applications under continuous compounding.

In cases involving annuities, it is also possible to envision that the amounts flow continuously rather than at discrete periods as in Equations 2–1' to 2–6'.

This changes the models still further, though it does not substantially affect the end results. Since the applicability of that refinement is not widespread, the relevant formulas will not be derived or used here.

Multiple Discrete Compounding

Frequently one encounters situations in which an interest rate is compounded discretely but more often than once per nominal period. The earlier section on nominal and effective rates explained the effect that more frequent compounding has on the rate. This section will explain how Equations 2–1 to 2–6 may be applied to specific problems.

   If the nominal annual interest rate i is compounded, say, quarterly, simply treat the number of periods as four times the number of years, and the periodic rate as one-fourth that of the nominal. Then proceed as before. For example, in computing the value of $100 in two years if the nominal annual rate is 12 percent and compounding is quarterly, the relevant rate per period is 3 percent and the number of periods is eight (quarters). Then it follows that,

$$F_{8 \text{ quarters}} = 100(1.03)^8 = 100(F/P \, 3\%, 8) = \$126.68$$

Note that this result is identical to what would be obtained if the effective annual rate, 12.551 percent, is applied to the principal for two years:

$$F_{2 \text{ years}} = 100(1.12551)^2 = 100(F/P \, 12.551\%, 2) = 126.68$$

Deferrals—Deferred Annuities

The derivation of the models described by Equations 2–3 to 2–6 assumed that the first amount, A, occurred at the end of the first period. In the event a situation does not correspond to such an assumption, the equivalence concept can be used to make the appropriate correction. Suppose, for example, that the first of n repayments of a loan occurs at the end of the second period rather than the first. This is depicted in Cash Flow Diagram D2–8.

D2–8

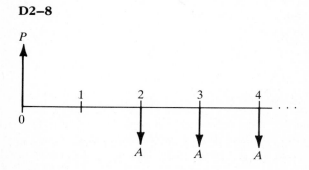

Clearly, any of the previously derived annuity formulas involving A implies a present value of the stream above at time 1. Thus, the correction called for is the computation of the equivalent value of P at time 1. This is what is referred to as a deferral, and it can be represented by the revised Cash Flow Diagram D2–9.

D2–9

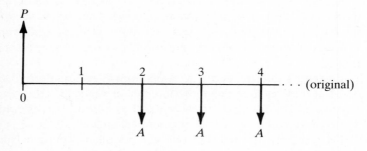

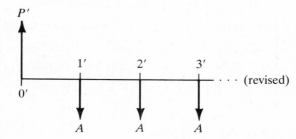

Now, the amount P' can easily be computed using Equation 2–1,

$$P' = P(1 + i)$$

and amount A can be computed from Equation 2–6 and P' as before.

EXAMPLE 2–8

Suppose that Example 2–6 is modified only so that the first repayment does not occur until two months after the loan is made. Thus, the new situation corresponds to a one-period deferral as depicted above. P' then becomes

$$P' = 75,000(F/P\ 1\%, 1) = 75,750$$

Then the revised A may be computed as

$$A = 75,750(A/P\ 1\%, 300) = \$797.82$$

Thus, it can be seen that the effect of deferring the repayment schedule by one period increases the periodic amount to be paid. This is due to the fact that the borrower has greater use of the funds (i.e., for a somewhat longer time).

Other types of deferrals or advances may be treated similarly as long as the proper equivalences are maintained. An example of an advance is Cash Flow Diagram D2–10.

D2–10

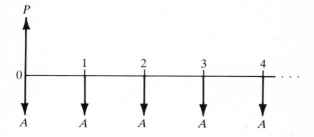

Here the series of annuity payments, A, begins coincident with P rather than at the end of the first period. This can be handled in a variety of ways. As an example, if P is a loan and it is desired to compute the requisite amount A to be repaid in n periods at i percent per period, given the description above, then the simplest solution is to combine the first A with P. This is easily done, since they occur at the same time. The result may be depicted as in Diagram D2–11.

D2–11

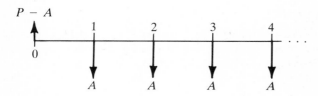

Here the reader should take care to verify that if n amounts are initially indicated, $n - 1$ are called for after the transformation. Then A may be computed from Equation 2–6 with $n - 1$ periods,

$$A = (P - A)\left[\frac{i(1 + i)^{n - 1}}{(1 + i)^{n - 1} - 1}\right]$$

Exercises

2–1 If $600 is deposited in a bank account now, how much would it be worth in 7 years if interest is credited at the rate of 8 percent per year?

2–2 Suppose you borrow $2,000 now. If you must repay the loan with interest in one lump sum in 3 years, how much would you repay if interest is at 10 percent per year?

2–3 A modification of equipment is scheduled for 4 years hence. If the cost is expected to be $8,000 then, what amount would need to be set aside now to provide for the change, if funds earn 8 percent interest per year?

2–4 You will need $5,000 in 3 years for a down payment for a new car. How much would have to be deposited now to provide for it, if the funds will earn 10 percent interest per year?

2–5 Your nephew has just been born. Suppose you resolve to deposit to a bank account $100 on each of your nephew's birthdays through his age 21. How much would be in the account just after the last deposit, if the account pays 8 percent interest per year?

2–6 A firm must repay a loan of $5 million in exactly 10 years. As part of the loan agreement, the firm must deposit an amount annually into a bank account earning 5 percent interest annually so as to accumulate the $5 million when the last deposit is made. How much should each deposit be?

2–7 In Exercise 2–5, suppose that you are unable to begin making the deposits until the third birthday. If everything else is the same, how much would be in the account after the deposit for the 21st birthday?

2–8 Again, returning to Exercise 2–5, suppose the deposits begin at the first birthday, but the last two deposits are missed, that is, those for the 20th and 21st. If all funds earn 8 percent, how much would be in the account at the 21st birthday?

2–9 Your firm must replace a machine costing $500,000 in four years. If it plans to accumulate the necessary funds by depositing into a bank account an equal amount each month, how much should the deposits be if the funds can earn 3/4 percent per month?

2–10 If you borrow $10,000 and agree to repay it in 5 equal annual installments starting one year from the date of the loan, how much would each payment be if the annual interest rate on the loan is 10 percent?

2–11 Suppose, in Exercise 2–10, the loan is repayable monthly. The monthly interest rate is 10/12 percent. How much would be repayable each month under these conditions?

2–12 If your firm leases a machine to a customer for $3,000 per year for 5 years, and money is worth 12 percent per annum, how much would the firm be indifferent to receiving in one lump sum now, instead of the annual lease payments? (Assume that the lease payments occur at the beginning of each period.)

2–13 In Exercise 2–11, what would the balance on the loan be after the tenth monthly payment?

2–14 Suppose you take out a 48-month auto loan. You borrow $5,000, agreeing to start monthly repayments in one month. The interest rate is 12 percent per year. How much is the monthly payment?

2–15 You buy a house, taking out a $50,000 conventional mortgage. The terms of repayment are 15 percent per year, and the loan will be repaid monthly over 25 years. How much would be paid each month?

2–16 Costs associated with a new process are projected to be $10,000 in the first year, $12,000 in the second, $15,000 in the third, and $20,000 in each of the next two years. If funds are worth 10 percent, what is the equivalent present worth of these future costs? What 5-year annuity is equivalent to these costs?

2–17 Suppose your firm borrows $25,000 for working capital and agrees to repay the loan on an installment basis over 60 months at a 15-percent interest rate beginning 6 months hence. How much would the monthly payments be? What would the balance on the loan be after the 8th payment?

2–18 In Exercise 2–17, just after the eighth payment, more funds are needed, and the loan is renegotiated as follows: an additional $50,000 is borrowed, the principal of which is added to the remaining balance on the old loan. The new balance will then be repayable on a monthly basis for 84 months, beginning the next month, at a rate of interest of 16 percent per year. How much is the monthly payment now?

2–19 If funds are worth 8 percent per year, what is the equivalent worth of $2,000 10 years hence?

2–20 Suppose one expects to receive $6,000 in 4 years. If funds are worth 10 percent per year, how much would those funds be worth now; that is, what is the current equivalent worth?

2–21 An annuity of $400 per month is paid into a fund earning 1 percent per month. How much would be in the fund after 1 year?

2–22 If funds are worth 1 percent per month and your firm expects to receive $10,000 in 4 months and an additional $15,000 after the 12th month, what is the present value of the expected receipts?

2–23 Suppose that instead of the two amounts in Exercise 2–22, the firm expects to get $2,000 per month for each of the next 12 months. What is the present worth under these circumstances, if funds are still worth 1 percent per month?

2–24 If you borrow $5,000 for school expenses and must pay it back over the next year on a monthly basis, starting the month after you receive the loan, how much would the monthly payment be if the interest rate were 1.25 percent per month?

2–25 Suppose your firm expects revenues of $4,000 in 3 years and $6,000 in 5 years. If funds are worth 12 percent per annum, what would be the present worth of those revenues? What annuity value would be equivalent to the above revenues (over 5 periods)?

2–26 Your bank lends a client $25,000, which he agrees to repay in one total lump sum in 6 months, at a rate of interest of 15 percent per year, compounded monthly. Just after the third month, the client indicates that more funds are needed and wishes to borrow $15,000 more. There is also some indication that there is added difficulty in repaying the loan. The bank agrees to the added loan but stipulates that the consolidated amount or balance must be repaid monthly over the next 12 months at an 18-percent annual rate. How much would the monthly payment be?

2–27 You borrow $60,000 in the form of a variable-rate mortgage to be repaid in monthly amounts over the next 30 years. The initial rate of interest is 12 percent per annum. How much is payable monthly during the first year? After the twelfth payment, the bank has the right to adjust the interest rate according to some externally set benchmark. Suppose that the rate is raised to 13 percent for the second year. How much would the monthly payments be now?

2–28 Your firm invests $200,000 in a new process. Revenues from the new process are projected at $300,000 in the first year, $350,000 in the second, and $400,000 in each of the next 5 years. Costs are estimated to be $250,000 in the first year, $290,000 in the second, and $330,000 in each of the next 5. If funds are worth 10 percent, what is the present worth of the anticipated flows?

2–29 A company borrows $100,000, agreeing to repay it monthly over 6 years, beginning next month. Interest is at 16 percent per year. After the 15th payment the firm seeks an additional loan, which is negotiated as follows: an additional $50,000 is borrowed at that time which will be combined with the balance remaining on the old loan. The consolidated balance will be repaid in monthly payments for 8 years (96 payments). The first repayment will not be made until 3 months from now, but interest will remain at 16 percent per year. How much must be repaid under the revised arrangement?

2–30 Your firm expects to receive 3 $5,000 payments at 3-year intervals, the first payment of which is scheduled in 3 years. If funds are worth 12 percent, what is the present worth of those amounts? What annual amount (over 9 years) is equivalent to that present worth?

2–31 *a*. A company has two loans outstanding. One involves an initial amount of $55,000 borrowed exactly one year ago at a rate of interest of 15 percent per annum, payable monthly over 3 years. The other involves a $25,000 principal, borrowed 2 years ago, at a rate of interest of 16 percent per annum payable monthly over 4 years. How much is the company paying per month on these loans?

   b. The company now finds it difficult to meet the payments specified in *a*. It seeks to refinance these debts. What would the balance on the loans be?

   c. If it can refinance this balance payable monthly at 18 percent per annum over the next 5 years, how much would the new monthly amount be?

2–32 Your firm undertook a loan of $300,000 2 years ago at an interest rate of 12 percent per year, to be repaid monthly over 5 years. One year ago it undertook an additional loan of $500,000 at an interest rate of 14 percent per annum, also to be repaid over 5 years. Currently interest rates have declined to 9 percent per year and the firm is thinking about refinancing the existing debt. Suppose there is a prepayment penalty (in addition to the balance) of 1 percent of the outstanding debt if the loan is to be prepaid. Furthermore the new, consolidated loan would not simply be an installment loan of the original type, but would be a balloon instrument, explained as follows: the new consolidated loan would be repaid monthly at 9 percent per annum over 4 years, at which time a balance of $100,000 would remain. At that time the firm could refinance the $100,000 at prevailing rates. How much would the monthly payment be on the original loans and on the refinanced loan?

3

Yardsticks for Evaluation of Alternatives

Introduction

The preceding chapter introduced the concepts used in time value models. This chapter will begin to explore the use of those models in decision-making.

Economic decision-making problems generally involve choosing between alternatives that offer superior performance for some attributes but inferior performance for others. A classic example is that of two competing types of equipment, one with greater acquisition cost but lower operating cost than the other. What is needed in order to evaluate the relative desirability of the two alternatives is some yardstick by which they can be compared.

This chapter will introduce tools useful for comparison or evaluation procedures applicable to an independent alternative. Such an alternative is one whose desirability can be judged by an objective criterion, without the presence of other alternatives to which the candidate may be compared. Thus, this procedure can be characterized as determining whether an alternative is acceptable with respect to some criterion. The next chapter will show how some of these tools may be used for comparing two or more alternatives. The time value models of Chapter 2 provide the basis for the yardsticks presented here, and what follows is a description of those used in this text, including:

1. Present worth,
2. Future worth,
3. Periodic worth,
4. Internal rate of return, and
5. External rate of return.

These five are, in fact, related. Often it is possible to compute each of them for a particular alternative. In other instances, the economic situation and the structure of the problem dictate that only one of the methods should be employed. (This may be due to the inapplicability of one or another of the methods.) The reader should alert himself to the characteristics of each, so that he may intelligently choose among them when conducting an analysis. (The benefit-cost ratio is another yardstick that can be used. It also utilizes the time-value concepts of Chapter 2. However, since its use is typically restricted to public projects, it will be presented separately in Chapter 14.)

In order to illustrate the methods above, a common situation will be presented in Example 3–1a. Examples 3–1b through 3–1e show how these yardsticks are computed. In this way their applications can be more easily compared.

EXAMPLE 3–1a

Consider a possible investment of $100,000 (assumed to occur at time zero). For simplicity, assume that revenues and expenses associated with the project are annuities which are projected to repeat for five years. At the end of that time the investment is terminated (by selling it, say) for a sum expected to be $30,000. If annual revenues exceed expenses by $20,000 (and are assumed to occur at the end of the periods), how can the desirability of that investment be measured? The projected cash flows for that situation can be summarized by Cash Flow Diagram D3–1, in which the amounts are labeled in thousands of dollars.

D3–1

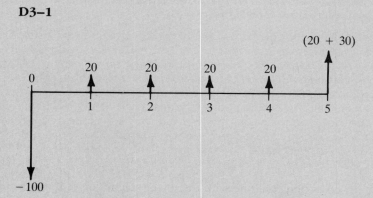

Present Worth

The present-worth method simply converts the entire set of flows projected to occur over a future time horizon to a single present value at time zero. This amount is variously referred to as present worth, present value, or net present value. Clearly, this can only be done given an assumption on the relevant earnings, discount, or interest rate. In other words, the decision analyst must use as a discount rate that which is applicable for the given economic situation. It represents a general estimate of worth, and it is often referred to as the minimum attractive rate of return, or MARR.

EXAMPLE 3–1b

Suppose the MARR is 12 percent per annum. Then the present worth of the flow is easily computed using Equations 2–2 and 2–5.

$$P_0 = -100 + 20\left[\frac{(1.12)^5 - 1}{0.12\,(1.12)^5}\right] + 30\left[\frac{1}{(1.12)^5}\right]$$

or

$$P_0 = -100 + 20(P/A\,12\%,\,5) + 30(P/F\,12\%,\,5)$$

$$= -100 + 72.096 + 17.023$$

$$= -\$10.882 \text{ thousand}$$

Here the present worth of the investment at 12 percent is less than zero. The interpretation of this result is important to an understanding of the method. This is equivalent to saying that if resources are valued at 12 percent, one would be indifferent between the flow depicted above and one in which one "lays out" $10,882 at time zero. The concept of equivalency makes the interpretation straightforward, for it is clear that a one-time outflow of $10,882 at time zero is an unacceptable alternative if resources are worth 12 percent. This example is generalizable, in that negative present values are indicative of undesirable alternatives if both revenues and costs are identifiable.

Conversely, if the present worth is greater than zero, the flow is equivalent to an immediate "gain," and the alternative is desirable. If the flow is just equal to zero the alternative is barely acceptable. (A further significance of a zero net present value will be discussed in detail in the section on internal rate of return.)

If only costs or only revenues are being considered for a cash flow, then all the signs are the same. Often in such cases, when the aggregate magnitude is of concern, the minus signs may be omitted. Thus, if two machines are being compared with respect to costs only, the one with the lower magnitude of the present worth of costs may be the more desirable one.

Future Worth

The future-worth method is very similar, in principle, to present worth. Both methods compute a lump sum quantity value which is equivalent to a stream of flows. The difference is that the future-worth method computes the equivalence in terms of value at a future time.

EXAMPLE 3–1c

Future worth could be computed in terms of period 5. Then,

$$F_5 = 30 + 20(F/A\ 12\%, 5) - 100(F/P\ 12\%, 5)$$

$$= 30 + 127.057 - 176.234$$

$$= -\$19.177 \text{ thousand}$$

where F_5 is the value of the flow at time period 5.

The reader should verify that this result is the same as that obtained by taking the present worth of $-\$10.882$ thousand and multiplying by $(F/P\ 12\%, 5)$. Thus, the present-worth and future-worth methods are related, differing only in the period for which the equivalent lump-sum value is calculated.

Similarly, the interpretation of future worth is the future value that one would be indifferent to receiving in place of the given set of flows. For the sample problem, a future worth of $-\$19,177$ indicates equivalency to "laying out" $\$19,177$ at the end of period 5. This clearly undesirable conclusion is identical to the one under present worth.

If the future worth is greater than zero when revenues and costs are jointly considered, the interpretation is a future "gain"—an acceptable alternative at the assumed interest rate. If the future worth is zero, the alternative is barely desirable at the given interest rate. Zero future worth has special significance which will be explored in the section on internal rate of return.

As with present worth, if only costs or revenues are considered, their magnitudes can be used for comparison with other alternatives.

Periodic Worth

Periodic worth is yet another "quantity" yardstick. It differs from the preceding two in that it computes an annuity for the horizon in question which is equivalent to the indicated flows at a given interest rate. Although the relevant period can be any length of time, most applications employ an annual period, because most decision analyses state costs and revenues in annual terms. In such cases, the periodic worth is often referred to as annual worth.

In general, periodic worth may be computed by taking the present worth (or future worth) of the entire set of flows and then converting that to an annuity. In instances in which a specific investment results in a series of subsequent flows, it is desirable to identify the periodic equivalence of the investment. This is illustrated in (*a*) through (*c*) in Example 3–1d. Of course, the periodic worth of the investment may then be combined with the periodic worth of the remainder of the flows.

EXAMPLE 3–1d

The following annual-worth analysis may be employed: First, it should be clear that the revenues of $20 thousand are already in annuity form and need no modification. (In other words, they already represent an annual worth of 20.) What remains is to convert the one-time capital investment of 100 (with expected salvage value of 30) to an annual amount. This will be referred to as the annual equivalent to the capital investment, or capital equivalent, *CE*. It can be done in a variety of ways, all of which are algebraically equivalent. They will be referred to as versions (*a*) through (*c*).

(*a*) From the point of view of equivalency concepts, the simplest procedure is to begin with the present worth of the salvage value and combine it with the investment outlay.

$$P_0 = -100 + 30 \, (P/F \, 12\%, 5)$$

$$= -100 + 17.023 = -\$82.977 \text{ thousand}$$

Since this is a "lump sum," it can be converted to an annuity as follows:

$$CE = P_0 \, (A/P \, 12\%, 5)$$

$$= -82.977 \, (0.27741)$$

$$= -\$23.019 \text{ thousand}$$

Note that this is equivalent to the more expeditious formulation given in general form as

$$CE = P(A/P \, i\%, n) + F \, (A/F \, i\%, n)$$

since it directly converts initial investment and subsequent salvage value to a periodic value. Here,

$$CE = -100(A/P \, 12\%, 5) + 30(A/F \, 12\%, 5)$$

$$= -100(0.2774) + 30(0.1574)$$

$$= -\$23.018 \text{ thousand}$$

as before.

(*b*) Alternatively, *CE* may be computed by algebraically converting

$P(A/P\ i\%, n)$ to $P(A/F\ i\%, n) + Pi$. (The reader should verify that this is an identity.) Then the capital equivalent becomes

$$CE = (P - F)(A/F\ i\%, n) + Pi$$

$$= (-100 + 30)(A/F\ 12\%, 5) + (-100)(0.12)$$

$$= -70(0.1574) + (-100)(0.12)$$

$$= -\$23.018\ \text{thousand}$$

as above. The interpretation of this approach is that of an annuity composed of the replacement amount $P - F$ at time n, and the periodic amount foregone, having invested P at $i\%$.

(c) Finally, the periodic value of the investment may be computed from

$$CE = (P - F)(A/P\ i\%, n) + Fi$$

which is also algebraically equivalent to (a), and which can be interpreted as the periodic worth of the net investment plus the opportunity cost of the salvage value to be received. Here,

$$CE = (-100 + 30)(A/P\ 12\%, 5) + (-30)(0.12)$$

$$= (-70)(0.2774) + (-30)(0.12)$$

$$= -\$23.018\ \text{thousand}$$

as before.

The total annual worth of the example can now be determined by adding the annual revenue and the annual equivalent of the capital investment.

$$AW = 20 + CE = 20 + (-23.019) = -\$3.019\ \text{thousand}$$

As with present-worth and future-worth analyses, the investment is seen to be undesirable at 12 percent, since it is equivalent to an annual "outlay" of $3.019 thousand. By contrast, if annual worth is positive when revenues and costs are jointly considered, the alternative is acceptable.

The periodic-worth approach is entirely consistent with the present- and future-worth methods. This may be seen by converting the present worth for the example, -10.882, to an annuity.

$$AW = -10.882(A/P\ 12\%, 5) = -10.882(0.2774) = -\$3.019\ \text{thousand}$$

as before.

If the alternative under evaluation has recurring amounts that do not constitute an annuity, which is the most general situation, then the periodic worth can be computed by converting the present worth of all the flows to periodic worth, using:

$$AW = PW(A/P\ i\%, n)$$

Rates of Return

The methods mentioned above—present worth, future worth, and periodic worth—all have something in common. That is, all are quantity measures equivalent to a set of flows. Although such measures are sufficient, theoretically, as tools to evaluate the relative economic desirability of alternatives, it is often thought to be useful to have rate measures for judging the acceptability of independent projects. In practice, rate measures are widely used because they provide a sense of comparability to easily comprehensible standards. However, they must be used with caution when comparing alternatives. There are pitfalls to their use, which will be described later on in the chapter.

A simple illustration will serve to distinguish between rate and quantity measures and to point out the reasons for their widespread use. Machines of a certain type may be compared with respect to their annual operating cost or energy consumption under specified operating loads and conditions. These are illustrative of quantity measures. By contrast, the energy consumption of the machines may also be compared by examining their efficiencies, stated as a rate in terms of output/input. These are illustrative of rate measures. Such rate measures suggest greater comparability because the conditions under which the machines operate, including the degree of use, need not be indicated. The use of rate measures in connection with the analysis of time value flows provides a sense of comparability with readily identifiable financial situations. For example, a rate of return of 10 percent can be identified with a bank account earning 10 percent interest per year.

Internal and external rates of return are examples of rate measures as applied to economic decision-making. Both are based on time value models, and are thus related to the methods described above. As with other tools, they must be applied and evaluated properly if correct inferences are to be made. As has been indicated, some pitfalls to the use of rate-of-return measures will be pointed out in this and subsequent chapters. However, at the very least, rate measures can safely be used to indicate whether an alternative, considered singly, is acceptable or not when a minimum acceptable rate of return is required.

Internal Rate of Return

Internal rate of return is a rate measure which is an extension of present- and future-worth methods. It exists when there are flows in opposite directions—that is, inflows and outflows—in which the nondiscounted sum of the positive and negative flows exceeds zero. (In this and subsequent descriptions, inflows will arbitrarily be designated as positive, with outflows as negative.)

Example 3–1 will help to develop the concept of internal rate of return. Recall that the present value of the flows at 12 percent was negative, indicating an undesirable alternative, assuming the earnings potential of funds is 12 percent. This is equivalent to saying that the flow generates an equivalent rate of return less than 12 percent. In contrast, the present worth of the flows of Example 3–1 at $i = 5$ percent is positive, $+10.095$. (The reader should verify this.) This is equivalent to saying that the flow generates an equivalent rate of return of more than 5 percent. To find that rate of return, one must compute a discount rate for which the present worth of outflows and inflows would just equal zero. The internal rate of return is that rate i for which the net present value is zero. Symbolically,

$$-I + \frac{R_1 - O_1}{(1 + i)} + \frac{R_2 - O_2}{(1 + i)^2} + \cdots + \frac{R_n - O_n}{(1 + i)^n} = 0 \qquad (3-1)$$

where R_j denotes inflow (or revenue) for period j, and O_j denotes outflow (or cost) for period j. Investment is designated as I. (For cases with multiple investments, the period associated with each investment can also be subscripted so that a set of investments, I_j, may be considered.)

Further recall that this is the situation referred to in the present-worth section as barely desirable (i.e., where present worth is zero at the criterion rate). In fact, this can be used to explain the meaning of internal rate of return. When the net present value of the flow, as computed from Equation 3–1 for a given value of i, is greater than zero, the flow is "desirable" with respect to i, or the real rate applicable to the flow is greater than i. Conversely, when the net present value is negative, the flow is "undesirable," or the real rate applicable to the flow is less than i. In other words, the internal rate of return is the value of the discount rate for which the net present worth is zero. This can generally be found only by trial and error. Such a procedure can be somewhat tedious when done by hand. However, the use of a computer can simplify this problem significantly.

EXAMPLE 3–1e

A trial-and-error solution done "by hand" is illustrated below. In addition, a computer-assisted method of solution will be described and contrasted in Chapter 5.

To solve for the internal rate of return for the sample problem, the following equation must be solved for i:

$$-100 + \frac{20}{(1 + i)} + \frac{20}{(1 + i)^2} + \cdots + \frac{20 + 30}{(1 + i)^5} = 0$$

A trial-and-error approach might begin with 12 percent for i. It should be noted that each iteration of the trial-and-error approach is equivalent to application of the present-worth method for the given rate, i. Recall that previously the net present worth of the flow above at 12 percent was computed to be $-\$10.882$ thousand, indicating that the internal rate of return is less than the 12 percent of the trial. (This is easy to visualize algebraically, since the net flow can equal zero only if the contribution of the discounted values is increased. In turn, this can be accomplished only if the discounting is redone with smaller denominators, or at a lower rate.) One might next try 8 percent for i, which yields

$$-100 + \frac{20}{(1.08)} + \frac{20}{(1.08)^2} + \cdots + \frac{20 + 30}{(1.08)^5} = + \$0.2717 \text{ thousand}$$

Since this is positive, it indicates that the solution lies above 8 percent. The rationale for this is the reverse for the case at 12 percent. With the knowledge that the net present worth is zero for i between 8 and 12 percent, one could continue to make additional trials to "home in on" the solution. Here, it is clear that the solution is close to 8 percent, and so one might be tempted to try 9 percent. However, if the problem is being done "by hand," at some point the analyst will wish to interpolate to approximate the solution. If one interpolates linearly between 8 and 12 percent, the solution for i is approximated by

$$i = 8\% + \left(\frac{0.2717}{0.2717 + 10.882}\right)4\% = 8\% + 0.097\% = 8.097 \text{ percent}$$

This result compares favorably with a computer solution based on many more trials, which is 8.090 percent. Once again, the substantive result is consistent with what was obtained under present-worth or periodic-worth analysis. That is, the investment is undesirable, since it yields a rate of return of 8.09 percent, whereas the minimum acceptable rate is 12 percent.

The appeal of measuring an investment's attractiveness by a rate should now be apparent. However, although the rate measure above can be compared to the absolute criterion of 12 percent, it cannot readily be compared to those of other alternatives whose rate measures have been computed. Such comparisons must be approached with caution. (Some of these issues will be addressed in more detail in the next section.)

Although the method described above will be referred to in this book as internal rate of return, it is often alluded to differently, depending on the field of application. Expressions such as "return on investment" and "yield" are employed by business and financial analysts.

Problems Associated with Internal Rate of Return

The most noteworthy difficulties associated with internal rate of return are that

1. Ranking of mutually exclusive alternatives in order of their measured rates may not be consistent with their rankings of desirability under present and annual worth, and,
2. Some cash flows have multiple solutions for Equation 3–1, thus confusing the possible interpretation of the computed rates.

Each of these is discussed separately below.

Inconsistency of IRR as Compared to Other Measures   This shortcoming will be explored with the aid of some examples.

EXAMPLE 3–2

Suppose two mutually exclusive cash flow proposals exist. (Mutually exclusive proposals are those for which the choice of one eliminates the others.) The alternatives in question may be described by Cash Flow Diagrams D3–2a and D3–2b.

D3–2*a*

Alternative *A*:

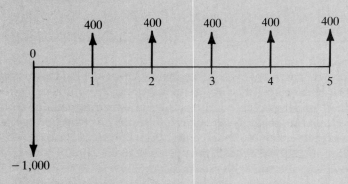

D3–2*b*

Alternative *B*:

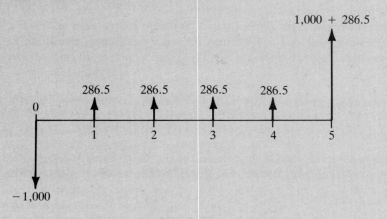

Alternative *A* may be described as a $1,000 investment which provides $400 inflows in each of the following five years. Alternative *B* also has a $1,000 investment, but is followed by five yearly returns of $286.50, and at the end of the last period the $1,000 investment is returned.

The analysis begins with the computation of the internal rate of return for each proposal. For A,

$$-1,000 + 400(P/A\ i\%, 5) = 0$$

must be solved for i. This yields an internal rate of return of 28.65 percent. For B, solving

$$-1,000 + 286.5(P/A\ i\%, 5) + 1,000(P/F\ i\%, 5) = 0$$

for i also gives an internal rate of return of 28.65 percent. Thus, examination of the internal rates of return alone would imply that the two sets of cash flows are equivalent. But are they equivalent? If, for example, capital is worth 10 percent, the equivalent present worth of each alternative may be calculated as

$$PW_A = -1,000 + 400(P/A\ 10\%, 5) = \$516.31$$

$$PW_B = -1,000 + 286.5(P/A\ 10\%, 5) + 1,000(P/F\ 10\%, 5) = \$706.98$$

Clearly, if funds are worth 10 percent, B is the more desirable alternative in that it has a higher present equivalent worth. If present worth is computed for various assumed discount rates, and the results are plotted, the curves would look like the ones in Figure 3–1.

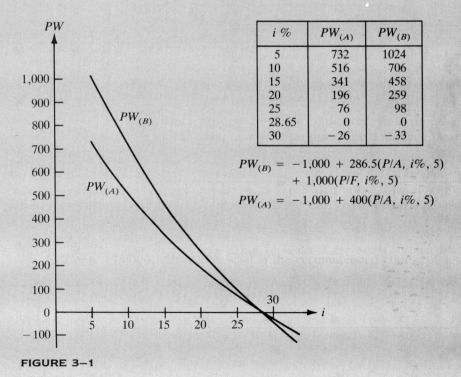

$i\ \%$	$PW_{(A)}$	$PW_{(B)}$
5	732	1024
10	516	706
15	341	458
20	196	259
25	76	98
28.65	0	0
30	-26	-33

$$PW_{(B)} = -1,000 + 286.5(P/A,\ i\%, 5) + 1,000(P/F,\ i\%, 5)$$

$$PW_{(A)} = -1,000 + 400(P/A,\ i\%, 5)$$

FIGURE 3–1

From the curves one can see that at discount rates other than 28.65 percent the two alternatives would not be viewed as equivalent. If funds are worth less than 28.65 percent, *B* has higher present worth, and if funds are worth more than 28.65 percent, *A*'s present worth is greater. Thus, the fact that both alternatives have the same internal rate of return should not lead to the conclusion that both alternatives are equivalent.

EXAMPLE 3–3

This example also serves to indicate the shortcomings associated with using the internal rate of return for ranking mutually exclusive alternatives. Consider alternatives *X*, *Y*, and *Z*, in Cash Flow Diagrams 3–3*a*, *b*, and *c*.

D3–3a

Alternative *X*:

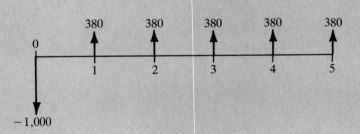

D3–3b

Alternative *Y*:

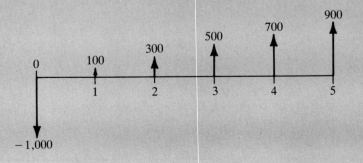

D3–3c

Alternative Z:

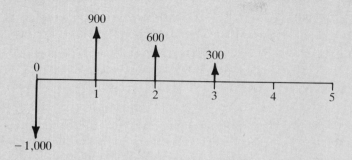

X can be compared to Y by computing the internal rate of return for each, as well as present worth at a variety of interest rates. The reader should verify, employing methods used previously, that the internal rates of return for X and Y are 26.07 and 28.83 percent, respectively. The reader should also verify that the present worth of X and Y at discount rates of 10, 15, 20, and 25 percent are as given in Table 3–1.

TABLE 3–1   Present Worth vs. i for Alternatives X and Y

	i	10	15	20	25
PW	X	$440.5	$273.82	$136.43	$ 21.93
	Y	751.44	490.24	280.29	109.63

These results, when plotted as in Figure 3–2, indicate that rankings of the alternatives under internal rate of return are the same as under present worth. That is, X has a lower internal rate of return than Y, and similarly, the present worth of X is always lower than Y.

However, it is important to note that rankings, as evidenced by comparing X to Y, are not always consistent. Consider the comparison of Y to Z. Here the reader should verify that the internal rate of return for Z is 45.44 percent, substantially higher than that for Y, previously reported to be 28.83 percent. Does this imply that Z is always ranked higher by the present-worth yardstick? This question can be answered by computing the present worth of the two alternatives, Y and Z, for various discount rates, as was done for X and Y. The results of these computations are shown in Table 3–2 and Figure 3–3.

These curves show that if funds are worth less than 17 percent, Y has higher present worth than Z, in spite of a higher internal rate of return for Z. Thus, if funds are valued at, say, 10 percent, Y would be favored, and the internal rate of return would have given an improper result.

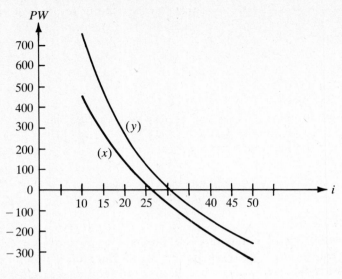

FIGURE 3–2

TABLE 3–2 Present Worth vs. *i* for Alternatives Y and Z

	i	10	15	20	25
PW	Y	$751.44	$490.24	$280.29	$109.63
	Z	539.4	433.55	340.28	257.6

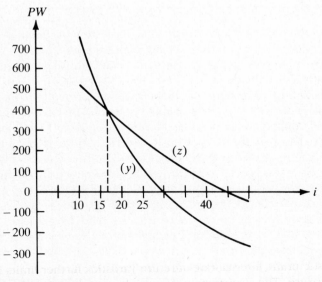

FIGURE 3–3

These examples indicate that, although internal rate of return may be used to judge the acceptability of a project in relation to a criterion, it is not safe to rely on it for ranking mutually exclusive alternatives. In the latter case, the correct approach would be to rely on other measures, such as periodic (annual) worth, or present worth. The next chapter will show how incremental internal rate of return may be safely used.

Multiple Solutions for IRR When the periodic cash flows in the discount equation for the internal rate of return (Equation 3–1) change in sign more than once, then there may be multiple roots for i. In other words, the graph of present worth versus i may look like Figure 3–4, as opposed to Figures 3–2 and 3–3, which characterize the examples presented thus far.

EXAMPLE 3–4

This example illustrates the possibility of multiple solutions for IRR. Suppose the following cash flows are forecasted:

D3–4

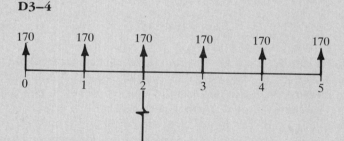

The reader should note that there are two changes in sign for the cash flows; going from time 1 to time 2, the cash flows change from $+170$ to -830, and from time 2 to time 3, the cash flows change from -830 to $+170$. (Recall that, by contrast, the previous examples involved cash flows that changed in sign only once.) The existence of two sign changes in the cash flows signals the possibility that multiple roots exist. If the present worth of the flows in this example is computed for various values of i, the graph shown in Figure 3–4 may be plotted. It is apparent from the graph that two roots exist. The reader may verify that $i = 4.7$ and 37.0 percent both satisfy Equation 3–1 for the given flows.

This characteristic of the internal-rate-of-return yardstick further limits its use as a comparative measure. The occurrence of multiple roots requires further as-

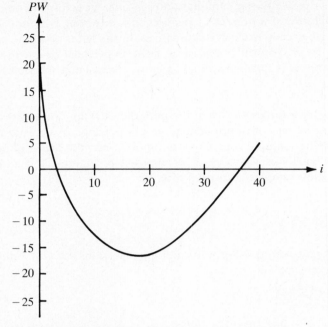

FIGURE 3—4 Present Worth vs. *i*, Example 3—4

sumptions in order to judge the flow in question when comparing it to others. These assumptions concern the value placed on capital. In general, if one can make an assumption concerning the relevant discount rate, the most direct analytical approach is to use present or annual worth as a yardstick.

It should be pointed out that, in practice, the existence of multiple roots is not the most likely pattern encountered by analysts. Most applications of internal rate of return give results described by Figures 3–1 to 3–3. However, as time value models become more widely used, more instances of multiple roots are encountered.

External Rate of Return

External rate of return is more easily computed than internal rate of return, but the assumptions underlying its use are different. Although the internal rate of return is independent of the value placed on capital, the external rate of return explicitly uses such an assumption in its rate measure computation.

The approach involves discounting or valuing the outflows and inflows at a criterion rate of interest, typically the MARR or the rate at which the organization's funds can alternatively be invested. The outflows are discounted to the present, and the inflows are projected to their future value at the end of the study period in question. Then the rate for which these present and future amounts are equivalent is computed. That rate is the external rate of return. Symbolically,

$$\left[-I - \frac{O_1}{1+r} - \cdots \frac{O_n}{(1+r)^n} \right] (1+i)^n$$

$$+ \left[R_1 (1+r)^{n-1} + \cdots + R_n (1+r)^0 \right] = 0 \tag{3-2}$$

where I, O_j, and R_j are as before, r is the given criterion interest rate, and, here, i is the external rate of return. Equation 3–2 is in a form that permits straightforward solution for i. The term $(1+i)^n$ can always be isolated, and then the nth root gives $1+i$, which in turn yields i. Note that this approach always yields a single positive real root (in contrast to the procedure for the internal rate of return, which may yield multiple roots).

This is the solution to Example 3–1 using the external rate of return, in which investment was 100, inflows were 20 per period with a salvage value of 30 at time 5, and zero periodic outflows:

$$\left[-100 \right] (1+i)^5 + 20(1.12)^4 + \cdots + 50(1.12)^0 = 0$$

which yields,

$$(-100)(1+i)^5 + 157.057 = 0$$

$$(1+i) = (1.57057)^{1/5}$$

$$\text{or } i = 9.445 \text{ percent}$$

Note that the external rate of return obtained is not identical to the internal. This is generally the case. When $r > i$ for the alternative, as it is in this example, then the external rate of return is higher than an internal rate computed for the same flows. This results from the fact that the generated flows are valued at a rate higher than i. When $r < i$, the reverse holds. The only instance in which internal and external rates of return would turn out to be equal is when the criterion rate (r) used is the same as the value obtained for the internal rate of return for that set of time value flows.

Once again, the substantive recommendation as applied to Example 3–1 as an independent project is the same. Since $i = 9.445$ percent, which is less than 12 percent, the investment is not acceptable for the required earnings rate.

The application of external rate of return to Example 3–4 shows how it may eliminate confusion associated with the incidence of multiple roots for internal rate of return. Using 3 percent for the MARR, one must solve the following for i:

$$1,000(P/F \; 3\%, 2)(1+i)^5 = 170(F/P \; 3\%, 5) + 170(F/A \; 3\%, 5)$$

This yields $i = 3.16$ percent. The interpretation is that if the proceeds of this cash flow are valued or reinvested at 3 percent, the set of flows may be thought of as earning 3.16 percent and is, therefore, acceptable under the given criterion. Of course, the same conclusion may be reached by using the present-worth yardstick if the flows are discounted at 3 percent. The reader should verify that the present worth of the indicated flows at 3 percent is positive and, thus, that the project would be viewed as acceptable.

Exercises

3–1 Engineers at a plant that manufactures wood furniture are considering the installation of a system to reclaim and utilize the sawdust that is a by-product of the process. It includes a preheater for the hot water system that will burn the dust safely, ductwork throughout the plant, and a set of blower motors to propel the dust and deliver it to the preheater. The system as currently designed requires investment of $225,000. It is expected to save $40,000 per year in conventional fuel, as well as reducing dust cleaning and disposal. The system is expected to serve for 15 years, after which it is considered to be worthless. What internal rate of return would these savings provide?

3–2 After landing a job, you consider buying a new car for $12,000, planning to hold it for 3 years. After the 3 years, you expect it would have a market value of $7,000. If annual operating costs are expected to be $1,500, $1,700, and $1,900, respectively, what would be the equivalent annual cost of owning and operating the vehicle? Assume that funds are worth 12 percent.

3–3 *a*. Suppose you purchase a $1,000-face-value industrial bond from your broker for $920. It bears a 10-percent coupon rate, which means that it will pay you $50 every 6 months. Assume that your first such interest payment will occur exactly 6 months from now. If the bond matures in exactly 5 years (10 interest payments), at which time you receive the last interest payment of $50 as well as the principal of $1,000, what would the internal rate of return be for the entire bond transaction? (Use 6 months as the relevant unit period.)

    b. Assuming you put the interest proceeds into your 5-percent checking account (treating that as 2.5 percent per 6 months), what is the external rate of return? (In other words, here the MARR is 2.5 percent per period.) Why does this differ from the result in *a*?

3–4 Having insufficient production capacity to satisfy the market for a newly developed device, your firm is considering a licensing agreement with another firm. It is expected that the terms of the agreement will generate revenues as follows:

Year	1	2	3	4	5	6	7	8	9	10
Revenue ($000)	50	60	70	100	100	100	90	80	70	60

If funds are worth 10 percent, what is the present worth of this flow of royalties?

3–5 Suppose, for simplicity, that a bank's mortgage department has a portfolio of 3 mortgages that it currently wishes to evaluate. For that purpose, the original interest rates on the mortgages are not relevant. What is important is the monthly amount being repaid on these mortgages, the number of payments remaining on each, and the current opportunity cost of equivalent capital. Suppose that the latter is 15 percent per annum, or 1.25 percent per month. The characteristics of the 3 mortgages are:

Number	Monthly Payments Remaining	Amount of Payment
1	80	$310
2	230	$620
3	189	$480

In order to establish the current market value of this portfolio, compute the present worth attributable to the 3 mortgages, assuming the MARR is 1.25 percent per month.

3–6   *a.* Your firm is considering the purchase of a new computer system. The total cost, including peripherals, would be $350,000. The maintenance contract would amount to $20,000 per year. A 5-year study period is being used, at the end of which time the system is thought to be worth $100,000. What would the equivalent annual cost of purchase and maintenance be? Assume that the MARR is 16 percent.

    b. A lease alternative is presented by the vendor in which maintenance is included. For the 5-year period the annual lease payment (prepaid) would be $110,000. What would the equivalent annual cost be for this alternative? (Remember that a prepaid lease means that the payments occur at times 0–4.)

3–7 The owner of an existing office building is investigating the prospects of conserving electricity. The building is heated electrically with thermostats in each leased area. A computer-controlled system is being examined which will permit the building manager to set and control the heat for a given temperature in each area, as opposed to giving the tenants the ability to set the heat levels themselves. It is felt that this will reduce waste associated with uncontrolled access to the thermostats. The system will cost $75,000 to install and will last for 10 years, at which time it will have a salvage value of $10,000. If the MARR is 10 percent, what annual saving is needed for the 10-year period if it is desired that the external rate of return be 12 percent?

3–8 Suppose a road project requires an initial investment of $6.5 million. Annual maintenance and repairs will amount to $25,000 for the first 5 years, $30,000 for the next 10 years, and $36,000 for the next 5 years. In addition a road resurfacing is anticipated 10 years after the road is constructed (not included in the figures for maintenance and repair) and is expected to cost $500,000. If money is worth 5 percent, what is the equivalent present worth of all road expenses? What would the equivalent annual annuity be, over 20 years?

3–9 Your firm developed a new product, investing $2 million. Its sales, less out-of-pocket expenses, for 5 years are: $300,000, $400,000, $500,000, $300,000, $250,000. At the end of the fifth year, the productive assets associated with the product are disposed of for $750,000. Compute the internal rate of return on the firm's investment.

3–10 Assuming the MARR is 10 percent, compute the external rate of return associated with the investment in Problem 3–9.

3–11 Your firm develops a new process, the investment in which is computed to be $40,000. Since it does not have the productive capacity to exploit the development now, it licenses another firm to do so. Revenues from the license would be $10,000 per year for the next 3 years, and $12,000 for each of the following 2 years. After that point the license is thought to be worthless since the process would need to be substantially modified to avoid obsolescence. What would the internal rate of return be for the flows above? If revenues are valued at 12 percent, what would the ERR be for those flows?

3–12 Your firm is planning to participate in an overseas development project. Participation appears to be profitable, but the timing of the required financial outflows is heavily weighted toward the beginning, while the returns are weighted more toward the end. These flows have been summarized and appear as follows:

Year	1	2	3	4	5	6	7
Amount ($000)	– $100	– $40	– $20	$30	$80	$100	$30

a. If funds are worth 16 percent per annum, what would be the annual worth of these flows (over the same time horizon)?
b. What is the internal rate of return for these flows?
c. What is the external rate of return?

3–13 Your group has developed a new process which requires an initial investment of $600,000. It will generate increasing net revenues, as follows: $200,000 in the first year, increasing by $50,000 per year for 4 years (5 years in total). In addition, the process will require substantial refitting in year 3, at a cost of $400,000. Draw a cash flow diagram for this project and compute its internal rate of return. If the criterion for adequacy of investment is 15 percent, is the project justifiable?

3–14 Compute the external rate of return for the project above (assuming an applicable discount rate of 15 percent).

3–15 Your firm is designing, fabricating, and assembling equipment for a new process that it will put into production. The cost of this investment will occur over the coming 2 years, but will be represented as incident at time 0 and time 1. The amounts are $175,000 at time 0 and $125,000 at time 1. The process is expected to generate revenue over the succeeding 5 periods as follows:

Period	2	3	4	5	6
Revenue ($000)	30	75	75	100	100

Finally, it is anticipated that the process will have a salvage value of $50,000 at the end of the 6-year period.

a. If funds are worth 15 percent per annum, what would the present worth of this set of flows be? What does this indicate about the economic attractiveness of the process?

b. What would the internal rate of return be for this set of flows? Does this IRR give a consistent interpretation with present worth?

c. What would the equivalent annual worth of these flows be if funds are still worth 15 percent and the annuity runs from time 1 to time 6? Is this result consistent with a?

d. Suppose, for argument's sake, that one wishes to compute an equivalent annuity that is deferred 1 period as compared to the one in (c) above (i.e., it runs from time 2 to time 6). What is the periodic annual amount now if funds are still worth 15 percent?

Selected References

Arrow, K. and D. Levhari, "Uniqueness of the Internal Rate of Return with Variable Life of Investment," *Journal of Economics*, Vol. 79, September, 1969, pp. 560–566.

Au, T. and T. Au, *Engineering Economics for Capital Investment Analysis*, Boston: Allyn and Bacon, 1983.

Bernhard, R. "'Modified' Rates of Return for Investment Project Evaluation: A Comparison and Critique," *The Engineering Economist*, Vol. 24, No. 3, Spring, 1979, pp. 161–167.

Fleischer, G. "Two Major Issues Associated with the Rate of Return Method for Capital Allocation: The 'Ranking Error' and 'Preliminary Selection,'" *The Journal of Industrial Engineering*, Vol. 17, No. 4, April, 1966, pp. 202–208.

4

Comparison of Alternatives

Introduction

The two preceding chapters have introduced time value concepts and yardsticks usable for measuring the worth of independent alternatives. In this chapter, those principles will be utilized to explore how mutually exclusive alternatives (defined in Chapter 3) may be compared. In some instances, the application of time value concepts is quite direct, almost "cook-bookish." In other cases, some imagination must be used to adapt the principles in a way that gives the analyst the insights he needs in order to make sound decisions.

Before proceeding, it is useful to review the general dilemma associated with economic decision-making. When two or more alternatives present themselves, the decision analyst often finds that some attributes of one alternative are superior to those of the others. However, other alternatives may have different attributes for which they exhibit superiority. It is even possible that an alternative may not have any one attribute for which it is viewed as most desirable; but it may be "almost best" for so many attributes that it may be found superior overall, a sort of "compromise candidate." It is these situations that pose a decision-making dilemma: how to choose? The response developed is a conceptually sound analytical treatment based upon time value concepts. Of course, it should be recalled that modeling the attributes of an alternative cannot capture all of its relevant details. Thus, modeling should be viewed as an aid to sound decision-making and should be thought of as just one part of the overall process of wise, human decision-making.

An exception to the foregoing conditions sometimes occurs but constitutes an obvious case not requiring much in the way of analysis—namely, when one alternative is superior in every respect. In such an instance, no effort at decision-making is required, only action!

Definitions

In order to proceed with economic modeling using time value concepts, certain commonly used attributes of alternatives will be defined. Once that has been done, the text will outline a variety of cases that can be used to describe many economic decision situations.

Investment Cost

Investment cost is the amount needed to acquire and install the asset. It generally occurs at the beginning or at least very early in the life of the asset or study period. Usually it is modeled as occurring at time zero and would be depicted as on Cash Flow Diagram D4–1.

D4–1

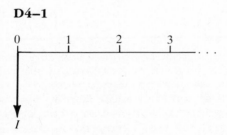

Clearly this is a simplification of reality in many cases, since actual investment outlays may take place over several periods. Construction expenses are a good example of this. In fact, major construction projects such as a nuclear power plant may take several years or more to complete. Again, although the pattern of construction (investment) expense may not be exactly as depicted, Diagram D4–2 may serve as a good model.

D4–2

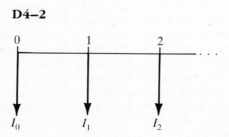

Here I_j symbolizes investment for period j. Sometimes supplemental investment may be anticipated at an intermediate stage of a project—for example, if the capacity of a manufacturing plant is to be expanded in the future, or if a major repair, modification, or subsystem replacement is normally anticipated. The latter might occur when planning the construction of a roadway, where a resurfacing could reasonably be expected during its future useful life. Such situations can be modeled as in Cash Flow Diagram D4–3, where the second investment occurs at period k.

D4–3

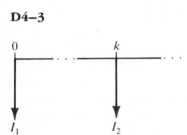

Algebraically, investment costs will be treated as negative amounts, to be consistent with the general treatment of outflows as negative values.

Salvage Value

Salvage value is a realistic estimate of the value to which an asset will decline after some period of use. It is almost universally true that productive assets, other than land, decline in value as time passes, through use or obsolescence. (Assets other than "productive," such as art works and other "collectibles," are not included here, because they are typically exceptions to this rule and are not of concern in this book.)

Salvage values may be estimated from sales data on similar assets, previous experience with trade-in values, historical patterns of decline in value for similar assets, or other sources relating to the anticipated market value of the asset at a future time.

Economic Life

This is a numerical value in units of time which represents the usable lifespan of the asset from the perspective of the analysis being conducted. It helps relate investment cost and salvage value over time to permit the time value model to be completed as in the Cash Flow Diagram D4–4.

D4–4

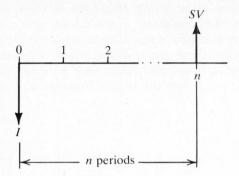

More than most other components of economic modeling, the estimate of economic life is somewhat arbitrary. In other words, although the assigned economic life should be realistic, it may be chosen to suit the analysis. This may seem contradictory, but it really is not. What is important is that economic life and salvage value bear a realistic relationship to one another. Taken together, the pair must make sense. Consider, for example, the purchase of a truck. Clearly its economic life may be estimated within broad limits rather than as a specific value. However, salvage value may be combined with other factors to assign economic life for the analysis. The shorter the value of economic life chosen, the greater the salvage value, and vice versa. In other words, a particular choice of economic life does not signify a guess at the maximum length of time the asset can be used. Nor does it necessarily coincide with the depreciable period (to be discussed later), although it certainly may do so.

Revenue

Revenue is an estimate of the periodic income associated with an alternative. In this book revenues will be treated like other inflows—namely, as positive numbers—although the reader should bear in mind that this approach is arbitrary. Revenue may be based on sales or rentals, and may take on any conceivable time pattern. In other words, it may be modeled as an annuity, a uniformly increasing or decreasing amount (gradient), uniformly increasing or decreasing percentage gradient, or any other pattern that can reasonably be expected. (More will be said of this in Chapter 11, where life cycles are discussed.)

The reader should bear in mind that revenue is not always readily identifiable in decision analysis. It tends to be true that the larger the scope of the entity being analyzed, the more likely it is that revenue can be identified. To appreciate this point, consider two illustrations at opposite extremes. One involves the evaluation of a planned manufacturing establishment. In conducting the analysis, the projections for salable output of the plant, along with the prices that are expected to prevail, yield revenue estimates. The other illustration is evaluation of the relative desirability of a new machine. In undertaking this analysis, it may be impractical or even impossible to identify the revenue associated with the use of the machine. This is often due to the fact that the output of the machine is quite variable with respect to the range of products, schedules, and other factors, all of which influence its performance.

Another issue contributing to the possible inability to identify revenue is the impracticality of imputing value to the output of an asset which is not a "final product," but is, instead, an intermediate one.

Finally, there are many instances where flows not normally thought of as revenues may be treated as such. For example, if an investment alternative is under consideration for the cost savings it can provide, the savings produced may be viewed as an economic revenue for purposes of analysis.

This chapter will explore several decision situations. In some of them revenue is estimatable in a straightforward way. Where that is impossible, the reader should carefully observe the approach taken to overcome this difficulty.

Costs

Costs will be treated here as real observable or projectable periodic outflows. They will be assigned negative values, although, as with revenues, this approach is purely arbitrary. The components to such costs can be categorized, in the main, as:

Labor   Wages and salaries of operative personnel that are directly related to the alternative.

Energy   The energy costs of the alternative.

Maintenance   The costs of preventive, or scheduled, maintenance, as well as the anticipated costs of remedial maintenance. The latter can often be projected from experience with similar assets. Maintenance costs include labor, materials, parts, and the like, and may include the cost of downtime or lost production, depending on the context of the study.

Taxes, insurance, misc.   Here, taxes do not include income taxes, which will be discussed in a subsequent chapter. They do include periodic taxes such as property taxes or "use" taxes related to the alternative in question. Property taxes are those assessed by a governmental entity, based on the value of the asset. Thus, they are usually estimated as a percentage of asset value. For study purposes, insurance costs are usually estimated in a similar way. Other secondary costs peculiar to an alternative that can be estimated as a percentage of first cost are also usually included here.

Overhead   Often there are many costs not directly attributable to the operation of the asset, such as heat, light, and managerial supervision. It is not the existence of such costs that is the issue, but rather the magnitude of such costs attributable to the alternative under study. Various means are commonly employed to estimate overhead. These will be discussed in another chapter. In this chapter, if overhead is to be considered, it will be expressed as a percentage of some other cost(s).

Alternatives with Different Economic Lives—Revenues and Costs Estimable

Given the definitions above, it is now possible to delineate the comparison of mutually exclusive alternatives. This will be done by categorizing the various cases that the analyst is likely to confront.

The first such case is the most general, with alternatives thought of as having different economic lives but revenues and costs that can be estimated. This may be referred to as the analysis of revenue-enhancing projects. Situations where revenue is estimatable tend to occur for larger projects, or where the project may be viewed as enclosed as a distinct economic entity, providing measurable and evaluatable service, while entailing measurable and evaluatable costs. Such is the case when a new factory, power plant, or process is being considered, perhaps in several different configurations. To analyze this case, consider Example 4–1.

Two machines are being considered for purchase. Each is deemed capable of performing the required functions but only one will be chosen. Their estimated economic parameters are given below. (Normally, the estimating process will yield figures with fairly consistent degrees of numerical rounding. The data in Table 4–1 contain what may appear to be an overly precise value for labor costs. This was done to create an example with certain numerical properties and for the purpose of simplifying subsequent computation.)

TABLE 4–1 Two Investment Alternatives

	A	B
Investment cost	$50,000.	$65,000.
Economic life	5 yrs.	6 yrs.
Annual revenue associated with machine operation	$20,000.	$20,000.
Annual costs associated with machine operation:		
Labor	$5,627.70	$5,190.35
Energy	$1,000.	$1,200.
Maintenance	$500.	$600.
Taxes, insurance, misc.	$500.	$650.
Total annual costs	$7,627.70	$7,640.35
Salvage value	$5,000.	$6,000.

The firm operates on the assumption that the opportunity cost of capital, or MARR, is 9 percent.

As the data indicate, B requires a greater investment than A. The equality of the annual revenue attributable to each indicates that their productive capacity is about equal. The economic life of B is longer than A, and its anticipated salvage value is a bit higher. Labor costs for B are lower than A, which is not an uncommon pattern. (The economic trade-offs involving special-purpose machinery often include higher purchase price for a concomitant reduction in labor costs, due to less actual labor required, or labor at a lower skill level and a consequently lower cost.)

Energy, maintenance, taxes, and insurance are higher for B, which is also frequently the case for more expensive and specialized equipment.

The overall result for this example is that B, the alternative requiring greater investment, yields about the same revenue and annual cost but has greater economic life. Thus, the economic decision on which should be chosen revolves about the issue of whether the greater economic life of B is worthwhile in relation to its higher investment requirement. These competing factors are what make this situation akin to a classical decision dilemma. (In general, the trade-off is most frequently between higher investment cost and lower operating cost, although there may be other competing factors.) The time value models of Chapter 2 and the yardsticks developed in Chapter 3 will help to shed light on the problem and aid in reaching a decision.

This problem will be analyzed using the periodic-worth yardstick. As indicated, Example 4–1 poses the difficulty of comparing situations encompassing different time horizons. The solution is to use a quantity measure which takes into account the difference in time periods. Such a measure is periodic worth. It accommodates alternatives with different lives because it allows the computation of an economic measure pertaining to a convenient unit period, which for this problem would be one year.

EXAMPLE 4–1

Thus, one may compute the annual worth for A, which is the annuity equivalent to all projected inflows and outflows over the study period.

$$AW_A = [-50 + (20 - 7.6277)(P/A\,9\%, 5)$$
$$+ 5(P/F\,9\%, 5)](A/P\,9\%, 5)(\text{in thousands})$$
$$= \$353.14$$

For B one may obtain the annual worth from

$$AW_B = [-65 + (20 - 7.64035)(P/A\,9\%, 6)$$
$$+ 6(P/F\,9\%, 6)](A/P\,9\%, 6)(\text{in thousands})$$
$$= -\$317.06$$

If the MARR is 9 percent, A is recommended since its annual worth is positive and higher than B.

The results for Example 4–1 may be generalized as follows: In cases where more than one revenue-enhancing alternative exhibits positive annual worth, they may be ranked in order, with the highest being the most desirable. Where no alternatives show positive annual worth, the substantive decision to be made revolves about other issues. In most instances, such a result would indicate that none of the alternatives is justified. In others it may be considered necessary to choose at least one of the available alternatives. In such cases, the alternative with highest (or least negative) annual worth may be indicated.

It should be pointed out that an implicit assumption in this analysis was that capital can be invested for at least a 9 percent return during year 6 if *A* is chosen. Or, stated alternatively, if *A* is chosen, service of a like sort can be purchased for year 6 which will provide an annual worth of $353.14 in that year. This assumption allows for comparability between *A* and *B*, since a common time horizon, namely 6 years, then exists.

Alternatives with Equal Economic Lives—Revenues and Costs Estimatable

This revenue-enhancing situation is clearly a special case of the preceding one. It is also simpler, owing to the more natural comparability provided by equal economic lives. It may be treated in exactly the same way as the preceding case—namely, by use of periodic (or annual) worth—but a present-worth or future-worth perspective may also be used. This is due to the fact that the "lump sum" equivalence provided by present and future worth are comparable when the alternatives in question have the same economic life. In addition, an incremental internal rate of return may be used. This will be explained later, in Table 4–2a and b. No illustration for this case will be explored here, since it is a relatively straightforward adaptation of the previous case. Ample opportunity for such analysis will be presented in the exercises at the end of the chapter.

Alternatives with Equal Economic Lives—Costs Estimatable but Revenues Not (or Not Applicable)

The reader may be curious about what sort of industrial situation can be described with revenues not estimatable. It is actually quite common, and tends to occur for relatively small projects that are subunits of larger entities. This is essentially a further special case of our original one but, before proceeding with its analysis, we will examine the conditions that typically give rise to it.

Consider, for instance, the purchase of a new machine in a large plant. The machine in question may be capable of performing a variety of jobs, at different rates, involving different products, etc. That variation alone might prevent a satisfactory estimate of revenue attributable to the project. However, further estimating difficulties are caused by the fact that such a machine typically performs service which is only a part of the "value added" process of the entire establishment. Thus, the revenue attributable to it is only a part of the revenue associated with the product as a whole. The impossibility or impracticality of estimating the value added attributable to the machine in question precludes any attempts to approximate revenue.

In other instances, revenue may simply not be a relevant issue, as where the approach is one of cost minimization. This may occur when revenues are deemed

TABLE 4–2   Three Investment Alternatives

	A	B	C
Investment cost	$10,000	$15,000	$21,000
Economic life	5 yrs.	5 yrs.	5 yrs.
Annual costs associated with machine operation:			
Labor	$32,000	$30,300	$27,900
Power	$2,000	$2,200	$2,500
Maintenance	$600	$700	$800
Taxes, insurance, misc.	$400	$600	$800
Total annual costs	$35,000	$33,800	$32,000
Salvage value	0	0	0

the same under different alternatives, or when different alternatives are deemed to provide essentially equivalent service or output. Cost minimization is thus often used as the viewpoint for analysis in design situations.

The following situation is presented as a basis for analysis of this case: Three machines are being considered for a certain function. Each can perform the required work, although their economic characteristics, given in Table 4–2, are different.

The alternatives are listed in order of increasing investment requirements. *B* and *C* require greater investment but have lower labor costs, which, even when combined with slightly higher taxes, maintenance, and power, result in lower aggregate operating costs per year. The question, then, is whether the increased investment is warranted, and, if so, should *B* or *C* be chosen? It is assumed that the service of only one of the machines is required and that the organization has available the $21,000 needed for the most expensive alternative.

These alternatives will be analyzed using present-, future-, and annual-worth yardsticks, along with an incremental-rate-of-return analysis. Before doing so, it should be pointed out that in this illustration the costs for each alternative over the five-year period form an annuity. This is a special case, not ordinarily occurring in practice. The reader should study the following analysis not only with the given example in mind but with a view to extending the same analysis to situations not described by annuities.

Present Worth

Since the economic lives of the alternatives are the same, a lump-sum equivalent value such as present worth can be computed for each of them and affords a comparable yardstick. Thus,

$$PW_A = -10,000 - 35,000(P/A\,9\%,5) = -\$146,138$$

to the nearest dollar. Similarly,

$$PW_B = -15,000 - 33,800(P/A\,9\%,5) = -\$146,470$$

$$PW_C = -21,000 - 32,000(P/A\,9\%,5) = -\$145,469$$

From this perspective, alternative C is best since the present worth of its costs is least negative. (Had only the magnitude of the costs been considered, the same conclusion would have been drawn, since the present worth of costs for C would have been least.) If a complete ranking is of interest, A is next in desirability and B the least desirable.

It should be pointed out that the arithmetical differences between the alternatives is not great in this example. The question of whether the differences are significant is of practical moment and is treated elsewhere.

Future Worth

Comparison of these alternatives using their future equivalent worths is conceptually similar to that for present worth. In most instances the future worth would be computed for the end of the economic lives, here period 5. Thus,

$$FW_A = -10,000(F/P\,9\%,5) - 35,000(F/A\,9\%,5) = -\$224,851$$

$$FW_B = -15,000(F/P\,9\%,5) - 33,800(F/A\,9\%,5) = -\$225,363$$

$$FW_C = -21,000(F/P\,9\%,5) - 32,000(F/A\,9\%,5) = -\$223,822$$

The rankings and conclusions are identical to those under present worth. In fact, for each of the alternatives

$$FW_i = PW_i(F/P\,9\%,5)$$

Annual Worth

Annual worth is the other lump-sum measure which can be used to compare these alternatives. Since only costs are involved, this is sometimes referred to as equivalent uniform annual costs. This may be computed in various ways, as described in Chapter 3. One way is to compute the present worth and then to convert that amount to an equivalent annuity. Since the economic life is 5 years for each of the alternatives, the approach would be to compute the annuity over that period. Thus,

$$AW_A = [-10,000 - 35,000(P/A\,9\%,5)](A/P\,9\%,5) = -\$37,571$$

$$AW_B = [-15,000 - 33,800(P/A\,9\%,5)](A/P\,9\%,5) = -\$37,656$$

$$AW_C = [-21,000 - 32,000(P/A\,9\%,5)](A/P\,9\%,5) = -\$37,399$$

Once again, the conclusions are the same as those reached previously, given the identical rankings under annual worth.

Incremental Internal Rate of Return

In general, an incremental analysis seeks to compare two or more mutually exclusive alternatives by examining the difference in investment, or capital, required and weighing the incremental benefits received for that incremental investment. This approach may be used with any of the yardsticks already developed as well as with the benefit-cost method. Accordingly, the alternatives should be ranked in order of increasing initial investment before beginning incremental analysis. The idea is that the investment of each discretionary unit of capital would be made only if it is warranted by comparison to the MARR, if capital is to be used profitably. This principle will be illustrated using the information in Table 4–2.

Since this case involves alternatives with no specified revenues, positive internal rates of return do not exist for each alternative per se. This results from the fact that the discount equation

$$-I - \frac{C_1}{(1 + i)} - \frac{C_2}{(1 + i)^2} - \cdots - \frac{C_n}{(1 + i)^n} = 0$$

has no solution for $i > 0$.

Nonetheless an incremental analysis using rate of return can be made, which yields important insights. An incremental analysis involves an examination of the added investment needed to acquire B as opposed to A, as compared to the cost reduction thereby obtained. From this perspective the cost reduction obtained can be conceptually treated as an implied revenue. This revenue can be measured against (incremental) investment by computing the internal rate of return on the incremental investment. Then the desirability of the incremental investment can be judged against the MARR criterion, as is demonstrated below. For the data in Table 4–2, the incremental values are as shown in Table 4–2a.

TABLE 4–2a

	A	B	C
Investment	$10,000	$15,000	$21,000
Annual cost	$35,000	$33,800	$32,000
		B − A	*C − B*
Incremental investment		$5,000	$6,000
Incremental cost (implied revenue)		$1,200	$1,800

Column $B - A$ indicates that if one chooses B rather than A, an incremental investment of $5,000 occurs. For that added investment, one obtains an economic savings or an implied incremental revenue of $1,200 per year for 5 years. Once again the question is whether this investment is warranted if the opportunity cost of funds, or MARR, is 9 percent. This can be answered by computing the internal rate of return on the incremental investment as follows: Solve the equation below for i:

$$-5,000 + 1,200(P/A\ i\%, 5) = 0$$

The reader should verify that the solution is $i = 6.4$ percent. Since this is below the MARR of 9 percent, B is not deemed desirable. Although this substantive result is the same as was obtained previously with lump-sum yardsticks, a somewhat different insight is achieved with an incremental-rate-of-return analysis. This shows that *B* is undesirable because the added capital required cannot generate sufficient cost savings to earn at least the MARR requirement. It earns only 6.4 percent; thus, capital can be put to better use elsewhere.

To continue, *C* must next be evaluated. The first impulse of the novice might be to compare the incremental rate of return of *C* vs. *B*, since previously *B* was compared to *A*. However, this would be incorrect here. The reason is that *B* has already been found undesirable, and thus, it makes no sense to compare *C* to it. Rather, *C* should now be compared to *A*, the last desirable alternative of those previously considered in order of increasing investment. To do so requires a modification of the incremental figures given before, as shown in Table 4–2b.

TABLE 4–2b

	A	B	C
Investment	$10,000	$15,000	$21,000
Annual cost	$35,000	$33,800	$32,000
		B − A	*C − A*
Incremental investment		$5,000	$11,000
Incremental cost (implied revenue)		$1,200	$3,000

The incremental investment associated with *C* as compared to *A* is $21,000 − $10,000, or $11,000. The annual savings associated with *C* as compared to *A* is $35,000 − $32,000 or $3,000. Now it remains to evaluate the merit of investing an additional $11,000 in order to achieve a $3,000 annual savings for five years. To do so, the following equation must be solved for *i*:

$$-11,000 + 3,000(P/A\ i\%, 5) = 0$$

This yields a rate of return of 11.32 percent, which is greater than the MARR. Thus, it would seem that here the incremental investment is justified.

The importance of comparing *C* to *A* in this case, as well as generally comparing an alternative only to other desirable ones in an incremental-rate-of-return analysis, may also be appreciated by the following argument: Suppose *B* is deemed undesirable from an analysis like the one undertaken above. It is possible that the incremental investment of *C* compared to *B* seems justified even though it may be that the incremental investment of *C* compared to *A* is not, in fact, desirable. In other words, an undesirable choice, *C*, might be erroneously recommended if compared to an inappropriate alternative.

In general, the root of the discount equation of the incremental flows between two alternatives, referred to as the incremental internal rate of return, is identical to the intersection discount rate for the two curves of present worth vs. the discount rate for each alternative cash flow.

Alternatives with Unequal Economic Lives—Costs Estimatable, Revenues Not (or Not Applicable)

At first glance, this might seem like the case in Example 4–1. In fact, the annual-worth procedure employed there can also be used here; the difference is that here only costs are relevant, and the annual-worth yardstick becomes an annual cost. However, the decision-maker has different analytical choices. Apart from annual cost, those available involve two other basic approaches: the "coterminated assumption," or "common study period," and the "repeatability assumption."

For the purpose of analyzing the case, consider the two alternatives for which the relevant data are displayed in Table 4–3. (Once again, the cost flows are expressed as annuities, for simplicity. The methods are easily extended for nonuniform cost flows.)

TABLE 4–3

	A	B
Investment cost	$8,000	$13,000
Economic life	3 yrs.	6 yrs.
Total annual costs	$12,000	$10,000
Salvage value	$1,000	$1,000

Again the classic decision dilemma presents itself. Is the added investment requirement of B warranted, given the advantage it provides in lower operating cost and longer life, if capital is worth 15 percent?

Repeatability Assumption

This approach involves comparison of the alternatives over a common time span at least as long as the longest-lived alternative. This may be accomplished by using the least common multiple of the economic lives of the alternatives. In this example, with economic lives of 3 and 6 years, the least common multiple would give a horizon study period of 6 years. (The use of the least common multiple makes the repeatability approach somewhat unwieldy when there are many alternatives, since it becomes likely that a very long time horizon may be needed.)

Using this approach, A and a replacement of it at the end of year 3 can be compared to B over the 6-year horizon, as in Cash Flow Diagram D4–5. (Data are in thousands.)

D4–5

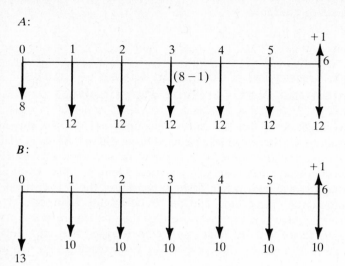

A comparison can now be made using any of the lump-sum yardsticks. Present worth is computable in a very straightforward way, as shown below. Again, data are in thousands, for simplicity.

$$PW_A = -8 - 7(P/F\,15\%, 3) + 1(P/F\,15\%, 6) - 12(P/A\,15\%, 6)$$

$$= -57.584, \text{ or } -\$57,584$$

$$PW_B = -13 + 1(P/F\,15\%, 6) - 10(P/A\,15\%, 6) = -50.412, \text{ or } -\$50,412$$

This suggests that B is indeed superior since its present worth is greater (or that the present worth of costs is less). The analysis indicates that the added investment provides benefits of disbursement reduction and longer life that is worthwhile if funds are valued at 15 percent.

Annual Worth (Equivalent Annual Cost)

Because this approach is the most direct, it has the greatest appeal and is widely used for this case. It simply converts the costs of each alternative to their equivalent annual amount, yielding a common basis for comparison. The advantage to the annual-worth approach is that it does not necessitate choosing a least common multiple time horizon. For the given data,

$$AW_A = [-8 - 12(P/A\,15\%, 3) + 1(P/F\,15\%, 3)](A/P\,15\%, 3)$$

$$= -15.216 \text{ or } -\$15,216/\text{year}$$

$$AW_B = [-13 - 10(P/A\ 15\%, 6) + 1(P/F\ 15\%, 6)](A/P\ 15\%, 6)$$

$$= -13.321 \text{ or } -\$13,321/\text{year}$$

Here this result is consistent with that obtained in the "repeatability" approach—namely, that B has lower cost and is thus more desirable. In a sense the implicit assumption of repeatability is the same here. The computation of annual worth converts the flows to annual amounts for the alternatives having unequal time horizons. By comparing the annual costs directly, one implicitly assumes that the annual flows for the shorter-lived alternative will be the same after the end of its economic life. This, in itself, is a sort of repeatability assumption.

Coterminated Assumption

This approach utilizes a time horizon shorter than the longer-lived alternatives, usually corresponding to the life of the shortest. In this example the time horizon would thus be 3 years. There are two variations to this approach, the most direct utilizing the annual worth of the longer-lived alternative(s). It may be formulated as follows: Compute a lump-sum measure of both alternatives, such as the present worth, over the 3-year horizon. For A this is direct, since that corresponds to its economic life.

D4–6
A:

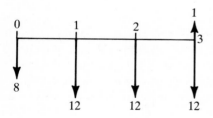

$$PW_{A,3} = -8 - 12(P/A\ 15\%, 3) + 1(P/F\ 15\%, 3) = -34.741, \text{ or } -\$34,741$$

For B the lack of comparability due to its 6-year life can be treated in the following way: First, convert its capital cost and salvage value to an equivalent annuity:

AW (Investment, Salvage value) $= [-13 + 1(P/F\ 15\%, 6)](A/P\ 15\%, 6)$

$$= -3.321, \text{ or } -\$3,321/\text{year}$$

Thus, the initial investment and salvage value can be depicted as in Cash Flow Diagram D4–7.

D4–7
B:

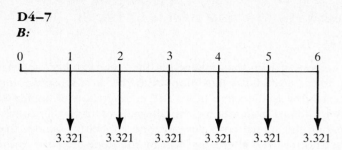

Next, find the present worth of the net capital cost annuity computed above for the 3-year study horizon. For convenience this may be combined with the disbursement annuity of $10,000 per year, yielding the equivalent present worth of alternative B over 3 years.

D4–8
B:

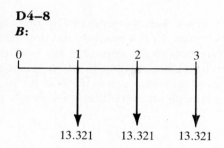

$$PW_B = (-3.321 - 10)(P/A\ 15\%, 3) = -30.415\ or\ -\$30,415$$

Once again, the result is consistent in that B is recommended as the lower-cost alternative.

The other variation on this method also utilizes the shorter time horizon, in this case, 3 years. However, in it one attempts to assign a realistic value for the longer-lived asset at the end of the study period, ignoring the rest of its flows, thus permitting a common time frame for comparison. This approach is practical for assets whose future value may be estimated. This is usually the case when an active market exists for used equipment of that sort, or where trade-ins are common.

In this example, if one estimates that B would be worth $5,000 at the end of the third year, then the analysis proceeds as in Diagram D4–9.

D4–9
A:

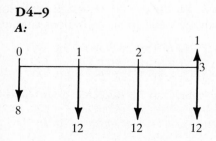

The present worth of A was found before to be $-\$34,741$. For B the capital investment, annual cost flow, and projected market or salvage value are depicted in Diagram D4–10.

D4–10
B:

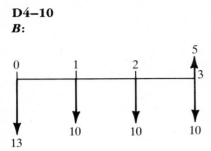

The present worth of B is then

$$PW_B = -13 - 10(P/A\,15\%,3) + 5(P/F\,15\%,3) = -32.545, \text{ or } -\$32,545$$

Once again the added investment required for B is seen to be desirable. However, the fact that the difference between the two alternatives, as measured this way, is not identical to that computed just before indicates the sensitivity of this method to the estimated market value for the longer-lived alternative. This characteristic should be borne in mind by the analyst.

Deferred-Investment Alternatives

Frequently decision-makers are faced with situations in which needed productive capability is not viewed as static over a time horizon but as increasing. Common examples abound, such as the need for increasing manufacturing capacity for a new product, electric power generating, or transportation capacity. Furthermore, the need for productive capability may be expected to increase in any conceivable fashion, such as rapidly at first, followed by a slower rate; slower at first, and then at an increasing rate; a constant rate of growth; or a constant absolute growth. Some of the possibilities are illustrated in Figure 4–1.

Clearly, a limitless variety of possibilities exists with respect to the pattern of growth which may be anticipated. One task of the analyst (or his working group) is to estimate the pattern of growth expected in a given design situation. As an example, suppose the pattern of growth expected for a certain project is as depicted in curve (a) of Figure 4–2.

Superimposed on the graph of the anticipated pattern of growth are representations of two possible design alternatives whose economic attractiveness must be compared and evaluated. The step function represents the immediate provision of capacity at level C_1, which is expected to suffice until time t_1. At that time the added capacity of the step is expected to be added, bringing total capacity to C_2, covering the rest of the planning horizon. This will be referred to in this section

Anticipated need
for productive capacity

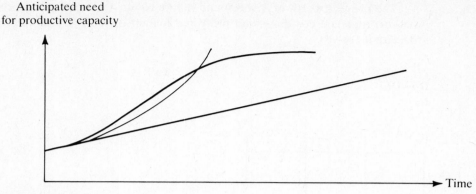

FIGURE 4—1

as the deferred-investment alternative. The other function, (*c*), represents one
alternative design, that of providing for all of the capacity that one can anticipate
needing over the entire time horizon of the analysis. It will be referred to here
as providing full capacity.

Before proceeding, it should be pointed out that the advantages associated
with the option of providing full capacity immediately are based on the potential
for achieving economies of scale, the potential for savings in overhead and setup
costs in construction, and the potential for savings due to simpler designs. The
disadvantages of providing full capacity immediately are more obvious and are
primarily attributable to higher operating disbursements toward the beginning than
would be experienced if some investment is deferred. Thus, the problem can be

Capacity needed,
provided

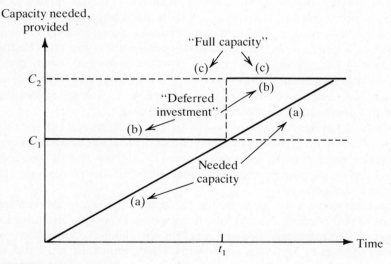

FIGURE 4—2

modeled by weighing these opposing economic forces in order to determine which alternative is superior, given certain assumptions on the timing of the need for additional capacity.

EXAMPLE 4–2

Illustrating the issues bearing on a deferred-investment situation, Figure 4–3 depicts the projected growth in required production for 15 years. It also indicates the nature of the two options under consideration: providing full capacity at once or deferring some investment for some of the capacity which won't be needed until later.

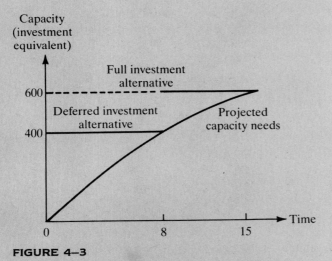

FIGURE 4–3

The deferred-investment option necessitates a $400,000 initial investment. Under this alternative, disbursements are estimated to be $40,000 per year for the first 8 years. At that time the addition to the physical plant necessitated by increasing requirements is expected to be made. After that the disbursements are expected to rise to $80,000 per year over the last 7 years. (This is a gross simplification, employed for illustrative purposes only. Most practical situations are more likely to exhibit a more gradual increase in operating disbursements. However, the method of analysis remains the same.) Finally, it is estimated that the plant will have a $150,000 market or salvage value by year 15.

By contrast, the full-capacity option is estimated to require $600,000 in initial investment, $50,000 per year in disbursements for the first 8 years, and $80,000 per year for the last 7 years. (This latter assumption is an even greater simplification than the preceding one.) Its market value is estimated to be $100,000 by year 15. Cash Flow Diagram D4–11 illustrates the two alternatives. (Data are in thousands.)

D4–11
Full Capacity:

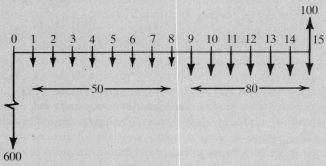

Deferred Investment:

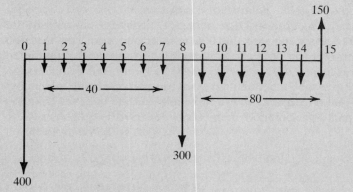

The most direct time value approach to this problem involves computing a lump-sum equivalent value for each of the two alternatives, such as present worth, and then comparing them. This is done under the assumption that capital is worth 10 percent.

$$PW_{Full} = -600 - 50(P/A\ 10\%, 8) - 80(P/A\ 10\%, 7)(P/F\ 10\%, 8)$$

$$+ 100(P/F\ 10\%, 15)$$

$$= -1024.5,\ \text{or} - \$1,024,500$$

$$PW_{Deferred} = -400 - 300(P/F\ 10\%, 8) - 40(P/A\ 10\%, 8)$$

$$-80(P/A\ 10\%, 7)(P/F\ 10\%, 8) + 150(P/F\ 10\%, 15)$$

$$= -899.1,\ \text{or} - \$899,100$$

Thus, in this instance the model suggests that it will be economic (i.e., less costly) to defer investment until it is needed, which is expected to be in period 8. The reader should note that in a real application one does not wait absolutely until then. Rather, the additional facility is provided when it is needed. Whether or not the correct decision has been made, therefore, depends partly on the accuracy of the original estimate concerning the timing of the growth of future needs. The response of the results of this analysis to such timing will be explored further in Chapter 6, "Break-Even Analysis." At that point it will become clear that the computer can be an extremely helpful tool, particularly when its graphic capabilities are put to use.

A final point on deferred-investment alternatives is that the analyst may wish to consider more than two options or a series of stepwise deferrals. In other words, one may analyze the relative desirability of providing full capacity immediately, as opposed to various deferral options. Clearly, the principles remain the same.

Exercises

4–1 *a.* Five design configurations for a metal-processing by-product recovery system are being evaluated for their economic characteristics. The alternative designs are listed in order of increasing investment requirements. If one design is to be chosen, if at least $300,000 is available, and if the MARR is 12 percent, which is most attractive?

	Yr.	A	B	C	D	E
Investment ($000)	0	− 75	− 100	− 180	− 250	− 280
Annual savings generated (revenue) ($000)	1–5	+ 25	+ 27	+ 43	+ 75	+ 82

b. If the capital available were only $200,000, which would now be most attractive?

4–2 Four proposals for a machining process are being considered. Each proposal will satisfy the demands for quantity and quality of product. The difference between the proposals is their cost structure, as follows:

	Yr.	A	B	C	D
Investment ($000)	0	− 50	− 60	− 100	− 150
Annual costs ($000)	1–4	− 130	− 125	− 113	− 90

a. If the opportunity cost of capital is 15 percent, which alternative is cost-minimizing?

b. If only $75,000 is available, what is the recommended course of action?

4–3 Four mutually exclusive cost-saving (or revenue-enhancing) projects are being studied. The data follow:

	Yr.	A	B	C	D
Investment ($000)	0	− 150	− 250	− 300	− 450
Annual cost savings (revenue) ($000)	1–4	+ 32	+ 70	+ 90	+ 125
Salvage value ($000)	4	+ 40	+ 60	+ 70	+ 95

a. If the MARR is 12 percent, which is superior, assuming $450,000 is available?

b. What if only $300,000 is available?

4–4 One of the four cost-minimizing proposals shown below is to be selected:

	Yr.	A	B	C	D
Investment ($000)	0	−125	−160	−190	−230
Annual costs ($000)	1–4	−140	−132	−110	−100
Salvage value ($000)	4	+35	+50	+60	+75

Which is best if the MARR is 15 percent?

4–5 Four cost-minimizing configurations need to be analyzed. The data follow:

	Yr.	A	B	C	D
Investment ($000)	0	−125	−145	−170	−200
Annual costs ($000):	1	−80	−80	−80	−65
	2	−80	−75	−70	−65
	3	−80	−70	−60	−65
	4	−80	−65	−50	−65
Salvage value ($000)	4	+20	+30	+40	+50

Which is best if the MARR is 10 percent?

4–6 The data for four cost-minimizing alternatives are as follows:

	Yr.	A	B	C	D
Investment ($000)	0	−95	−130	−150	−175
Annual costs ($000):	1	−60	−50	−40	−38
	2	−60	−50	−40	−38
	3	−60	−50	−40	−38
Salvage value ($000)	3	+50			
Annual costs ($000)	4		−50	−40	−38
Salvage value ($000)	4		+50	+60	
Annual costs ($000)	5				−38
Salvage value ($000)	5				+60

Analyze the data, assuming the MARR is 15 percent. How does this problem differ from the preceding ones?

4–7 Two processes are being considered to fulfill the same function. The first requires an immediate investment of $100,000 and will provide revenues of $300,000 and $360,000 in each of the first two years, $510,000 for the third year, $505,000 for each of the next two years, $400,000 in the sixth year, and $300,000 in the seventh. Costs will be $310,000 in the first year, $350,000 in the second, $490,000 in the third, $475,000 in each of the next two years, $360,000 in the sixth, and $250,000 in the seventh. In addition, the salvage value of the process at the end of the seventh year is expected to be $30,000.

The other alternative requires a $125,000 initial investment. Revenues would be $400,000 per year for five years and costs would be $350,000 per year. The salvage value at the end of the fifth year would be $40,000.

If funds are worth 12 percent, which investment is superior? Use annual worth as a yardstick.

4–8 Repeat Exercise 4–1 using incremental internal rate of return as a yardstick.

4–9 Two minicomputers are being compared for use by an engineering consulting firm. Although other characteristics of the machines will be considered, an economic analysis yielded the following data: System A would cost $10,000 to acquire and $3,000 per year to operate, including power, materials, and a maintenance contract. It is expected to be used for 6 years, after which it would have a salvage value of $2,000.

System B would require a $12,000 investment but would cost only $2,000 per year to operate, including maintenance. Its economic life is also felt to be realistically estimated at 6 years, after which its salvage value would be $3,000. If funds are valued at 15 percent and these are viewed as mutually exclusive alternatives, which is superior?

4–10 A machine may be purchased for $20,000. Operating costs are expected to be $3,500 per year with maintenance being $500 per year. Its anticipated economic life is 7 years, after which it is projected that salvage value would be $5,000. Under a leasing alternative to the purchase arrangement, the machine may be leased for 7 years at an annual cost of $3,900, the lease payments to be prepaid for each year. Operating expenses are expected to be the same for the leased machine as for the purchased one; however, maintenance under the lease will be the responsibility of the manufacturer.

If capital is worth 16 percent, which alternative is superior? What assumptions must be made in order to undertake the analysis with the given information? What other considerations should be borne in mind when conducting such a comparison in practice? (At this point, the analysis can be done only on a pretax basis (i.e., not considering the effects of income taxes). This sort of problem will be re-examined in a subsequent chapter after the effect of income taxes on economy studies has been outlined.)

4–11 Energy costs for a certain process are now $12,000 per year. They are expected to increase at an average of 5 percent per year over the next decade, due to increasing activity. A microprocessor-based control system can be obtained and installed for $8,000, its expected economic life being 10 years. It is designed to cut down on the energy required by 10 percent. If the salvage value of the system after 10 years is $2,000 and funds are valued at 15 percent, can the system be justified?

4–12 Your company is considering one of two forklift trucks. A costs $11,000, requires $2,000 per year in operating expense, and is expected to have a 3-year economic life. Its salvage value at that time is projected to be $4,000. B costs $15,000 but requires $1,500 per year to operate. It is expected that its economic life will be 5 years, with a salvage value of $4,000. If funds are worth 12 percent, which is superior?

4–13 Data for a deferred-investment vs. a full-investment alternative are depicted in the following cash flow diagrams:

Full ($000):

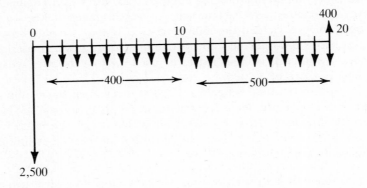

The full-investment diagram may be interpreted to mean that the investment is $2.5 million, operating and maintenance expenses are $400,000 per year for the first 10 years, $500,000 per year for the last 10 years, with a salvage value of $400,000 at year 20.

Deferred ($000):

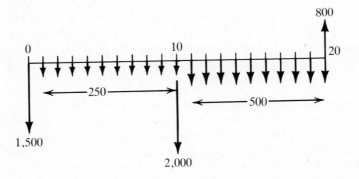

The deferred-investment diagram shows a subsequent investment of $2 million in year 10. The other parameters of initial investment, operating expense, and salvage value also differ from the preceding configuration. If the MARR is 12 percent, which has a more desirable present worth?

4–14 A bridge project is being considered under the following two configurations: Under the first, a four-lane bridge can be constructed for $8 million. It is expected to serve the anticipated needs for the next 20 years. Maintenance would be $80,000 per year for the first 5 years, going up by $25,000 for the next 5 years and similarly for the remaining 10 years. The salvage value at year 20, to be used for studying this alternative, is $2 million.

The other alternative is to construct a two-lane bridge now at a cost of $5 million. Maintenance would be $55,000 per year for the first 5 years, going up to $70,000 for the next 5. At the tenth year two more lanes would need to be added to accommodate the anticipated growth in traffic. This would cost an additional $6 million. Maintenance of the combined four-lane facility would then be $130,000 in years 11 to 15, and $155,000 in years 16 to 20. Salvage value for the second alternative would be $2.5 million at year 20. If funds are worth 10 percent, which alternative is superior?

4–15 You have been asked to evaluate the following two alternatives. One requires an immediate investment of $10,000 and is expected to generate revenues in excess of costs over the next 3 years of $3,000, $4,000, and $5,000 respectively. The other alternative requires a $12,000 investment and will provide net returns of $4,000 per year for the next 4 years. If capital is valued at 12 percent, and only one of the two alternatives can be chosen, which should it be? To proceed,

a. Compute the annual worth of each alternative.
b. Compute the external rate of return for both alternatives, assuming a 12-percent discount rate.
c. Interpret all your results.

4–16 Three cost-saving machines are being considered for an existing function. Data follow:

	A	**B**	**C**
Investment ($)	60,000	80,000	90,000
Annual costs ($)	142,000	136,000	128,000
Economic life (yrs.)	5	5	5

If funds are worth 20 percent before income taxes, which alternative is superior, assuming that the firm has at least $90,000 of investable capital? (Do this problem two ways: incremental-rate-of-return method; and either the periodic (annual-worth) method or present-worth method.

Selected References

Au, T. and T. Au, *Engineering Economics for Capital Investment Analysis.* Boston: Allyn and Bacon, 1983.

Ward, T. and W. Sullivan, "Equivalence of the Present Worth and Revenue Requirement Methods of Capital Investment Analysis." *AIIE Transactions,* Vol. 13, No. 1, March, 1981, pp. 29–40.

5

Computational Considerations

Introduction

Chapters 2 through 4 have delineated the exponential equations that govern time value concepts and computations. These computations may be undertaken in various ways. In this chapter the use of computers will be explored and compared to "traditional" computational procedures. Before proceeding with a discussion of computer capabilities for time value applications, a brief review of common computational practice is presented, in the following order:

1. General-purpose calculator and/or tables,
2. Special-purpose calculator, and
3. Computer, using specialty software.

General-Purpose Calculator and/or Tables

The use of calculators, with or without tables, has been the dominant mode of computation for time value concepts. Until the 1960s, only electromechanical or simple electronic calculators were available. These machines did little more than add, subtract, multiply, and divide. The advent of electronic machines permitted the storage of results in memory. However, this was still a rather primitive aid for time value calculations.

As in most disciplines, computational constraints had a noticeable bearing on the way work was carried out, as well as on the way new practitioners were trained.

As a result, most authors and practitioners resorted to the use of tables of values for the exponential factors derived in Chapter 2. This approach still receives the dominant emphasis in the text and reference books in the field, even though there is now greater computational power available. Although it will not be discussed extensively in this book, the basic use of tables is outlined below, and tables for that purpose are provided for reference in Appendixes A and B.

Example 2–1, in Chapter 2, will serve to illustrate the use of the tables. It was desired to compute the value, four periods hence, of \$2,000, given that $i =$ 12 percent per period. Thus, one must compute

$$F_4 = 2,000(1.12)^4$$

Using the notation introduced in Chapter 2, this may be written as:

$$F_4 = 2,000(F/P\ 12\%, 4)$$

If 1.12^4, or $(F/P\ 12\%, 4)$, were given in a table, then all that would be necessary would be to look up that value and multiply by 2,000 to get the desired result. That, in essence, is the service the tables provide.

Most tables of the time value factors are organized like Table 5–1. On each page are found the values of the factors for a variety of commonly used time periods for a given interest or discount rate. The set of tables then consists of a page for each commonly used rate between 0.5 and, say, 50 percent, such as 5, 6, 8, and 10 percent. In addition, a duplicate set of tables is usually included for continuous compounding. Although very useful, such a set of tables is clearly not helpful for other than round values for rates or for commonly used time periods, unless one is willing to interpolate.

Table 5–1, for discrete compounding at 12 percent, indicates that $(F/P\ 12\%, 4)$ $= 1.57352$ (to 5 places). This allows completion of our problem,

$$F_4 = 2,000\,(1.5735) = \$3,147$$

which agrees, within reasonable rounding difference, with the result obtained previously in Chapter 2.

In summary, tables can aid in the computation of time values even when only the simplest of machines is available. In such instances, table reference is required for each occurrence of a different time value factor.

Modern hand-held calculators have eclipsed much of the usefulness of the tables, however. Even typical "bottom of the line" models of today have exponential functions as well as at least one separately addressable memory. The use of such calculators for the purpose at hand can be illustrated using the same problem. Again, what is needed is to compute an exponential like $(1.12)^4$. For calculators that use an "algebraic" format, one enters 1.12, the exponentiation key (frequently labeled y^x), the exponent, which is here 4, and then the " $=$ " key. At that point the result is shown. If results are displayed in floating point format, one would get 1.57351936, assuming a 10-character display. One would then complete the task by pressing the "x" key for multiplication, then 2,000, followed by the " $=$ " key again, for the product of 3147.03872, which agrees with our original result. For calculators using "reverse polish format," the procedure differs somewhat, with the same results.

TABLE 5–1 Discrete Compounding: *i* = 12.0%

N	(F/P)	(P/F)	(F/A)	(P/A)	(A/F)	(A/P)
1	1.12000	.89286	1.00000	.89286	1.00000	1.12000
2	1.25440	.79719	2.11999	1.69005	.47170	.59170
3	1.40493	.71178	3.37439	2.40183	.29635	.41635
4	1.57352	.63552	4.77931	3.03734	.20924	.32924
5	1.76234	.56743	6.35283	3.60477	.15741	.27741
6	1.97382	.50663	8.11516	4.11140	.12323	.24323
7	2.21068	.45235	10.08898	4.56375	.09912	.21912
8	2.47596	.40388	12.29965	4.96763	.08130	.20130
9	2.77307	.36061	14.77560	5.32824	.06768	.18768
10	3.10584	.32197	17.54866	5.65022	.05698	.17698
11	3.47854	.28748	20.65450	5.93769	.04842	.16842
12	3.89596	.25668	24.13303	6.19437	.04144	.16144
13	4.36348	.22917	28.02899	6.42354	.03568	.15568
14	4.88709	.20462	32.39245	6.62816	.03087	.15087
15	5.47355	.18270	37.27955	6.81086	.02682	.14682
16	6.13037	.16312	42.75306	6.97398	.02339	.14339
17	6.86601	.14564	48.88342	7.11962	.02046	.14046
18	7.68993	.13004	55.74940	7.24966	.01794	.13794
19	8.61272	.11611	63.43933	7.36577	.01576	.13576
20	9.64624	.10367	72.05201	7.46944	.01388	.13388
21	10.80379	.09256	81.69824	7.56200	.01224	.13224
22	12.10024	.08264	92.50199	7.64464	.01081	.13081
23	13.55227	.07379	104.60224	7.71843	.00956	.12956
24	15.17853	.06588	118.15442	7.78431	.00846	.12846
25	16.99995	.05882	133.33295	7.84314	.00750	.12750
26	19.03993	.05252	150.33276	7.89566	.00665	.12665
27	21.32474	.04689	169.37281	7.94255	.00590	.12590
28	23.88368	.04187	190.69734	7.98442	.00524	.12524
29	26.74973	.03738	214.58104	8.02180	.00466	.12466
30	29.95969	.03338	241.33071	8.05518	.00414	.12414
35	52.79913	.01894	431.65943	8.17550	.00232	.12232
40	93.04997	.01075	767.08307	8.24378	.00130	.12130
45	163.98566	.00610	1358.21375	8.28252	.00074	.12074
50	288.99829	.00346	2399.98566	8.30450	.00042	.12042
55	509.31323	.00196	4235.94344	8.31697	.00024	.12024
60	897.58228	.00111	7471.51866	8.32405	.00013	.12013
65	1581.84521	.00063	13173.70960	8.32806	.00008	.12008
70	2787.74805	.00036	23222.89947	8.33034	.00004	.12004
75	4912.95703	.00020	40932.97363	8.33164	.00002	.12002
80	8658.29687	.00012	72144.13776	8.33237	.00001	.12001
85	.15259E+05	.00007	.12715E+06	8.33279	.00001	.12001
90	.26891E+05	.00004	.22409E+06	8.33302	.00000	.12001
95	.47392E+05	.00002	.39492E+06	8.33316	.00000	.12000
100	.83520E+05	.00001	.69599E+06	8.33323	.00000	.12000
120	.80565E+06	.00000	.67138E+07	8.33332	.00000	.12000
150	.24137E+08	.00000	.20114E+09	8.33333	.00000	.12000
180	.72314E+09	.00000	.60262E+10	8.33333	.00000	.12000
200	.69756E+10	.00000	.58130E+11	8.33333	.00000	.12000
240	.64908E+12	.00000	.54090E+13	8.33333	.00000	.12000
250	.20159E+13	.00000	.16799E+14	8.33333	.00000	.12000
300	.58260E+15	.00000	.48550E+16	8.33333	.00000	.12000
360	.52293E+18	.00000	.43578E+19	8.33333	.00000	.12000

It should be emphasized that the separately addressable memories of modern calculators offer further capabilities usuable in time value calculations. These memories can cumulate terms that are entered successively. In discounted cash flows, for instance, such as in Example 2–3, one can compute the present worth of each term (using the exponentiation function discussed above) and then cumulate them in memory. This obviates the need to record and sum the present worth of each of the terms after they have been individually computed. For the given problem, one can compute the present worth of 10 one period hence (using what was discussed above), put it in memory, compute the present worth of 15 two periods hence, put that in memory, and so on. After the last amount has been discounted and put in memory, the memory contains the entire sum desired. At that point all that needs to be done is to recall the contents of memory.

This sort of procedure is useful for computing internal rate of return when using a hand calculator. In such cases one can conceptualize a table to facilitate the computation of net present worth for each i in the trial-and-error approach, as illustrated in Table 5–2 for Example 3–1.

TABLE 5–2

Year	Flow	$\dfrac{1}{(1 + i)^n}$ at 8%	Present Worth at 8%	$\dfrac{1}{(1 + i)^n}$ at 12%	Present Worth at 12%
0	−100	1.0	−100.	1.0	−100.
1	20	.9259	18.518	.8929	17.858
2	20	.8573	17.146	.7972	15.944
3	20	.7938	15.876	.7118	14.236
4	20	.7350	14.700	.6355	12.710
5	50	.6806	34.030	.5674	28.370
NPV (sum)			+0.2717		−10.882

For each iteration, the net present worth can be arrived at by cumulatively storing the products of the discount factor, $1/(1 + i)^n$, and periodic flow. Also, it should be noted that the word "conceptualize" was used above, because the intermediate values of the table, such as the discount values, need not be stored.

To arrive at the internal rate of return, the analyst must repeat the present-worth computations implied by Table 5–2 until the solution is found. This requires a present worth (at one rate) greater than zero, and another (at a different rate) less than zero, in order to use linear interpolation. This procedure is the same as was described in Chapter 3.

Another use of the calculator memory is for computing the time value factors without having to record intermediate values. For instance, in the mortgage annuity problem of Chapter 2, Example 2–6,

$$(A/P\ 1\%, 300) = \frac{.01(1.01)^{300}}{(1.01)^{300} - 1}$$

is needed. Here, one can compute the denominator, using the exponential function, and put it in memory; then one can compute the numerator, press the " ÷ " key, recall memory, which brings forth the denominator previously computed, and, finally, press the " = " key, which yields the factor, 0.01053224, with a small amount of effort and no need to write down intermediate values. Note that this result, when multiplied by the $75,000 of the problem, gives the same amount as before.

Many practitioners combine the use of modern calculators with tables. For example, the discounted cash flow above could have been done about as expeditiously by using the tables to discount each term. Then the products of the values and their discount factors can be cumulated into memory as before.

Special-Purpose Calculators

Since the late 1970s, several manufacturers (such as Texas Instruments and Hewlett Packard) have introduced calculators—for example, Texas Instruments Model BA–II—that directly compute the time value exponentials $(F/P\ i\%, n)$, $(F/A\ i\%, n)$, etc., of Equations 2–1 to 2–6. These machines expedite time value calculations by reducing the number of keystrokes needed to compute the exponentials.

The TI machines use one key for each of the variables defined in Chapter 2. The correspondence in notation is as follows:

F	FV
P	PV
A	PMT
$i\%$	$i\%$
n	N

The procedure is to input the value of each of the variables above which constitute the given information, followed by the key indicating what that value represents; (for example, -100 may be entered on the keyboard, to be followed by PV, indicating that it represents a present value); after all the data are entered, as indicated above, then one presses "2nd" (for uppercase key), followed by the variable representing the unknown quantity desired. Then the machine computes and displays the required quantity. For example, to compute the future-value solution to Example 2–1, one would enter 10, then the $i\%$ key; 2,000, followed by the PV key; 4, then the N key; finally, the 2nd and FV key; and the now familiar result of 3147.04 would be displayed.

These calculators offer greater speed for computation of simple time value quantities than their predecessors, but they do not bring to bear the great power of the modern computer for sophisticated and repetitive analysis for time value models. It is the aim of this book to point toward what can be achieved using the modern computer for time value analysis. This chapter only scratches the surface; what follows is a list and explanation of the basic characteristics that modern computer software should possess, along with an illustrative program in flow chart and code form. The program discussed in this chapter is listed in Appendix C. In

Appendix D the scope of computer application to time value problems will be broadened to include after-tax and sensitivity analysis. They are also available from the publisher in disk form for the IBM PC and Apple II series.

Computers

In this section a computer program written in BASIC will be introduced which is capable of solving all of the time value problems presented in the previous chapters, including the storage of the individual time value measures for the comparison of mutually exclusive alternatives.

The experienced time value analyst will find the program very easy to use, since it is written interactively. However, the novice may need more time to learn to use it. The following is not a detailed explanation of each step of use but is, instead, an overview of the structure of the program, which also will serve as an illustration of the desirable user characteristics that software should possess for this purpose. These characteristics are listed and explained in conjunction with the description of the program. (Figure 5–1 represents a program flow chart.) Later on in the chapter, a more detailed discussion of the logical procedures contained in the blocks will be presented.

Computing All Time Value Quantities

Minimally a program should be able to compute present, future, or periodic values, as well as internal or external rates of return, where applicable. Such software is easy to prepare and exists in most computer-center libraries. In addition, it should be capable of making different computations, such as present value and equivalent annual worth, for the same set of data (flows).

Examination of the flow chart shows that block A1 contains the time value Equations 2–1 through 2–6 in the form of BASIC functions used for computation in the program. In block B1 the user chooses which time value measure is to be computed for a particular iteration. Blocks D3, D4, C5, and G5 contain the logic for the time value computations. Note that the program permits the user to return to B1 in order to compute a different time value measure for the same cash flow data.

Amalgamating Factors

This refers to the ability of the program to accept "elemental" data on flows and automatically combine them for the purpose of time value analysis. Consider Example 4–1, which contains data on investment cost, annual revenue, annual costs associated with labor, energy, maintenance, taxes, insurance, and miscellaneous expense, as well as salvage value. At first glance, the ability to combine these flows for any

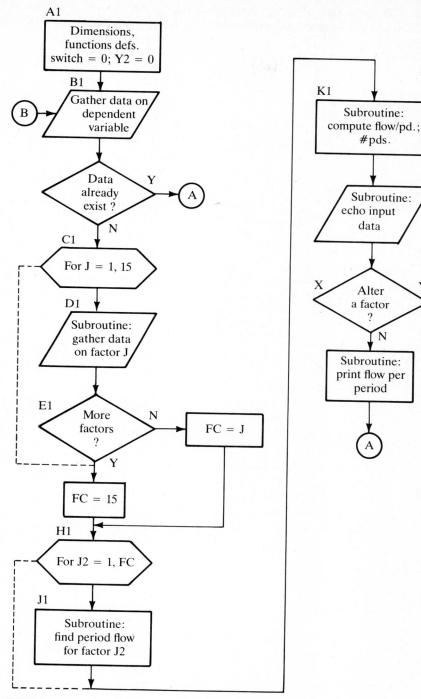

FIGURE 5–1

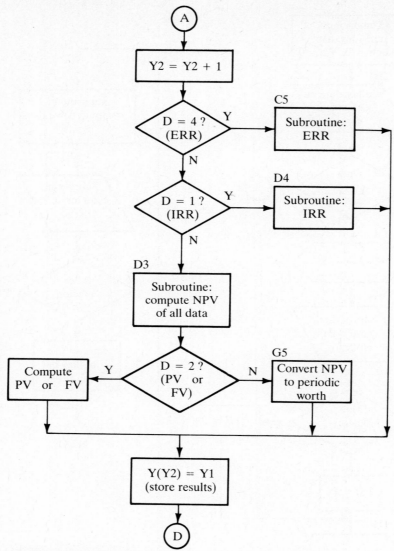

FIGURE 5–1 *(Continued)*

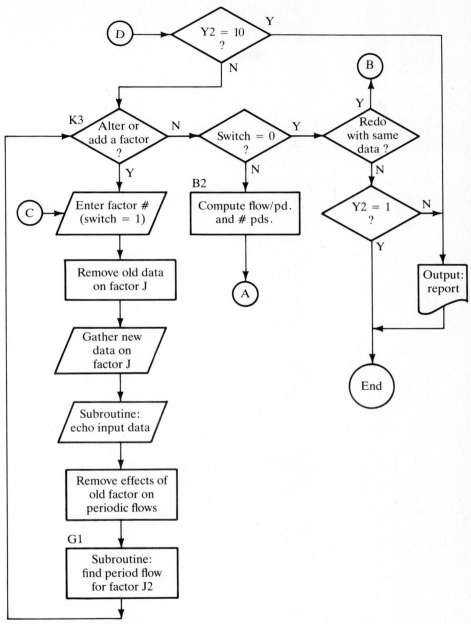

FIGURE 5–1 (Continued)

period may seem like an inconsequential advantage. In fact, the table of data for Example 4–1 does show the total annual cost (of labor, power, maintenance, and taxes) as a simple sum. However, sophisticated application of the models often requires different projected rates of increase over time of the separate "elemental" values. For instance, it may be anticipated that labor expense will remain the same, while maintenance might increase at a rate of 4 percent per year. Or the analyst may wish to conduct a sensitivity analysis, varying energy costs, in order to see how the time values vary in response to such assumed changes. Recall that such an analysis was discussed in Chapter 1. Clearly both of these practical requirements illustrate the desirability of treating the data "elementally," while relying on the computer to "amalgamate" the flows for each period before proceeding with the time value computations. Most available software does not perform this amalgamation function. Instead, it requires the analyst to prepare the data by hand in advance of time value analysis by computer. Often this preparation can be as much or more work as the time value computation itself.

The following illustration will serve as a clarification. Consider the three "elemental" flows shown as Cash Flow Diagram 5–1a:

D5–1a

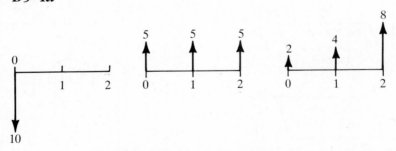

The analyst would need to combine the data for D5–1a "by hand" to obtain D5–1b. A good program, such as the one provided in this chapter, will perform

D5–1b

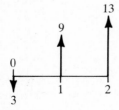

this function for the analyst.

On the program flow chart, blocks C1 through E1 describe a process of gathering data, interactively from the user, on each of the "elemental" flows, called factors. Here a factor may take on one of three forms: single-period amount; uniform series (annuity), for which this program permits other than single-period increments;

or gradient series, either uniform or percent, also with other than single-period increments allowed. Blocks H1 through K1 perform the amalgamation function, so that a single net cash flow is computed.

Treating Gradients—Arithmetic and Percent

Often it is necessary to treat a flow which increases or decreases according to a consistent pattern. Two such cases are uniform (arithmetic) and percent gradients. An arithmetic gradient is one in which the flows increase or decrease by the same absolute amount each period. Two examples are given in Cash Flow Diagram 5–2. The uniform gradient in the first case is +3; in the second case it is −1.

D5–2

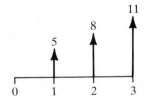 

A percent gradient is one in which the flows increase or decrease by the same percentage amount each period. Two examples are shown in Cash Flow Diagram D5–3:

D5–3

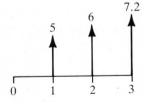 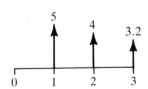

In the first case the percentage increase after period 1 is +20 percent, and for the second it is −20 percent. Such gradients are very useful for projection of some elemental flows. It is highly desirable that software incorporate the capability for computing time values for them, since simple algebraic formulations don't exist for this case, as they do for uniform gradients. Thus, the analyst wishing to treat percentage gradients must tediously do them "by hand." With computer software, the gradient values for a particular factor can be enumerated for each period, using Equation 2–1 for the percentage gradient series, or uniform increases for the uniform gradient. Then the period values can be stored for amalgamation just as for the other factors.

Facilitating Sensitivity Analysis

Sensitivity analysis is another important capability frequently ignored in existing software for time value analysis. The process of modeling for economic purposes necessitates numerous estimates concerning the values of relevant variables. The applicability or predictive value of the model used partly depends upon the extent to which these turn out to be correct later on. Although sensitivity analysis does not help to predict the future any better, it does help to gauge the risks involved in the estimating process. In other words, if the computations of the models are repeated allowing one of the variables of the problem, say labor costs, to change within realistic limits in either direction from the best estimate, and this does not alter the substantive recommendation of the model, then the analyst has a much greater degree of confidence in the model's results for decision-making.

Clearly, the converse effect also holds. Nevertheless, sensitivity analysis contributes to the decision-making process in such instances, since this initial lack of confidence may motivate a more thorough investigation regarding the variable in question. It may be possible, after such a redoubled effort, to estimate the variable more precisely, yielding more certain conclusions. At worst, the decision-makers get a better insight into which factors hold the greatest danger with respect to the outcome of their recommendations.

Chapter 1 and Example 1–2 indicate that sensitivity analyses can be computationally demanding. Computer software should be designed to ameliorate this difficulty. Sensitivity analysis is provided for in the program in the following way (see Figure 5–1): Blocks K3 through G1 show the logic for varying one factor at a time. Each time a factor is altered, the program gets new data for the factor from the user, removes the effect of the old factor, replacing it with the new, and then recomputes the period flows (by amalgamation). Finally, it performs the time value computation after the change in the factor has been accounted for and stores the value for future use (inspection, data file creation, and/or graphing).*

An additional benefit of such logic in the program is that it readily permits change in the data for any factors which may have been erroneously input by the user. This is particularly desirable for interactive-type programs, such as the one in question. Thus, after the data have been gathered for all the factors, they are "echoed" or displayed, consistent with their input format. At that point, the user has the opportunity to alter or add factors, as seen from block X of the flow chart.

Facilitating Comparison of Alternatives

The same program logic described above for sensitivity analysis may also be profitably employed to facilitate comparison of two or more alternatives. The analysis of the first alternative, say A, proceeds as has already been described. For example, one might begin by computing the present worth for A, repeating execution of the program for, say, annual worth at a given interest or discount rate. At that point

---

*The after-tax program described in Appendix D "automates" the sensitivity analysis. It allows the user to specify in advance the values of the factor to be varied.

two values would be stored, one for the present worth for *A*, the other for its equivalent annual worth.

To analyze one or more additional alternatives, say *B*, the program permits alteration of the factors constituting *A*, one at a time, until all of the input data for *A* has been replaced with those for *B*. Then the program will compute the time value measure for *B* previously specified by the user. Of course, the user may seek to compute different time value measures for *B* as he did for *A*. The program will store all computed values for future use or display.

Program Use/Program Summary

This section will summarize the use of the program described in Figure 5–1. In addition, it will elaborate some of its logic.

1. The user must choose the type of dependent variable to be computed.

    ENTER DEPENDENT VARIABLE:
      1 - INTERNAL RATE OF RETURN
      2 - PRESENT OR FUTURE VALUE
      3 - PERIODIC WORTH
      4 - EXTERNAL RATE OF RETURN

2. The user then inputs data on the first factor, assigning it a name and indicating its type in response to the following:

    ENTER FACTOR TYPE FOR FACTOR 1
      1 - SINGLE PERIOD AMOUNT
      2 - UNIFORM SERIES
      3 - UNIFORM GRADIENT SERIES

In addition, the characteristics of the factor, such as its value(s) and period(s) of incidence, are indicated by the user. These are illustrated with Cash Flow Diagrams D5–4 to D5–9.

    A. Single-period Amount. Required of the user are
      i. The value or amount, and
      ii. The period in which the value or amount occurs.

      e.g., value = +1,000, in period 0:

D5–4

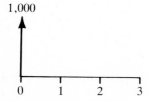

e.g., value = − 500, in period 3:

D5–5

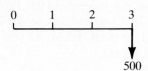

B. Uniform Series. Required of the user are
    i. The amount or value per period,
    ii. The starting period,
    iii. The period increment, and
    iv. The number of occurrences of the value in the series.

e.g., annuity amount = + 1,000, starting in period 2, increment = 1 period, for 3 occurrences.

D5–6

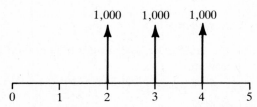

e.g., annuity amount = − 1,000, beginning in period 1, period increment of 2, 3 occurrences.

D5–7

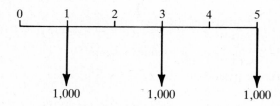

C. Gradient Series. The user must supply
    i. The initial value or amount of the series,
    ii. The starting period,
    iii. The period increment,
    iv. The number of occurrences,
    v. The type of increase or decrease
      (arithmetic or percent), and
    vi. The value of the increase or decrease.

e.g., starting value = 500, beginning in period 1, period increment of 2, 3 occurrences, and an arithmetic change of − 50.

D5–8

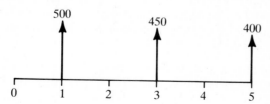

e.g., starting value = − 500, beginning in period 1, period increment of 1, 5 occurrences, with a percent change of − 10 percent.

D5–9

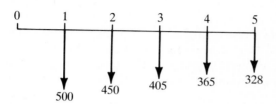

3. If there are more factors to be input (the sample program is dimensioned for 15 factors) the user is returned to step 2. If not, the program proceeds to step 4.

4. A. The program calls for computation of the period flow for each of the factors input in 2.
   B. It then computes the aggregate flow per period and the number of periods in the horizon (dimensioned in the sample program for 52 periods).
   C. Then it echoes the inputs (to permit any changes due to keyboard errors or changes of heart by the user before proceeding with computation).
   D. If the user wishes to alter (or add) a factor, program control transfers to the section for that purpose. If not, the flow per period is displayed for each period before proceeding to the next step.

5. The periodic flows are evaluated with respect to the dependent variable chosen in step 1, and the result is displayed.

6. At this point the user has the option to alter or add a factor (either for sensitivity analysis or to begin evaluation of a different alternative). The process requires:
   A. Entry of the factor number to be altered or added;
   B. Removal of data for the old factor;
   C. Gathering of data for the new factor (This is the same procedure as in step 2.);
   D. Echoing the data for user approval;
   E. Removal of the effects of the old factor on the aggregate periodic flow;
   F. Computation of the period flows for the new factor; and,

G. Finally, if no more changes are desired, computation and display of the new aggregate flow per period and the number of periods in the horizon.

7. Here, the user can opt to compute a different time value measure for the same set of factors. For example, having computed a present worth, he may then desire to compute periodic (say, annual) worth. The program permits return to step 1 for that purpose. (If that happens, the data need not be re-entered.) If no more computations are needed, the program ends with the output displayed and/or sent to a file for future use.

Computer Logic—Procedures, Subroutines This section describes in detail the logic of the key blocks of the program flow chart of Figure 5–1. Thus, the more casual reader, or the one who simply wishes to use the program without becoming acquainted with the details of its logic, may prefer to omit it. The reader wishing to adapt the program to a particular need will find reading it a necessity.

Each of the subsections that follow corresponds to a block of the main program. In most cases a flow chart is presented to represent the logic. In each instance, a list is included, defining those variables appearing in the procedure that have not been previously presented. In addition, the fully coded program is in Appendix C.

The reader should note that loops covering the different periods were begun with an index of 1, not 0. Thus, when period data are input (permitted to begin at 0), those data are transformed by adding 1. The reverse transformation is then performed before output is displayed.

Function Definitions In this part Equations 2–1 through 2–6, for discrete compounding, are defined as functions for subsequent use in the program. This avoids having to specify the computation in detail each time it is called for. (If the user wishes continuous compounding, he may replace these functional definitions with those for continuous compounding. Or, if he wishes a choice of either, he may put both into place along with a switch indicating which is to be used as the program is executed.)

Variables:
D3: the interest or discount rate
N: the number of periods

Functions:
FN A (N) = $(1 + D3)^N$
  (This gives F/P.)
FN B (N) = $1/(1 + D3)^N$
  (This gives P/F.)
FN C (N) = $[(1 + D3)^N - 1]/D3$
  (This gives F/A.)
FN D (N) = $D3/[(1 + D3)^N - 1]$
  (This gives A/F.)
FN E (N) = $[(1 + D3)^N - 1]/[D3(1 + D3)^N]$
  (This gives A/P.)
FN F (N) = $[D3(1 + D3)^N]/[(1 + D3)^N - 1]$
  (This gives P/A.)

Compute Periodic Flow for a Factor (Block G1 or J1) This section is a prerequisite for the amalgamation function. It computes the appropriate values attributable to the factor being considered for each time period and stores them in the array $A(i, j)$. To do so, it must distinguish among the three types of factors: single-period amount, uniform series, and gradient series. The flow chart (Figure 5–2) indicates how each is treated.

Variables:
J2:   the factor number
F(i,1):   the starting period
F(i,3):   the type of factor
> 1: single-period amount
> 2: uniform series
> 3: gradient series

F(i,5):   the value (for single-period factor) or the starting value (for series factors)
F(i,6):   the incremental amount (for gradient series)
F(i,7):   the type of increase or decrease for the gradient
> 1: percent
> 2: arithmetic

A(i,j):   the aggregate net flow due to factor i in period j
T1:   the beginning period
T2:   the number of periods
T3:   the period increment
T4:   the starting value
J3:   an index for periods

Compute Flow Per Period, Number of Periods in Horizon (Blocks K1, B2) This is the amalgamation function in which the values of array $A(i, j)$ for each of the i factors is summed to give the aggregate flow for each time period j. The period flows are stored in $S(j)$.

Then the number of periods in the horizon is calculated, so as to minimize future computation. This is done by scanning, in order, the array $S(j)$ beginning with the highest values of j, until a nonzero value of $S(j)$ is found. This is stored in P.

Variables:
FC: the number of factors
S(j):   the sum of all flows for period j
P: the last period to contain a nonzero flow (i.e., the number of periods in the horizon)

Compute ERR (Block C5) Here the negative flows are discounted to period 0, using the given interest or discount rate, and the positive flows are evaluated in terms of the last period. Then the rate equating those flows is computed, giving the ERR.

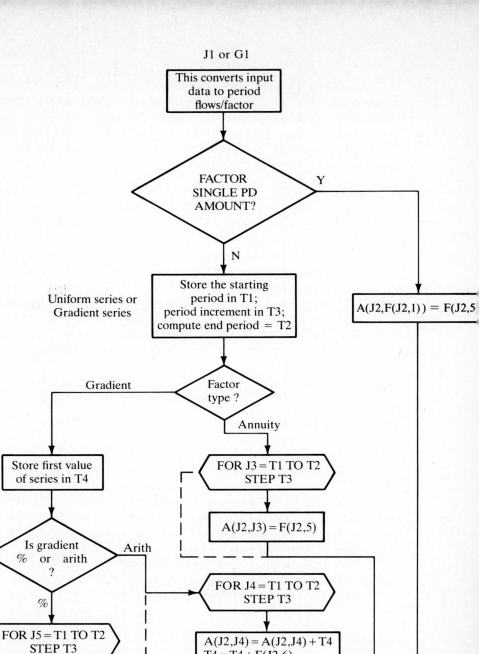

FIGURE 5–2

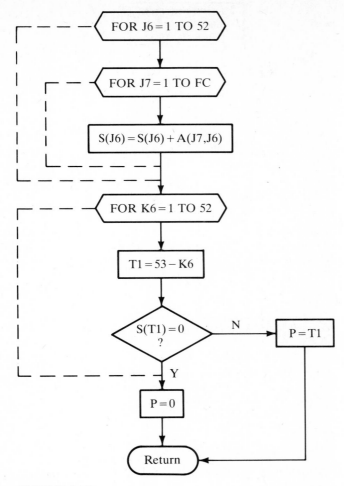

FIGURE 5–3

Variables:
E1: the sum of the negative flows evaluated at time 0
E2: the sum of the positive flows evaluated at the end of the horizon

   Compute NPV or FV of Flows (Block D3)   In this section, the present
worth of the flows is computed, one period at a time, while accumulating. Then
that amount can be expressed in terms of any other future period if such a value
is desired.

Variables:
N1: holds the value for the desired period
T1, Z9: temporary storage
D1: the starting period for the dependent variable (for *NPV* and periodic worth
     only)

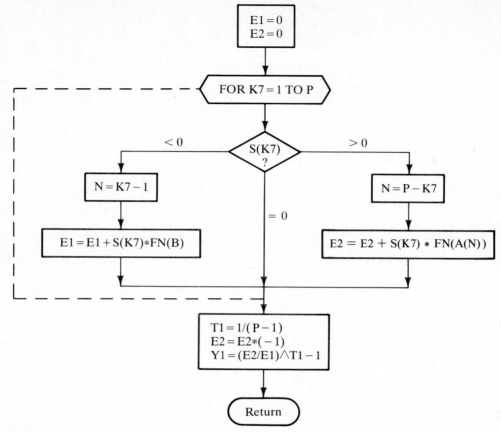

FIGURE 5—4

Compute IRR (Block D4) This subroutine calls others during its execution. It is a search algorithm seeking the root of Equation 3—1. It does so by calculating the *NPV* of flows, using that subroutine for successively higher values of *i*, where I1 is the increment. When *i* changes from positive to negative, the root has been crossed. The next step is then to backtrack, reducing the increment by a factor of 10 and repeating. The root is then successively crossed with smaller increments. When it is crossed and the increment is negligible, the current value of D3 is used as the solution.

Variables:

V1: temporary storage for the *NPV*
M1: used for "direction of search" in IRR subroutine
    0 - IRR moving in "right" direction
    1 - IRR moving in "wrong" direction
    2 - *NPV* has crossed 0
D3: holds the IRR
I1: the increment for IRR trials
Y1: the rounded solution

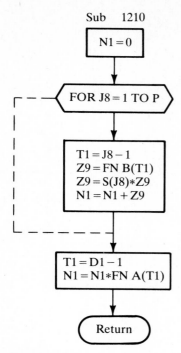

FIGURE 5–5

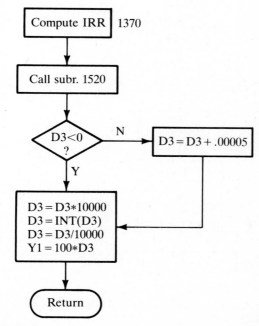

FIGURE 5–6

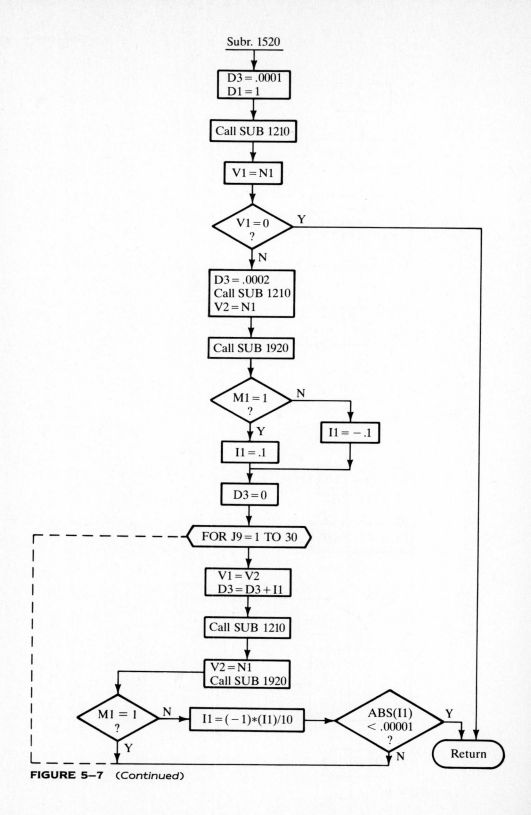

FIGURE 5–7 (Continued)

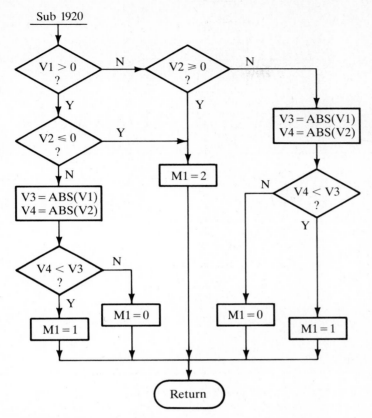

FIGURE 5–7 *(Continued)*

Conclusion The program described in this chapter and appearing in Appendix C will prove very useful to new students of this subject. It will permit them to perform pre-tax computations much more quickly than if done by hand. They will have the opportunity to try more exercises and to play more readily "what if?", thus enhancing the learning process.

6

Break-Even Analysis

Introduction

The preceding chapters have dealt with the use of time value models as aids in economic decision-making. Initially in this chapter a different viewpoint is taken, yielding what at first seems to be an alternative class of models useful for economic analysis. However, by the end of the chapter the reader will see how the two perspectives may be combined, offering the analyst a broader and more representative set of tools with which to attack investment decision-making problems.

Classical Break-Even Analysis

Break-even models were introduced roughly 50 years ago. Their basic concepts have become quite popular and useful, finding their way into everyday business practice and economic reporting. Essentially a break-even model attempts to characterize a relatively complete economic unit like a factory *at one point in time* with respect to the effect of alternative rates of operation on revenues and costs. It is a short-run model. (By short run is meant a period during which the essential nature of the establishment cannot be radically changed; this is generally assumed to be less than a year.) The term "break even" is used typically to denote the operating rate at which cost and revenue are equal in the short run. This model permits the comparison of process design alternatives with respect to their anticipated effects on operating conditions, the overall evaluation of current operations with respect to standards or plans, and sounder decision-making on operational matters. Its usefulness as a tool for planning and design as well as for operating purposes contributes to its popularity.

The difference in the approach, as compared to time value models already discussed, stems from the description of costs and revenues as a function of operating rates. Implicit in these models, then, is the recognition that most productive units, whether they be factories, chemical processing plants, power-generating stations, or service or transportation facilities, operate at varying rates of output. The models attempt to shed light on the effects of operating at such different rates.

Because break-even models specify functional relationships between costs and another variable (here, output rate), they provide an enhanced analytical viewpoint when combined with the perspective provided by the preceding treatment of time value models, in that they point the way to generalizing them. This functional approach can and should be applied to time value models where feasible. The analysis of such generalized cost functions yields more sophisticated and realistic time value analyses and can be enhanced by the use of the computer for computational and graphical purposes. These issues will be explored in the latter part of the chapter, after the basic concepts have been introduced.

Operating Rate

The first task in break-even analysis is to describe total cost as a function of the output rate of the establishment. If one visualizes a single-product establishment such as a factory, one can conceive of the output rate as measurable in units produced per unit time: e.g., tires per day, cartons per week, or gallons per month. This notion of output per unit time will be designated by Q/t.

Although the output rate is conceptually simple, it is, like many economic variables, more difficult to pin down and estimate in practice. The primary complicating factor is that most modern productive entities have a varied repertoire of productive output; it is an extremely rare establishment that produces a completely homogeneous product. The degree of variation in the product line differs quite considerably among establishments: some manufacturers produce a small number of different models in the same plant; some plants are designed for product variation within the same model designation (as is the case in auto assembly, where the buyer may specify optional equipment); and, as an extreme, some manufacturers produce a veritable cornucopia of various products in one plant (such as certain electrical parts manufacturers or commercial bakeries).

Where variation in product occurs, there are two major approaches to describing output rates, both involving the use of a common reference measure. The first is in the case of a limited number of models, limited variation within models, or limited variation in the physical characteristics of the product. The task of measuring output in such an establishment generally entails establishing a standard physical unit of output to which each actual type may be related, with respect to its demand on the productive facilities.

In some industries it is convenient to use a standard model as the basic unit of output. Then, each actual model or variant thereof can be expressed in terms of the standard model, and the output rate is the cumulative sum of production as expressed in terms of the standard model. This is done in auto manufacturing. In other industries the standard unit of output may be expressed in more physical terms. For example, in lumber milling, output may be expressed in board feet; in steel production it may be expressed in tons. In this book all physical output rates will also be denoted by Q/t.

The second approach may be applicable when the variation in product is so great that there does not exist a practical way to express an equivalent model or physical unit of output. In the case of a commercial bakery or electrical parts manufacturer, for instance, one may express output in the form of a common factor, the one most frequently used being money. Then it is possible to express output as its revenue equivalent. This will be designated as $\$Q/t$.

Finally, it is possible to express output rates as a percentage of practical capacity. This sort of measure is becoming quite common because it can be interpreted more easily. (Operation at 85 percent of capacity is information of a different sort from operation at 200,000 tons per week.) Percent of capacity can be computed for any of the foregoing cases, no matter what degree of variability exists in the final product. There are practical difficulties associated with this refinement, however, because there is an inherent degree of vagueness in specifying practical capacity and in estimating it. For example, is one or more shifts meant, or what degree of overtime is assumed? Nonetheless, when this measure is used, it will be denoted by $\%Q/t$.

Fixed vs. Variable Costs

In characterizing the relationship between costs and the rate of output of a facility, the first step is to distinguish among the various classes of costs. The simplest way to do this, and the most sensible for a first consideration, is to divide costs into two major categories: purely fixed and purely variable. Purely fixed costs are those that do not change at all in response to short-run variations in the output rate (those that can be implemented in less than one year). Examples are various types of prepaid insurance; supervisory, managerial, or staff salaries; planned research and development expenses; and rental payments, or expenses associated with investment in equipment, land, or buildings. Purely variable costs are those that do vary in response to output changes according to some functional rule. Examples are raw materials costs and energy costs associated with process or machine operations.

Operative labor costs have been treated in most texts as variable, because historically labor in free-market economies has been viewed as a variable factor. In other words, as activity increased and the need for labor increased, labor would be added, with a corresponding increase in costs. The reverse was also true. As time went on, however, this description became less universally valid because of a gradual change in the way labor was utilized in industry.

First of all, specialization in labor functions increased along with increases in labor productivity. This resulted in a concomitant increase in training required for workers. Thus, it took longer to bring workers onstream, and there was an implicit penalty in reducing such labor when activity declined in the short run. In addition, the gradual increase in labor productivity resulted in a decline in labor's share of total production costs. This provided a disincentive to varying the use of labor in order to "absorb" or respond to the changes in the rate of output. Finally, union-management collective bargaining often led to restrictions on management's discretionary ability to lay off workers. Even in some firms with no unions, management's aim to dissuade workers from organizing for unionization frequently inhibited downside labor changes. The upshot of all these developments is that it is not now always accurate to describe operative labor costs as purely variable.

The segmentation of total costs into fixed and variable components is depicted in Figure 6–1. Note that the simplest sort of functional form for variable costs is

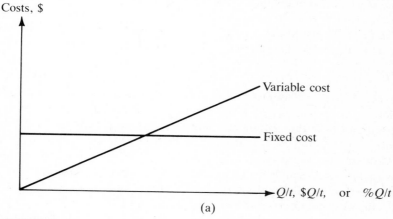

FIGURE 6–1a

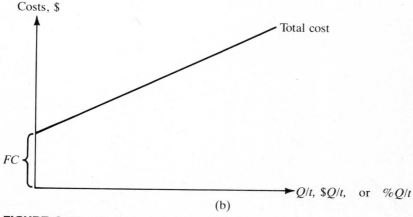

FIGURE 6–1b

used (i.e., linear). Each unit of additional output results in a proportionate increase in variable cost. Figure 6–1a shows fixed and variable costs separately, and Figure 6–1b shows their sum.

An algebraic expression for total cost as a function of output would then be as follows,

$$TC_Q = FC + VC \cdot Q$$

where TC_Q is total cost as a function of output and is in units of dollars per unit time; FC is the total fixed cost, a constant, and is in the same units as TC_Q; VC is the variable cost per unit of output, a coefficient, and is in units of dollars per unit; Q is the output rate, and is in units of output per unit time.

Revenue

Revenue as a function of output may be thought of most simply as directly proportional to output. This is easy to visualize in the case of a single-product plant with a constant price for its output. In that event, revenue as a function of output would appear as in Figure 6–2. Algebraically, this would be expressed as,

$$TR_Q = P \cdot Q$$

where TR_Q is total revenue as a function of output and is in units of dollars per unit time; P is the (constant) price per unit of output, and is in dollars per unit of output; and Q is the output rate and is in the same units as in the total cost function.

Although this is a good demonstrative model for revenue and, in fact, is used widely because of its simplicity, its limitations should be recognized. Strictly speaking, the instances of absolutely constant prices are as rare as those of single-product plants. Quantity discounts, differential pricing, and discounting during periods of slack demand are commonplace in the real world. Also, the preponderance of

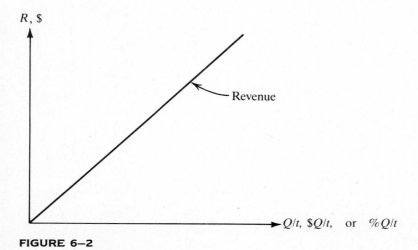

FIGURE 6–2

multimodel and multiproduct plants further complicate the task of constructing revenue curves. A good example of a plant producing a homogeneous product without constant prices is that of an electric utility. Industrial and residential customers typically pay different rates, and within the industrial class of customers there are quantity discounts for large users. Sometimes there are also differential rates for off-peak use.

Several approaches may be taken with respect to deviations in practice from constant prices. One is that within the operating range foreseeable in the short run, the expected average price may be used. This often gives good results for modeling and has the advantage that, in terms of algebraic treatment, it is the same as the revenue model illustrated above. Modeling the anticipated nonlinear behavior of price is more difficult and will be illustrated in a subsequent section.

Break-Even Analysis

Given the cost and revenue functions described above, it is possible to analyze the operating characteristics of a facility so described. The most salient characteristics of general interest include the break-even point (the output rate at which revenues and costs are equal); the amount of revenue, cost, and thus profit, at a target rate of output; and the effect on these two characteristics of a change in the operating rate or operating design of the facility. Each of these three will be examined in turn.

Break-Even Point In a free-market environment, the break-even point is of vital interest to analysts and planners because it indicates the rate of operation that needs to be maintained on the average, over the short run, in order to attain long-run survival. It is shown graphically in Figure 6–3 as the intersection of the total cost and revenue functions. Thus, at Q_0 revenue and cost are equal, amounting to TC_0 or TR_0.

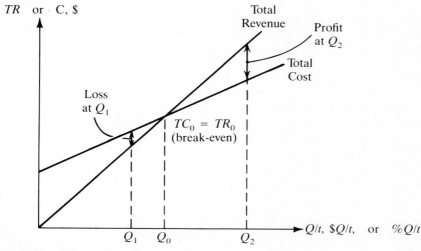

FIGURE 6–3

Algebraically, the break-even point may be evaluated as

$$TC_Q = TR_Q$$

$$FC + VC \cdot Q = P \cdot Q$$

$$FC = P \cdot Q - VC \cdot Q = Q(P - VC)$$

and finally,

$$Q_{BE} = \frac{FC}{P - VC}$$

where the units are the same as before.

Example 6–1 illustrates how cost and revenue curves may be depicted, as well as how to compute their break-even point.

EXAMPLE 6–1

A manufacturer estimates that fixed costs at his plant amount to about $600,000 per year, while variable costs are approximately $14 per unit. The price per unit of output averages $20 over the year. What is the break-even point? If practical annual capacity is estimated to be 160,000 units, how does the break-even point compare to it?

The information above points to linear revenue and cost curves:

$$TR_Q = P \cdot Q = 20Q$$

$$TC_Q = FC + VC \cdot Q = 600,000 + 14Q$$

and thus,

$$TC_Q = TR_Q$$

$$600,000 + 14Q = 20Q$$

so Q_{BE} = 100,000 units per year for break-even. If capacity is 160,000 units, break-even is equivalent to 100,000/160,000 or 62.5 percent of capacity. These are illustrated in Figure 6–4.

Other-than-break-even Point Break-even analysis is useful in giving insight into the effect of operating at rates other than the break-even point. Figure 6–3 also indicates the revenues, costs, and profit (or loss) at hypothetical points above and below break-even.

For Example 6–1, one can compute the revenue, cost, and profit at the output rates of, say, 90,000 and 120,000 units per year. At 90,000 units,

$$TR = 20(90,000) = 1,800,000$$

$$TC = 600,000 + 14(90,000) = 1,860,000, \text{ and}$$

$$PROFIT = TR - TC = -\$60,000$$

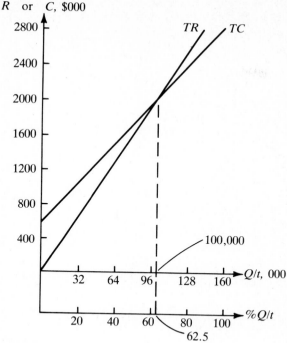

R or *C*, $000

FIGURE 6–4

or a $60,000 loss. At 120,000 units,

$$TR = 20(120,000) = 2,400,000$$

$$TC = 600,000 + 14(120,000) = 2,280,000, \text{ and}$$

$$\text{PROFIT} = TR - TC = \$120,000$$

Effects of Changes The usefulness of break-even analysis stems also from the aid it provides in anticipating the results of operating at various rates or of operational or design changes in the establishment.

Operating Changes Suppose the plant in Example 6–1 is operating at 85 percent of capacity, as shown in Figure 6–5, which shows a narrower range of operating rates than the preceding figure. The effect of reducing the rate of activity to 75 percent of capacity because of a slump in business caused by a recession is a reduction in profit from $216,000 to $120,000. (Note that revenues and costs are changing simultaneously to produce the indicated profit change.)

Another possible change which could be evaluated by break-even analysis is that of a price modification. If one begins by assuming perfect inelasticity (a small price increase causes no effect on the quantity demanded), then a price increase of, say, 5 percent in Example 6–1, when the plant is operating at an output rate of 136,000 units per year, can be depicted as in Figure 6–6. Note the change in the revenue function, which increases profit by $136,000.

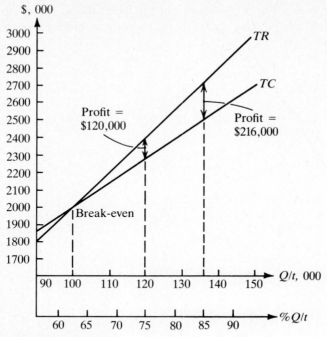

FIGURE 6–5

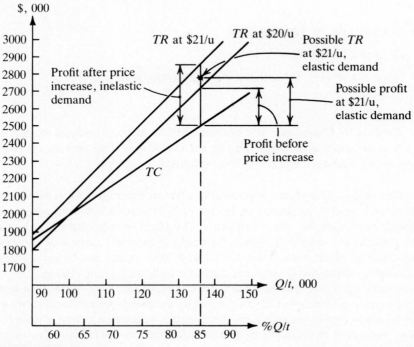

FIGURE 6–6

However, an assumption of perfectly inelastic demand is rather unrealistic for most market behavior. Usually, price increases with no underlying changes in demand behavior result in a percent reduction in quantity demanded which is less than the percentage price increase. This combination of increased price and somewhat reduced demand is referred to as relatively inelastic demand and yields a percentage increase in revenue somewhat less than the percentage increase in price. Such a change is depicted as a single point in Figure 6–6. (Only this single point is shown because the entire revenue function would not be linear. Nonlinear revenue curves will be treated in a subsequent section.)

Other causes for operating changes, sometimes involving combinations of factors, are illustrated by the exercises at the end of the chapter.

Design Changes Break-even analysis can be used to shed light on the change in short-term operating characteristics that would result from new equipment or a new *modus operandi*. Figure 6–7 shows the cost function for the existing operation of Example 6–1 along with that for a proposed one. The proposed operation differs from the existing one in that it requires increasing the investment

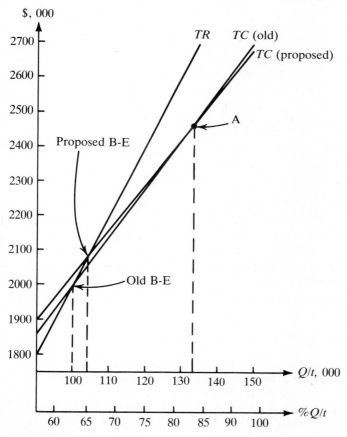

FIGURE 6–7

in fixed plant (which would then show a larger short-run fixed cost) to a level of $700,000 per year. However, this new investment would increase labor productivity, resulting in lower unit labor costs and, thus, a more gradually sloped variable cost function.

Suppose that the variable cost per unit is reduced to $13.25. The original revenue function (at $20 per unit), which is independent of these two alternatives, is superimposed. The break-even analysis shows that the new investment would yield greater operating profits if a high level of production can be maintained, say 85 percent of capacity or 136,000 units per year. (The reader should verify all of the following computations.) At the 85-percent output rate, profit would be $218,000 per year (as compared to $216,000 under the old method). If the operating rate can be pushed up to 93.75 percent, or 150,000 units per year, then profits would be increased to $312,500, as opposed to $300,000 for the old method. However, this is accomplished at the risk of greater losses if such relatively high operating rates cannot be achieved. It can be seen that the increased investment is accompanied by a higher break-even point, to 103,704, which means that a greater activity rate is needed to break even. By the same token, losses would be greater than would be experienced with the existing plant if activity is low. For example, if the operating rate were reduced to 80,000 units per year, losses under the proposed method would be $160,000 per year, as compared to $120,000 under the existing method.

In addition, one can examine the point of indifference between the two methods—i.e., that operating rate at which the costs of the two methods are the same. (In a sense, this is another form of break-even.) This is labeled as point A in Figure 6–7. It may be computed as follows:

$$600,000 + 14\,Q_A = 700,000 + 13.25\,Q_A$$

$$0.75\,Q_A = 100,000$$

$$Q_A = 133,333$$

At rates higher than 133,333, costs for the new method are lower, and vice versa.

The simple break-even model above, showing the effects of increasing investment, illustrates a classic pattern that has been in effect since the beginning of the industrial revolution: greater levels of fixed investment in manufacturing, accompanied by greater labor productivity, along with a strong incentive for the manufacturer to maintain high rates of activity (in order to be to the right of break-even), often increasing plant capacity.

Unit Costs

The break-even analysis presented in the preceding sections was based on the functions of total cost and revenue with respect to output rates. It is also possible to depict the same relationships by indicating costs per unit (as opposed to total costs) as they relate to output rates. Although the subject of estimating unit costs is a very important one in other contexts, this sort of analysis may be of supplementary interest to analysts using break-even. (The estimation of standard costs is the chief arena for a unit-cost viewpoint and is widely used for target pricing, short-term planning and scheduling, and analysis of performance as compared to plan. That subject is covered in Chapter 11.)

In the unit-cost approach, the cost curves of the preceding sections must be transformed. Thus, at every output rate interest is now focused on the ratio of cost to the number of units produced.

Whereas the curve of total fixed costs as a function of output was simply

$$FC_Q = FC$$

and appeared as in Figure 6–8a, if unit fixed costs are desired it now becomes

$$UFC_Q = \frac{FC}{Q}$$

which now appears as in Figure 6–8b. Note that unit fixed cost declines with increases in the output rate and is hyperbolic and asymptotic to the horizontal axis at $FC = 0$. The interpretation of this is straightforward, in that fixed costs are defined as a lump sum so that the greater the operating or production rate, the less fixed cost there is to be absorbed by each unit.

Unit variable costs as a function of output may be similarly generated. If one begins with a total variable-cost curve that is linear, as was done previously, then it may be expressed as

$$VC_Q = VC \cdot Q$$

where VC is the variable cost per unit. This appears as in Figure 6–9a. Dividing by Q as before yields the unit variable cost curve, or

$$UVC_Q = VC$$

and is shown in Figure 6–9b. The result should be as expected—namely, that variable cost per unit is constant over the feasible operating rates.

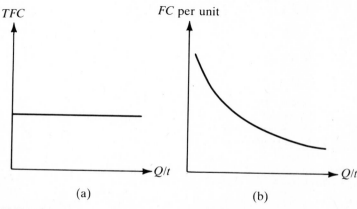

(a)

(b)

FIGURE 6–8a

FIGURE 6–8b

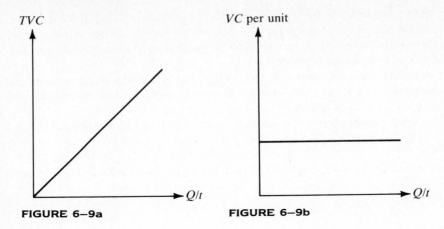

FIGURE 6—9a   **FIGURE 6—9b**

If the fixed and variable unit cost curves are combined, one obtains the cost curve depicted in Figure 6–10. Depending on what one uses for a unit revenue curve, a variety of unit break-even curves may be obtained. Assuming the same total revenue function as before, namely

$$TR_Q = P \cdot Q$$

gives a unit revenue curve of

$$UR_Q = P$$

This is depicted with the unit costs in Figure 6–10.

Although both total and unit curves give the same information, there is sometimes a preference for one or another, depending on the application. For example, if an analyst is interested in the effect of an action on total profit, total curves may be preferred, whereas if the focus is on pricing, unit curves may be preferred.

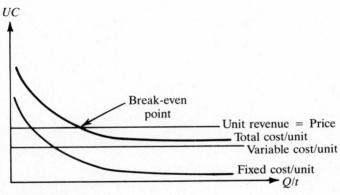

FIGURE 6—10

Nonlinear Break-Even Curves

The use of linear functions for total fixed cost, variable cost, and revenue represents a great simplification of reality. Where the analyst perceives that oversimplification and the risk of an invalid model would result from using linear functions, then an alternative is to employ nonlinear curves. To illustrate the possibilities that exist in this regard, fixed costs can be represented as an absolute constant, as before, while allowing the function of short-run variable costs to describe some of the nonlinearities commonly thought to exist. For example, it is often observed that variable cost per unit increases as the establishment approaches practical capacity. This is referred to as "diminishing returns" and may be due to less coherent and efficient use of the facility as such use becomes concentrated. These two descriptions of fixed and variable cost may be combined simply in a parabolic curve such as the one in Figure 6–11. There total cost may be expressed as

$$TC_Q = FC + VC \cdot Q^2$$

where FC is the fixed-cost component and corresponds to the y-intercept and $VC \cdot Q^2$ gives the variable cost component, the unit increment of which increases as plant capacity is approached.

Turning attention to the total revenue function, one may perform a break-even analysis under the same revenue function as before, namely

$$TR_Q = P \cdot Q$$

This appears in Figure 6–12a. However, if this revenue assumption is also viewed as oversimplified and invalid, one may seek nonlinear extensions as was done for

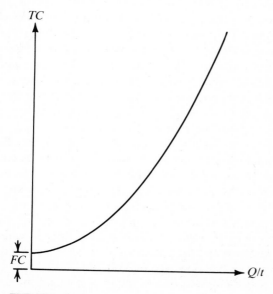

FIGURE 6–11

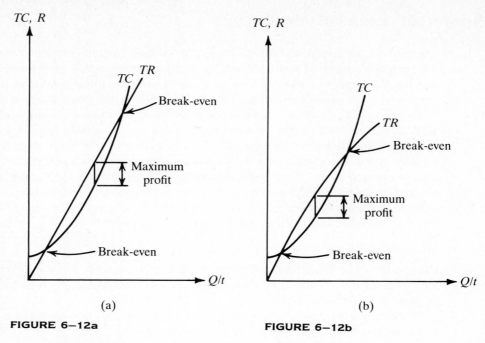

FIGURE 6–12a   **FIGURE 6–12b**

costs. One possible approach is to use a function of the form

$$TR_Q = AQ - BQ^2$$

which appears in Figure 6–12b along with the same cost curve as before. Such a revenue curve is more realistic in that it is based on a downward-sloping demand curve for the establishment's output. Thus, it takes into account the likely condition that higher sales can be achieved by lowering unit prices. Such being the case, revenue would not rise linearly as in Figure 6–12a, since higher sales are accompanied by somewhat lower prices. The combined effect is that of increasing revenue, but at a decreasing rate. (The demand curve used for the purpose of generating the preceding revenue function was a simple linear one. It is left for the reader as an exercise at the end of the chapter to show how such a linear, downward-sloping demand curve may be shown to yield the revenue function above.)

In general, nonlinear functions such as those illustrated have certain characteristics which distinguish them from their linear counterparts. To begin with, there may be more than one break-even point. The reason for this is that the optimal point of operation, from the point of view of maximizing profit, is not simply "as far as possible to the right of break-even," or at full capacity, as was implied by linear break-even analysis. Figure 6–12 indicates that optimal profit is attained between the two break-even points. Mathematically, this is where the difference between the cost and revenue functions is greatest (or where the first derivatives of the cost and revenue functions are equivalent).

Although this is a more realistic description of the typical establishment than that provided by linear functions, in practice, users of linear models typically apply

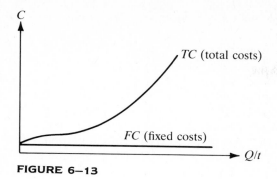

FIGURE 6–13

them in a range surrounding the break-even point, not approaching extreme values such as full capacity. Thus, the disparity in conclusions reached through using different models is less than would otherwise be expected.

Finally, a special form of nonlinear cost curve is used by economists to describe the behavior of the firm. It appears in most microeconomic textbooks as a graph like Figure 6–13. This analysis is entirely consistent with break-even analysis. In fact, break-even analysis was initially developed as a simplification of the classic microeconomic model in order to facilitate its application in an industrial setting.

Once the nonlinear revenue and cost functions have been generated, their use and analysis follow along the same lines as for the linear models. Of course, in practice the analyst must spend a good deal of time in estimating the functional forms and coefficients that best fit the behavior of the establishment that is the focus of his work. This is a process not dealt with in this book. (For a comprehensive treatment, see the references for this chapter.)

Break-Even Extensions: Indifference Analysis Applied to Time Value Models

Break-even analysis has been shown to be a practical and valuable tool for studying the short-run effect on an economic entity of changes in the operating rate. Although the short-run viewpoint employed is in marked contrast to the long-term view facilitated by time value analysis, nonetheless there are abstract features of the break-even analysis which may be profitably utilized for time value analyses. However, in order to avoid confusion, this extension will be referred to in this book as indifference analysis.

Indifference analysis can best be explained by referring to Example 4–2, from the section on deferred-investment decisions in Chapter 4. There it was assumed that the future date at which the increase in capacity would be necessary would be year 8. That sort of assumption may now be viewed as either an elementary pedagogical prop or a preliminary estimate in the course of a modeling effort. In any case, from a practical perspective, such an assumption is likely to turn out to be erroneous because it is generally not possible to estimate future events quantitatively with such precision.

An alternative approach using sensitivity analysis postpones the question of when the added capacity will be needed. Instead, it begins by treating that event as a variable and seeks to answer the question, "At what time period would a provision of increased capacity just equate the present worth of costs for the two design alternatives?" That question will be explored using indifference analysis as a prerequisite for a more detailed discussion of the design dilemma.

Before proceeding with the analysis, however, it should be pointed out that the "step" in annual expenses—that is, from 50 to 80—will continue to be treated as coinciding with the act of increasing capacity. Thus, not only is the time of the deferred investment an unknown, but so is the timing of the change in the disbursement pattern. This assumption is used because the increase in annual expense is one of the results of increasing the capacity.

Table 6–1 gives the values of present-worth calculations (done with the computer program described in Chapter 5) for the full- and deferred-investment alternatives of Example 4–2 under varying assumptions for the time of installation of the additional capacity. As an illustration, for year 5,

$$PW_{full,\ 5} = -600 - 50(P/A\ 10\%, 5)$$
$$- 80(P/A\ 10\%, 10)(P/F\ 10\%, 5)$$
$$+ 100(P/F\ 10\%, 15)$$
$$= -1,070.8$$

$$PW_{deferred,\ 5} = -400 - 300(P/F\ 10\%, 5) - 40(P/A\ 10\%, 5)$$
$$- 80(P/A\ 10\%, 10)(P/F\ 10\%, 5)$$
$$+ 150(P/F\ 10\%, 15)$$
$$= -1,007.2$$

If the times in Table 6–1 (and several others beyond year 5) were graphed, they would appear as Figure 6–14. The point of indifference (analogous to the break-even point), where the present worth of costs is the same for both alternatives, is between 2 and 3 years. If one uses linear interpolation as an approximation, the indifference point is 2.62 years. To the left of this indifference point the present worth of costs for the full-investment alternative is lower, while to the right the

TABLE 6–1 **Present Worth of Costs for Both Alternatives under Varying Assumptions for Timing of Increased Capacity**

	Time When Capacity Is Increased (yrs.)				
	1	**2**	**3**	**4**	**5**
PW_{full}	$-1157.3	$-1132.5	$-1109.9	$-1089.5	$-1070.8
$PW_{deferred}$	-1208.9	-1151.1	-1098.5	-1050.7	-1007.2

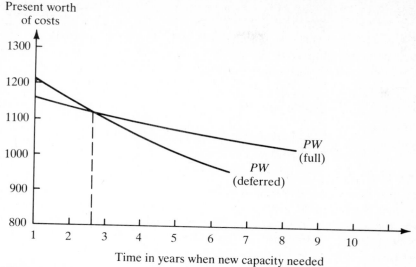

FIGURE 6-14

deferred-investment alternative exhibits lower present worth of costs. In other words, if one expects the additional capacity to be needed before 2.62 years, then the full-investment alternative would be more desirable; if the added capacity is not expected to be needed until after 2.62 years, then the deferred-investment alternative would be superior.

Although this may seem to beg the original question of which alternative to recommend, in fact it helps the professional analyst to focus on an important aspect of the problem. What this provides is the basis for a sensitivity analysis regarding the influence of the important variable, time when the new capacity is needed, rather than the overly simplistic assumption of a particular time, as was used in Chapter 4. The analyst, therefore, sees in stronger terms the value of examining the relative likelihoods of when the additional capacity will be needed. If such empirical studies give weight to a range of likely times greater than 3 years, then he may be confident in recommending a deferred investment.

The reverse also holds. If, as sometimes happens, the best estimates span the indifference point in a roughly uniform way, then no immediate guidance is provided. However, in that case, the analyst is no worse off than before and has the advantage of knowing what pitfalls accompany either choice. In addition, he may devise other means of analysis to supplement what he has already learned, in order to clarify the situation somewhat.

The joint use of indifference and sensitivity analysis forms an important complement to the time value models already presented. These will be used in combination in the rest of this book. Of course such an approach requires much more computation, which could obviously be facilitated by using a computer. This issue will be addressed further in Chapter 12.

Exercises

6–1 *a.* A firm has fixed costs of $350,000 per month and variable costs of $32 per unit. If the average price to distributors is $47, what is the break-even volume?
b. What would the profit (loss) be if sales were at 24,000 per month?

6–2 *a.* A firm has fixed costs of $475,000 per month. Variable costs are $18 per unit. The product can be sold for $25 per unit. What is the break-even point for the firm?
b. Suppose that the variable costs can be reduced by $1 per unit by increasing fixed costs by $25,000 per month. Is this desirable? Explain. The firm's "standard volume" has been around 75,000 units per month.

6–3 Your firm is considering changes at its roller-bearing plant. Fixed costs are $500,000 per month. Variable costs are $0.85 per unit (standard dozen). Currently the average price is $1.24 per standard dozen. Sales and production are now at a rate of 1.3 million units per month. The changes being considered are as follows:

a. Raise the price by $.05 per standard dozen. This is expected to reduce unit sales by 2 percent.
b. Make suggested improvements adding $50,000 per month to fixed costs. These will reduce variable costs by $.06 per unit.

What do you recommend and why? If these two proposals are deemed independent of one another, in that both may be implemented, what would you recommend?

6–4 *a.* A new plant has fixed costs of $450,000 per month. Variable costs are $21 per unit. The average price to distributors is $33 per unit. What sales volume must the firm maintain to break even?
b. If a short-term contract can be obtained to sell 2,500 units per month to an overseas customer at $26 per unit, what effect would this have on the firm's profitability, other things remaining the same? What would you then say about the desirability of such price-cutting?

6–5 An improvement for a post-assembly inspection operation is being studied. It will require investment in new machinery at the inspection station. The initial configuration is expected to generate savings in inspection labor and reductions in subquality goods worth $27,000 per year. The additional operating expense of this configuration, including added property taxes, insurance, and energy consumption, is about $2,000 per year. The system is expected to last 6 years, with no salvage value. At most, how much could the investment cost be for this system, if funds are worth 16 percent per annum?

6–6 A designer is considering the use of asphalt or slate shingles as roofing materials. Maintenance costs are negligible, and so are not considered. The roof of asphalt shingles is expected to give 20 years of service, while the slate would be serviceable for 30 years. If the cost of asphalt shingles for a particular installation is computed to be $4,000, and the MARR is 12 percent, at most how much could slate cost before asphalt becomes more desirable?

6–7 A production engineering group is performing a value engineering analysis of a manufactured part. The part is currently made of aluminum, costs $20, and, based on previous experience, lasts 10 years, on average. It is proposed to change to Teflon, which would reduce the unit part cost to $10. However, the useful life of the Teflon part is in question. What would the average useful life of the Teflon part have to be for the equivalent uniform annual costs of both to be the same? Here, only acquisition cost is pertinent. Assume that the MARR is 16 percent. What material would be favored if the life of the Teflon part could safely be estimated as 4 to 6 years?

6–8 A firm is considering investment in a new process to recover a by-product of its primary manufacturing activity. The equipment would require a $105,000 investment and serve for 8 years, with an expected salvage value at that time of $10,000. Although the idea seems attractive, it is less clear how much revenue the processed by-product would yield over the study period. In conducting a sensitivity analysis of the problem, the engineer in charge computed the minimum annuity income necessary to make the investment attractive, assuming funds were worth 15 percent. What is that amount?

6–9 A manufacturing firm faces a make-or-buy decision with respect to a certain subcomponent it uses. It can purchase the part from a reliable vendor for $23 per unit. Alternatively, it can produce the part in-house with no major change to its productive facility. However, it would have to invest $40,000 in new equipment, which is expected to serve for 8 years, with a salvage value of $5,000. In addition, production of the part would entail an increase in overhead (fixed cost) of $2,000 per year. The direct (variable) cost of manufacturing would be $9 per unit. At what volume of parts per year would the annual costs be the same for making or buying? Assume that the MARR is 16 percent.

6–10 An electronics manufacturer has just purchased a large number of new soldering machines for its plant, and it is studying the economic feasibility of a service contract with a local agency, as compared to doing the maintenance itself. In order for the maintenance department to do the work, $15,000 of new equipment will be needed, which will be useful for 10 years, after which it will have a salvage value of $5,000. Other fixed costs associated with internal maintenance amount to $8,000 per year. The average direct (variable) cost of repair per unit is estimated to be $50. If the number of such repairs per year is expected to be about 150, what annual flat price for a service contract would have the equivalent annual cost?

6–11 This situation was described in Chapter 1. A manufacturing company requires the service of a new 50 h.p., totally enclosed, fan-cooled electric motor for which its best estimate is that it will be run an average of 7 hours per day for 250 days per year. Past experience indicates that (1) its annual costs for property taxes and insurance can be estimated at 5 percent of investment cost, (2) it must earn 18 percent on invested capital before income tax considerations, and (3) it must recover capital invested in such equipment within 3 years. Two motors are offered to the company. Motor *A* costs $1,990 and has a projected efficiency of 90.5 percent at the indicated operating load. Motor *B*, a high-efficiency version, costs $2,429 and has a guaranteed efficiency of 94.5 percent under the same operating load. One h.p. = 0.746 kw. At what price of electric power would the equivalent annual costs of the two motors be the same? If power were expected to cost less than 7 cents per kilowatt-hour over the 3-year period, what conclusion can be drawn?

6–12 Reconsider Exercise 4–14. At what time would adding the second two lanes of the deferred-investment alternative give the same present worth of costs as the initial full-investment alternative? What is the significance of this time?

6–13 Show that a linear demand curve of the form

$$Q = a - bP$$

where *Q* is demand per unit time, *P* is the price per unit, and *a* and *b* are constants, gives a total revenue function that is curvilinear, of the form

$$TR = \frac{aQ - Q^2}{b}$$

Selected References

Benston, G. "Multiple Regression Analysis of Cost Behavior," *Accounting Review*, October, 1966, pp. 657–672.

Brigham, E. and J. Pappas, *Managerial Economics*, 2nd ed., Hinsdale, IL: Holt, Rinehart and Winston, 1976.

Draper, N. and H. Smith, *Applied Regression Analysis*. New York: John Wiley & Sons, 1966.

Johnston, J. *Statistical Cost Analysis*. New York: McGraw-Hill, 1960.

7

Depreciation

Introduction

In the broadest sense, depreciation refers to the inevitable decline in value through time of long-term assets such as machinery, other equipment, and buildings—that is, productive assets, as opposed to assets of increasing value, such as rare art works and other collectibles. Obviously, this is a subject of great importance to decision-makers concerned with capital investments. However, depreciation also has a more particular meaning in the context of the necessarily more arbitrary methods used in accounting practice for that decline in value.

Sound investment decision-making requires not only an estimate of the anticipated depreciation meant in the first sense but also an understanding of the treatment of depreciation for accounting purposes. The reasons for this are twofold. For one, decision-makers are often dependent on depreciation accounting records as part of the organizational data base at their disposal. It behooves them to know the definitions implicit in such records. Secondly, the actual depreciation schedule to be employed will have a bearing on the corporate income tax stream to be paid out over time. As will be seen more clearly in a subsequent chapter, such taxes are real economic flows on a par with labor, material, energy, and so on, which need to be reckoned with if valid decision models are to be constructed.

In order to grasp the significance of the methods used to account for depreciation of capital assets in a modern organization, one must have at least a rudimentary understanding of the way such investments are treated as compared with expenses such as labor, materials, and energy. The next section will delineate that distinction and provide an elementary view of the accounting function. This latter subject is also important for the decision-maker because much of the data available for analytical purposes are recorded and reported as part of the modern organization's accounting activity.

The Function of Depreciation Accounting

Productive organizations exist to transform resources such as labor, material, energy, and technology into goods or services desired by others. For a profit-seeking organization to remain in business, the revenues received from the sale or lease of that output over the long run must be sufficient to defray all the expenses incurred, as well as to provide for reinvestment in new productive assets, and still generate a satisfactory profit for the owners. (When this status is achieved, the expression "going concern" is used to describe such an organization.)

Revenues and expenses are accounted for as they are incurred. Investments, however, are not treated as expenses when they occur, even though they necessitate relatively large outlays. Instead, they are viewed as changes in the asset (or liability) structure. For example, a firm with assets of $2 million in cash that makes a $1 million investment in equipment is then still accounted to have $2 million worth of assets; now, $1 million is in cash and $1 million is in the form of equipment. If a firm that starts with assets of $500 thousand in cash and no liabilities borrows $1 million and uses it in conjunction with its cash to acquire $1.5 million worth of equipment, it would then have assets of $1.5 million, all in equipment, and liabilities of $1 million.

Depreciation is an accounting mechanism employed to describe an approximation to the decline in value of assets as they are used. It represents an accounting entry only, in that it does not correspond to any actual monetary or resource transaction. It is treated as an expense, identical in form to "real" ones such as wages or materials and, thus, it has an effect on the firm's net income (revenue less expense) and the income taxes that are based on it. In short, it is a procedure for allocating a time series of expenses to a relatively large investment outlay occurring at a single time. Sometimes the expression "amortize" is used to describe this function.

Depreciation also has an effect on the accounting value of the firm's assets as time passes. The firm's fixed assets are valued at first cost less their accumulated depreciation. Thus, an asset purchased for $10,000 two years ago, for which the depreciation charge is $1,000 in each of those years, would be valued at $8,000 at the end of that time.

With these definitions in mind, attention will now be directed to the most basic accounting statements encountered in the modern organization.

The Income Statement

Also known as "Statement of Profit and Loss," or "Statement of Income and Expense," the income statement summarizes the major revenue and expense categories for a particular time interval. The income statement may be thought of as indicating the highlights of the financial flows for the period it covers. The difference between revenue and expense is referred to as net income before income taxes. Then the income taxes are shown. The residual of net income before tax and income taxes is net profit, a part of which is often paid out to owners in the form of dividends and the balance of which is available for the firm to reinvest. When available profit is divided between dividends and corporate reinvestment, that segment of profit not directly paid out to owners as dividends is called retained earnings. A sample income statement is shown as Figure 7–1.

Note that depreciation is part of the cost of goods sold. Thus, to the extent that revenues are targeted to expenses plus profit, it is clear that depreciation helps to provide for a healthy capital recovery.

The Balance Sheet

The balance sheet may be thought of as a photograph of the firm's financial condition at a particular point in time. In contrast to the income statement, it depicts levels rather than flows. It uses the fundamental accounting identity,

Assets = Liabilities + Equity

FIGURE 7–1 Income Statement (period ended Dec. 31, 19—)

(1) Income
    net sales
    rentals or leases

(2) Expenses
    cost of goods sold (includes depreciation)
    research and development
    selling, general, and administrative

(3) Net income before interest, other income or losses, and taxes
    (1) − (2)

(4) Interest and other income (loss)

(5) Income before income taxes
    (3) − (4)

(6) Income taxes

(7) Net income
    (5) − (6)

as the basis for the structure of the report. Thus, it shows the assets of the organization on one side, and what the assets are attributable to on the other. This can be better understood by examining Figure 7–2.

Definitions for the Balance Sheet   Succinct definitions for the common entries of the balance sheet are as follows:

Current assets   Assets that turn over quickly, or may be converted to cash within one year.

Cash and short-term investments   Cash, checking, and other demand deposits, and marketable securities.

Accounts receivable   Amounts owed to the firm for goods or services rendered on account, but not yet paid for.

Inventories   The value of raw materials, work-in-process, and finished goods.

Prepaid expenses and taxes   Amounts paid in advance for goods or services not yet used.

Fixed assets   Assets that do not turn over quickly as they are used, typically lasting longer than one year.

Property, plant, and equipment   Land, buildings, machinery, and other tangible

FIGURE 7–2   Balance Sheet   (Dec. 31, 19—)

(1) Current Assets
        Cash and short-term investments
        Accounts receivable
        Inventories
        Prepaid expenses and taxes

(2) Fixed Assets
        Property, plant, and equipment at first cost,
            less accumulated depreciation = net book value

(3) Other Assets

Total Assets (1) + (2) + (3)

(1) Current Liabilities
        Notes and loans payable
        Accounts Payable
        Income taxes payable
        Other accrued liabilities

(2) Long-term Liabilities
        Long-term debt

(3) Other Liabilities

(4) Equity or Net Worth
        Capital stock
        Retained earnings (surplus)

Total Liabilities and Equity (1) + (2) + (3) + (4)

assets used for the firm's productive purposes, valued at original cost as amended by IRS regulations.

Accumulated depreciation   The total value of depreciation charged on the assets above, from the date of acquisition to the time of the balance-sheet statement.

Net book value   The difference between the two entries above.

Other assets   Other tangible assets not included above, such as investments in other corporations or subsidiaries, or intangible assets such as the value of patents, trademarks, copyrights, and customer goodwill.

Current liabilities   Obligations payable in the near future (i.e., in the next year).

Notes and loans payable   Borrowed funds that must be repaid in the next year.

Accounts payable   Amounts owed to others for goods or services received on account but not yet paid for.

Income taxes payable   Income taxes incurred but not yet paid.

Other accrued liabilities   Short-term liabilities not included above, such as wages and salaries accrued but not yet paid.

Long-term liabilities   Obligations payable over the long-term future (i.e., after one year).

Long-term debt   Borrowed money scheduled to be repaid over a period more than one year in the future.

Other liabilities   Amounts not included above that must be repaid in the future, such as deferred income taxes.

Equity or net worth   The value of the owners' investment in the enterprise.

Capital stock   The representation of the owners' investment outlays.

Retained earnings (surplus)   The value of the accumulated residual of profits less the dividends paid to shareholders; used for capital formation.

As can be seen from the balance sheet, depreciation plays an important role in accounting for the value of fixed assets in the firm. Note, however, that such "book value" is not designed to correspond to the actual market value of those assets at the time of the statement.

The Treatment of Depreciation

Accounting for the depreciation of a capital asset requires a procedure for allocating a depreciation allowance for the asset for each period, typically one year. This necessitates agreement on a variety of factors, which include the following:

The depreciable life of the asset   This is a time period, set a priori, over which the periodic depreciation allowances will be taken. It must conform to federal Internal Revenue Service guidelines. Table 7–1 lists some of these guidelines. It is important to note that the actual useful life of the asset is not likely to match the depreciable life. The actual life may be more or less, depending on the intensity of use, maintenance experienced, and other factors which determine its relative deterioration, as well as its relative obsolescence.

Investment cost   This refers to the total acquisition cost of the asset. It includes

TABLE 7-1 Asset Guideline Periods for Depreciation Purposes

Asset Description	Asset Guideline Period (yrs.)
Office Furniture, Fixtures and Equipment	10
Information Systems (includes computers and their peripheral equipment used for business activities)	6
Automobiles	3
Light General-Purpose Trucks	4
Heavy General-Purpose Trucks	6
Tractor Units for Over the Road	4
Mining and Quarrying Equipment	10
Oil and Gas Well Drilling Equipment, Onshore	6
Oil and Gas Exploration and Production Equipment	14
Petroleum Refining Equipment	16
Construction Equipment	6
Equipment for Manufacture of Apparel	9
Equipment for Manufacture of Wood Products and Furniture	10
Equipment for Manufacture of Chemicals and Allied Products	9.5
Equipment for Manufacture of Primary Nonferrous Metals	14
Equipment for Manufacture of Foundry Products	14
Equipment for Manufacture of Primary Steel Mill Products	15
Equipment for Manufacture of Fabricated Metal Products	12
Equipment for Manufacture of Electrical and Non-Electrical Machinery	10
Equipment for Manufacture of Electronic Components, Products, and Systems	6
Equipment for Manufacture of Motor Vehicles	12
Equipment for Manufacture of Aerospace Products	10
Ship and Boat Building Equipment	12

Source: Internal Revenue Service Publication 534, Rev. Dec. 1984.

the purchase price as well as delivery, installation, and other miscellaneous costs.

Salvage value   When used for depreciation purposes, it is an a priori estimate of the sale or trade-in value of the asset expected at the end of its depreciable life. Such estimates are most often based on historical experience with similar items. Thus, the reliability of such estimates is quite variable and depends partly on the degree of specialization of the asset. The more specialized the asset, the less guidance can be obtained from historical experience and, thus the greater risk in estimating the salvage value. Of course, the established salvage value may be zero, currently a common practice.

The reader should bear in mind that the expression "salvage value" may also be used in another context—namely, that for market value in an ex-post analysis, or for replacement analysis.

Once the depreciable life, investment cost, and salvage value of an asset have been agreed upon, a depreciation method can be used to determine the depreciation allowance for each year. However, before such depreciation methods are examined, several other aspects of depreciation accounting need to be explored. In this context, a simple example will be useful.

EXAMPLE 7–1

Suppose the investment cost of an asset is $10,000, the salvage value is $1,000, and the depreciable life is 3 years. Further suppose that the depreciation method generates allowances of $4,500, $3,000, and $1,500 in the first, second, and third years, respectively. One property common to all depreciation methods is illustrated by the example; the sum of the depreciation allowances over the depreciable life is equal to the depreciable amount (i.e., the investment cost less the salvage value).

Another important element of depreciation accounting is the concept of book value. Book value at any point in time is the investment cost less the accumulated depreciation allowances up to that point. For this example, book value at the end of the first year would be $10,000 − $4,500, or $5,500. At the end of the second year it would be $10,000 − ($4,500 + $3,000), or $2,500. Note that the book value, having been determined from the investment cost, the a priori depreciable life, and salvage value, is not likely to correspond to the actual market value at the moment.

Depreciation Methods

Although the principles of depreciation accounting, as delineated in the previous sections, have not changed in modern times, depreciation methods acceptable to the federal government for income-tax purposes have frequently been modified. Thus, current and future practice should be viewed as changing in response to public policy.

The following symbols will be used to facilitate an explanation of the various depreciation methods:

P = The investment cost of the asset;
n = Its depreciable life;
SV = Its salvage value;
d_t = The depreciation allowance for period t (d when the allowance is the same in all years, as in straight-line depreciation);
D_t = The accumulated depreciation through the end of period t;
BV_t = The book value at the end of period $t = P − D_t$.

Here, too, an example will illustrate each of the methods.

EXAMPLE 7–2

The investment cost of an asset is $24,000, the depreciable life is 5 years, and the expected salvage value is $3,000. For each of the following methods, a graph of book value through time is provided in Figure 7–3. Table 7–2 gives the numerical values used for the graphs, and may also be used to compare the various depreciation methods.

TABLE 7–2 Depreciation Methods Compared* (in dollars)

yr	SL		SYD		DB(k = 0.34)		ACRS		SF	
	d_t	BV_t	d_t	BV_t	d_t	BV_t	d_t	BV_t	d_t	BV_t
1	4,200	19,800	7,000	17,000	8,160	15,840	4,800	19,200	3,114	20,886
2	4,200	15,600	5,600	11,400	5,386	10,454	7,680	11,520	3,114	17,305
3	4,200	11,400	4,200	7,200	3,555	6,900	5,760	5,760	3,114	13,186
4	4,200	7,200	2,800	4,400	2,346	4,454	3,840	1,920	3,114	8,449
5	4,200	3,000	1,400	3,000	1,548	3,006†	1,920	0	3,114	3,002†

SL: straight-line; SYD: sum-of-the-years'-digits; DB: declining balance; ACRS: accelerated cost recovery system; SF: sinking fund.

*The data in this table are shown plotted in Figure 7–3.

†Does not equal 3,000 because of rounding.

Straight-line Depreciation

This is the simplest method of depreciation. Although it is not now used universally, it is still important, not only for the use that it still enjoys but also as a standard against which other methods may be compared. This will become clear later in the chapter.

The straight-line method allocates an equal proportion of the depreciable amount to each period of the depreciable life.

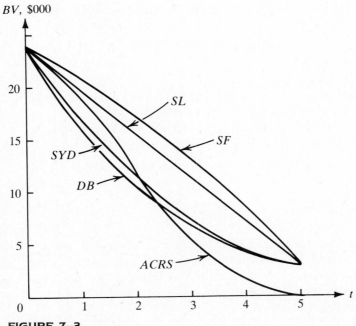

FIGURE 7–3

EXAMPLE 7–2a *Straight-line Depreciation.*

$$d = \frac{(P - SV)}{n}$$

Thus, for Example 7–2,

$$d = \frac{(24{,}000 - 3{,}000)}{5} = \$4{,}200 \text{ per year}$$

The accumulated depreciation is given by

$$D_t = \frac{t(P - SV)}{n}$$

After year 2 of example 7–2, book value would be

$$BV_t = 24{,}000 - \frac{2(24{,}000 - 3{,}000)}{5} = \$15{,}600$$

A graph of book value through time may be seen in Figure 7–3, and is based on data in Table 7–2. That book value declines linearly each year is the rationale for the name of the method: straight-line.

Sum-of-the-years'-digits Depreciation

Sum-of-the-years'-digits depreciation differs from straight-line in allocating larger depreciation allowances in earlier years, with smaller allowances toward the end of the depreciable life. It accomplishes this by using the sum of the digits of the years of depreciable life as the denominator of the ratio apportioning the depreciable amount and using as the numerator the year number in reverse order, highest first.

EXAMPLE 7–2b *Sum-of-the-years'-digits Depreciation.*

For this example the depreciation period is 5 years. Thus, the denominator of the ratio would be the sum of the digits, or $(1 + 2 + 3 + 4 + 5) = 15$. The fraction of the depreciable amount for the first year would then be $5/15$, for the second year $4/15$, for the third year $3/15$, and so on until $1/15$ is allocated for the last year. Since $(5/15 + 4/15 + \cdots + 1/15) = 1$, the entire depreciable amount is ultimately allocated. (The reader can verify that just as $(5/15 + 4/15 + \cdots + 1/15) = 1$ in this case, so it does in general for this procedure.) The annual and accumulated depreciation for this example are shown in Table 7–3 for each year.

Algebraically, the depreciation allowance for any year may be computed using,

$$d_t = (P - SV) \frac{2(n - t + 1)}{n(n + 1)}$$

TABLE 7–3 Sum-of-the-years'-digits Depreciation *(in dollars)*

Year	d_t	D_t	BV_t
1	$(5/15)(24{,}000 - 3{,}000) = 7{,}000$	7,000	17,000
2	$(4/15)(24{,}000 - 3{,}000) = 5{,}600$	12,600	11,400
3	$(3/15)(24{,}000 - 3{,}000) = 4{,}200$	16,800	7,200
4	$(2/15)(24{,}000 - 3{,}000) = 2{,}800$	19,600	4,400
5	$(1/15)(24{,}000 - 3{,}000) = 1{,}400$	21,000	3,000

Figure 7–3, which shows book value through time for several methods, facilitates comparison of the different depreciation methods. For the sum-of-the-years'-digits procedure, the depreciation allowance is greater at the outset. As a consequence, the book value declines more rapidly, as is evident from the figure. Such behavior, as compared to the straight-line procedure, is referred to as accelerated depreciation. Other types of accelerated-depreciation methods follow.

Declining-balance Depreciation

This procedure is based on a constant rate of decline in the book value of the asset, which will be denoted by k. It is analogous to the time value models of the preceding chapters, the difference being that here a constant rate of decrease is described. Thus, the book value after period t may be denoted by

$$BV_t = P(1 - k)^t$$

The depreciation during any period is the rate, k, applied to the book value remaining at the start of the period. Thus, the depreciation allowance in period t may be denoted by

$$d_t = (k)BV_{t-1}$$

where BV_{t-1} refers to the book value at the end of period $t - 1$.

The main weakness of this method is that book value cannot directly go to zero, as is sometimes necessary. (Clearly zero is the limit approached by book value for this method, given an infinite time span; however, real assets are depreciated over a finite period. When a zero salvage value is necessary, one can still use the declining-balance method initially, switching to another method, such as straight-line depreciation, in order to bring book value to zero.) Nonzero salvage values pose no difficulty, however, and are, in fact, one of the boundary values used for determining the value of the rate, k, which will depreciate an investment to a given positive salvage value in n periods. Since a schedule is desired that takes book value to salvage value,

$$BV_n = P(1 - k)^n = SV$$

This may be solved for k to yield,

$$k = 1 - \left(\frac{SV}{P}\right)^{1/n}$$

(7–1)

EXAMPLE 7–2c *Declining-balance Depreciation.*

Equation 7–1 can now be applied to Example 7–2 to give a rate of decline of

$$k = 1 - \left(\frac{3,000}{24,000}\right)^{0.2} = 0.34$$

The reader should verify that, when applied to the data of the example, it yields the results depicted in Table 7–4.

That the final book value did not decline to exactly 3,000 is due to rounding errors, principally rounding the applicable rate, k, to 0.34. The curve of book value through time is depicted in Figure 7–3. Both the figure and the table of data indicate that the declining-balance method is an even more accelerated procedure than sum-of-the-years'-digits.

A widely used variation of the procedure above is the double-declining-balance method. It calls for a rate independent of the investment cost and salvage value. Nominally, that rate is just $2/n$. However, IRS guidelines prohibit the depreciation allowance in any year to exceed twice the value of what would be applicable under the straight-line method. In such circumstances the usable rate may be reduced to $1.5/n$ or $1.25/n$. Also, the use of the arbitrary double-declining rate is likely to cause the book value to fall below the nominal salvage value by the end of year n. Since it is not permissible to depreciate an asset more than the depreciable amount, the solution is to cease depreciation charges when the salvage value is attained. Clearly, this would occur before period n.

Switching from One Depreciation Method to Another

The preceding section described an instance where it was necessary to switch from one depreciation method to another (i.e., from declining-balance to straight-

TABLE 7–4 **Declining-balance Depreciation** *(in dollars)*

Year	d_t	D_t	BV_t
1	.34(24,000) = 8,160	8,160	15,840
2	.34(15,840) = 5,385.6	13,545.6	10,454.4
3	.34(10,454.4) = 3,554.5	17,100.1	6,899.9
4	.34(6,899.9) = 2,346.0	19,446.1	4,553.9
5	.34(4,553.9) = 1,548.3	20,994.4	3,005.6

line) in order to accommodate a zero salvage value. Prior to 1981, such switching was generally permissible and could be done if certain rules were obeyed. It might have been advantageous to do so if, by switching, the aggregate depreciation schedule could be accelerated. In such instances the present worth, at time zero, of the stream of depreciation charges would be larger, thus offering real tax advantages. (A detailed analysis of the rationale of the last statement will be presented in the next chapter, where tax issues will be explored in more detail. This is a "chicken or egg" dilemma in that, although depreciation concepts are generally a prerequisite for the analysis of income taxes, an appreciation of the distinctions between depreciation methods depends upon some understanding of those taxes.)

The technique of switching between methods has more significance than just its occasional use before 1981, however. The next depreciation method to be presented, the accelerated cost recovery system, or ACRS (pronounced as "acres"), enacted in 1981 for the purpose of stimulating aggregate productive investment in the economy, utilized the technique of switching in order to arrive at its depreciation schedule. The rationale and technique of switching will be illustrated by an example.

EXAMPLE 7–3

Example 7–2 is modified in the following way and numbered 7–3 to keep the distinctions explicit: The salvage value is changed to zero, with the investment and depreciation period the same as before, $24,000 and 5 years respectively. Here the double-declining-balance method (DDB) will be contrasted with the combination of DDB and straight-line. The DDB method for a 5-year depreciation period results in a value of k of 2/5, or 0.4. Recall that the declining-balance methods do not accommodate a zero salvage value; hence the need for switching.

The timing for the switch from DDB to straight-line is indicated when the depreciable charge under straight-line for the then undepreciated book value exceeds the allowance indicated by the DDB method. The example will show that timing the switch in that way produces not only a feasible depreciation schedule, in the sense of satisfying the zero salvage value requirement, but also provides a schedule that is more accelerated. (See Table 7–5 and Figure 7–4.)

Note that the DDB method, unadjusted, would bring the book value down to 1,244.2 as opposed to zero, as required. Column 4 indicates the straight-line depreciation that would be used for the balance of the depreciable period if a switch is made at that point in time. In years 1–3, DDB is used since its depreciation charges are larger than for straight-line on the remainder of the book value. In year 4, however, straight-line on the remainder would be 2,592 per year (for each of the last two years), as opposed to 2,073.6 and 1,244.2, respectively, under DDB. Thus, the switch to straight-line in year 4 gives both a decline to zero salvage value and a more accelerated decline in book value in year 4. The resulting depreciation charges, which yield the book values shown in column 5, are 9,600, 5,760, 3,456, 2,592, and 2,592 for years 1–5 respectively.

TABLE 7–5 Switching Between DDB and Straight-line *(in dollars)*

Year	d_t DDB	BV_t DDB	d_t SL on remaining BV	d_t DDB→SL	BV_t DDB→SL
1	.4(24,000) = 9,600	14,400	1/5(24,000) = 4,800	9,600	14,400
2	.4(14,400) = 5,760	8,640	1/4(14,400) = 3,600	5,760	8,640
3	.4(8,640) = 3,456	5,184	1/3(8,640) = 2,880	3,456	5,184
4	.4(5,184) = 2,073.6	3,110.4	1/2(5,184) = 2,592*	2,592	2,592
5	.4(3,110.4) = 1,244.2	1,866.2	2,592	2,592	0

*Indicates switch to SL.

Depreciation under ACRS

The accelerated cost recovery system, commonly referred to as ACRS, was enacted in 1981 as part of the Economic Recovery Tax Act of that year. It replaced, for the most part, the methods previously presented with a simplified set of percentages to be used for computing the depreciation allowances. These appear in Table 7–6.

There are several important novel features about the system. An examination

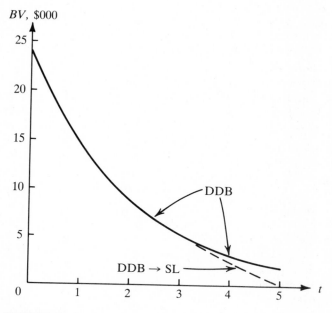

FIGURE 7–4

TABLE 7–6 Depreciation Rates under ACRS *(in percent)*

	Recovery Period											
	3			5			10			15		
	Year Asset Placed in Service											
Year	81 to 84	85	After 85	81 to 84	85	After 85	81 to 84	85	After 85	81 to 84	85	After 85
1	25	29	33	15	18	20	8	9	10	5	6	7
2	38	47	45	22	33	32	14	19	18	10	12	12
3	37	24	22	21	25	24	12	16	16	9	12	12
4				21	16	16	10	14	14	8	11	11
5				21	8	8	10	12	12	7	10	10
6							10	10	10	7	9	9
7							9	8	8	6	8	8
8							9	6	6	6	7	7
9							9	4	4	6	6	6
10							9	2	2	6	5	5
11										6	4	4
12										6	4	3
13										6	3	3
14										6	2	2
15										6	1	1

of Table 7–6 indicates that only four possibilities exist for the depreciable life, referred to under ACRS as the recovery period: 3, 5, 10, and 15 years. Thus, ACRS provided a general shortening of the depreciable period as compared with the asset depreciation guidelines of Table 7–1. For example, assets previously depreciable over 8 years would now be depreciable in 5. The effect of this change was to accelerate the depreciation process, allowing for faster capital recovery.

Also, ACRS differed from previous practice in that the entire first-year allowance called for in the schedule could be claimed in the year the asset was installed. Prior to ACRS the depreciation charge for the first year was prorated, based on the actual portion of that year the asset came into use. By the same token, under ACRS no depreciation is permitted in the year the asset is removed from use, whereas previous practice called for prorating the depreciation charge in the final year, based on when the asset left service or was fully depreciated. This aspect of the law also served to accelerate the depreciation process.

Another innovation of the system was its uniform treatment of salvage value as zero. Note that the percentages of annual depreciation allowances in Table 7–6 sum to 100 in each column. Again, the aggregate effect of such a change was to provide for greater capital recovery during the depreciable period and, thus, was an incentive to investment.

Clearly, the intent behind the aforementioned changes was to provide a public policy stimulus to investment in new plant and equipment, which had previously been stagnant at undesirable levels. (The next chapter will indicate in greater detail why shortening the depreciable period and increasing the permissible depreciable amount would have such an effect.)

For assets placed in service before 1981, the classic depreciation procedures discussed in the preceding sections apply. For those installed from 1981 to 1985, ACRS provided transitional rates. For assets installed after 1985 the final ACRS percentages hold. As an illustration, consider the following example:

EXAMPLE 7–4

This example uses the data of Example 7–2, where $P = \$24,000$, $n = 5$, and $SV = \$3,000$. Application of the post-1985 ACRS formula yields the following:

TABLE 7–7   ACRS Depreciation

Year	ACRS (percent)	d_t (dollars)	BV_t (dollars)
1	20	.20(24,000) = 4,800	19,200
2	32	.32(24,000) = 7,680	11,520
3	24	.24(24,000) = 5,760	5,760
4	16	.16(24,000) = 3,840	1,920
5	8	.08(24,000) = 1,920	0
	100	24,000	

Figure 7–3 provides a comparison of book value through time under ACRS, straight-line, sum-of-the-years'-digits, and declining-balance depreciation. Note that only the ACRS procedure reduces the book value to zero.

The ACRS percentages were based on the principles of switching introduced in the preceding section. For the schedule for 1985 and beyond, the switch is between double-declining-balance and sum-of-the-years'-digits rates. Specifically, after 1985 the first year's allowance is one-half the double-declining rate, or $2/2n$. In subsequent years the switch is made to sum-of-the-years'-digits, applied to the remaining book value, for the remaining depreciable period. Thus, for five-year assets, the percentage for the first year would be $1/n$ or 20 percent. That leaves 80 percent to be depreciated over the last four years, using sum-of-the-years'-digits. (For the four remaining years the sum of the digits is $4 + 3 + 2 + 1 = 10$). The five-year schedule is generated as follows:

Year	Percent	
1	1/5 =	20
2	(4/10)(80) =	32
3	(3/10)(80) =	24
4	(2/10)(80) =	16
5	(1/10)(80) =	8
		100

Sinking-fund Method

The sinking-fund method is rarely used in industry because it is decelerated by comparison with all of the others. (See Figure 7–3.) However, it is presented here for the sake of completeness. It may be encountered in the noncorporate or public domain, typically when the asset is long-lived and the behavior of value through time is expected to exhibit a gradual decline at first, with a more rapid decrease toward the end of the economic life. The sinking-fund procedure is based directly on time value concepts, specifying an annuity amount for the periodic depreciation charge. The accumulated allowances are valued at an opportunity cost rate, i, such that the entire depreciable amount, $P - SV$, is recovered by year n. Thus, the depreciation allowance is given by,

$$d_t = (P - SV)(A/F\ i\%, n) \text{ each period}$$

and the book value is the difference between the investment cost and the accumulated time value of the depreciation charges up to that point. Therefore, book value may be given by

$$BV_t = P - d\,(F/A\ i\%, t) = P - (P - SV)\,(A/F\ i\%, n)\,(F/A\ i\%, t)$$

EXAMPLE 7–5

To illustrate the application of the sinking-fund method, the same data are used as in the preceding examples, with an opportunity cost rate of 15 percent.

$$d_t = (24{,}000 - 3{,}000)\,(A/F\ 15\%, 5) = (21{,}000)\,(0.1483) = 3{,}114.3$$

which is considerably less than the allowance under the straight-line method. The book value after the third year is computed as follows:

$$BV_3 = 24{,}000 - 3{,}114.3\,(F/A\ 15\%, 3) = 24{,}000 - 3{,}114.3\,(3.4725) = 13{,}186$$

and is indicated in Figure 7–3 along with the curve of book value over the entire depreciable life.

Depreciation Based on Units of Production

In some cases it may be sensible to base depreciation on the amount of activity associated with the asset in a given period rather than simply on the amount of time that has elapsed since its acquisition. Procedures based on this notion are referred to as "units of production" methods. Necessary for implementing them are a valid measure of the rates of activity and deterioration of the asset; the ability to estimate the units of production available over the entire depreciable life; and the ability to keep track of production over the interval corresponding to the depreciation period in question.

The physical units of production employed may vary, depending on the application. Commonly, these are actual recorded production units, weights, or volumes. In other cases, units of production may be expressed in terms of operating days. Also, the units-of-production procedure may be employed for rental real estate assets by using income or projected income for the period as the unit of production, as compared to projected total income over the useful life of the asset.

The units-of-production method allocates the depreciation charge for a given period by forming the ratio of the units of production experienced in that period to the anticipated total for the entire depreciable life. That proportion of the depreciable amount becomes the charge for the period. Algebraically,

$$d_t = \frac{U_t}{U_{\text{life}}} (P - SV)$$

where U_t represents the units for the period in question and U_{life} represents the units for the entire depreciable life. Book value is computed as the investment cost less the accumulated depreciation, as before.

The units-of-production method cannot be classified a priori with respect to whether or not it is accelerated, since the pattern of depreciation charges corresponds to the nature of use of the asset. If the asset is used more intensively at first, then more gradually, the depreciation pattern will be accelerated. Clearly the reverse also holds.

Current Depreciation Practices

In conjunction with the aforementioned principles and methods, a variety of industrial practices have evolved with which the reader should have some familiarity.

The Treatment of Land

Land is one of the fixed assets appearing in the balance sheet because it is a necessary factor of production. However, it is generally not depreciated because its value does not typically decline with use.

Depreciation of Multiple Assets

Heretofore it has been implied that each asset is treated independently. For large organizations with abundant assets, this may prove to be an unwieldy and wasteful practice, since it may be possible to treat groups of assets in a coherent way. The collection of assets into a single account for depreciation purposes is often clerically desirable. The actual useful lives of the assets may, of course, vary. However, if a reasonable estimate of central tendency can be made among members of the collection, treating the set with that life is a practical and acceptable solution. Although some assets will last longer and others less, the aggregate effect will not differ in substance from treating the assets individually.

There are various ways to group assets into a single account, including what are called group, classified, or composite accounts. Group accounts are typically made up of assets with similar lives, such as motor vehicles. The depreciation charge can be easily arrived at by applying whatever depreciation method is being used to the aggregate depreciable amount and the common depreciable life. Classified accounts contain similar assets according to criteria of use, but with different lives. An example of such an account might include a variety of machine tools in a plant. The task is to arrive at a depreciation period to use for the set. Finally, composite accounts are those which are set up for assets that are different in nature, with possibly different economic lives. Here, too, the task is to arrive at a depreciable life useful for the entire account. This may be done by computing the aggregate depreciable amount for all the assets, and dividing it by the total depreciation charge for the next period.

Widespread Use of Accelerated Methods

As the business and engineering community has become more familiar with time value models, there has been a corresponding greater recognition of the desirability of accelerated depreciation methods. Although the various depreciation methods recover the same depreciable amount, they differ with respect to their timing, and greater depreciation charges toward the outset allow for faster recovery of invested capital. This is analogous to comparing two time series with equal totals but different payment schedules. Just as the present worth of a series with larger payments at the beginning has a greater present value, so, too, is the present worth of capital recovery greater under a more rapid depreciation schedule. This will be explored in more detail in the next chapter.

The introduction of ACRS in 1981 was entirely consistent with this trend but reduced the discretion available for choosing depreciation procedures. As was seen in this chapter, ACRS is an accelerated method and was, in fact, designed to provide faster capital recovery as a stimulus to investment. (In this era, government manipulates fiscal and monetary policy to accomplish its economic goals. The need for stimulation of investment has resulted in various forms of governmental acceleration of the depreciation process. One well-known business leader has even advocated that investment be treated as an expense in the year of the outlay. This is just about the ultimate in acceleration of capital recovery.)

Depletion

Depreciation is a capital recovery device for fixed assets that are replaceable with new or improved ones. An analogous procedure exists for accounting for the decline in value of natural productive properties that are not, strictly speaking, renewable but, rather, exhaustible. Examples of such properties are gas and oil wells, mines, quarries, and, in some sense, forests. The federal code permits a depletion allowance to be charged as an expense against revenue in computing income. This is intended to permit the owners to recoup the capital invested in the property in order to facilitate further exploration and development of like properties.

There are two recognized methods for computing depletion allowances. One is analogous to the units-of-production depreciation method in that it is based on the production rate rather than on the elapsed time of ownership. It is referred to as the factor depletion approach. The actual depletion charge for a year is computed as a proportion of the total resources available for exploitation, as follows:

$$\text{dep}_t = \frac{U_t}{U_{\text{total}}}(P)$$

where dep_t is the depletion charge for period t, U_t is the quantity of the nonrenewable resource removed in period t, U_{total} is the estimated total resource capacity of the property, and P is its investment cost.

The use of this method requires keeping track of the accumulated depletion charges over the useful life of the property because that amount cannot exceed the investment cost. It is this feature that generally makes the method less beneficial to the owners and, hence, less widely used, than the percentage depletion allowance alternative.

The use of the latter procedure permits as a depletion allowance a fixed percentage of the property's gross income for each period, with the proviso that no more than 50 percent of gross income may be deducted in any such period. However, the cumulative allowance has no limit, so that properties that have been extensively exploited over many years may have experienced total depletion allowances in excess of the original investment cost. Table 7–8 shows the applicable percentages for various types of properties. It should be noted that these percentages are adjusted by the federal government from time to time, as the need to provide incentive for those activities changes.

As an example, an oil well producing a gross income of $500,000 in a year would be permitted a depletion allowance of $0.15(500,000) = \$75,000$, irrespective of the cost of the property or of the exploration or exploitation expenses actually incurred during that year.

TABLE 7–8   Permitted Depletion Percentages

Resource	Percentage
Oil, gas	15
Coal	10
Gold, silver, copper, iron	15

Exercises

7–1 A piece of equipment is acquired for $12,000. Its depreciable life is 3 years, with a salvage value of $3,000. Prepare a table of the depreciation charges using the straight-line method. Include in the table the accumulated depreciation charges after each year and the book value.

7–2 *a.* For the asset in Exercise 7–1, compute the depreciation charges, accumulated depreciation, and book value under the declining-balance method, with $k = 0.2$.
b. Repeat part *a,* with $k = 0.37$.
c. Repeat again with the double-declining method (i.e., $k = 2(1/3) = 0.67$).

7–3 For the asset in Exercise 7–1, compute the depreciation charges, accumulated depreciation, and book value under the sum-of-the-years'-digits method.

7–4 For the asset in Exercise 7–1, compute the depreciation charges, accumulated depreciation, and book value under the sinking-fund method. Assume that funds are worth 12 percent.

7–5 For the asset in Exercise 7–1, compute the depreciation charges, accumulated depreciation, and book value under the ACRS method. Use ACRS percentages from Table 7–6.

7–6 A new machine is purchased for $245,000, F.O.B. the vendor. Delivery and installation amount to $5,000. The depreciable period is 10 years. The salvage value, although nominally $25,000, is really less, because dismantling and removal expenses amount to $5,000. Use this information to prepare a table of the depreciation charges by the straight-line method. Include in the table the accumulated depreciation charges after each year and the book value.

7–7 *a.* For the asset in Exercise 7–6, compute the depreciation charges, accumulated depreciation, and book value under the declining-balance method, with $k = 0.1$.
b. Repeat part *a* with $k = 0.2232$.
c. Repeat part *a* with the double-declining method (i.e., $k = 2(1/10) = 0.2$).

7–8 For the asset in Exercise 7–6, compute the depreciation charges, accumulated depreciation, and book value under the sum-of-the-years'-digits method.

7–9 For the asset in Exercise 7–6, compute the depreciation charges, accumulated depreciation, and book value under the sinking-fund method. Assume that funds are worth 12 percent.

7–10 For the asset in Exercise 7–6, compute the depreciation charges, accumulated depreciation, and book value under the ACRS method. Use ACRS percentages from Table 7–6.

7–11 *a.* If an asset is purchased for $18,000 and has a depreciable period of 6 years (under the IRS depreciation guidelines shown in Table 7–1), how much would the depreciation allowance be in the second year under straight-line, sum-of-the-years'-digits, declining-balance, and ACRS procedures? The salvage value after 6 years is expected to be $2,000.
b. What would the book value be after the second year for each of the procedures above?
Note: Use ACRS percentages for 5-year properties of 20, 32, 24, 16, 8 (after 1985).

7–12 The following three assets are included in a group for depreciation purposes:

	Investment Cost	Depreciable Period (yrs.)	Salvage Value
A	$39,000	6	$7,000
B	$25,000	5	$5,000
C	$42,000	7	$6,000

Compute the group depreciation charge per year under the straight-line method. Hint: Compute the depreciable amount for the group (i.e., the sum of $P - SV$ for each element in the group). Then compute the average depreciable period for the group, $(6 + 5 + 7)/3$. Then the ratio of the group depreciable amount and the average depreciable period is the annual straight-line charge for the group.

7–13 A property is acquired for copper mining at a cost of $2.5 million. It is expected to yield ore amounting to 5 million tons over its usable life. In the first year of operation 200,000 tons are removed, providing gross income of $5,000,000.

a. Compute a depletion allowance based on the factor depletion approach.
b. Compute a depletion allowance based on the percentage depletion approach.

7–14 Your firm installs a new press costing $85,000. Its depreciable period is 10 years with a salvage value of $5,000. Suppose you wish to investigate the depreciation schedule that would be applicable with the double-declining-balance method, with a switch to straight-line, if applicable. Determine such a schedule. Is a switch to straight-line called for and, if so, in which year? What are the depreciation charge and book value after each year?

Selected References

Bernstein, L. *Financial Statement Analysis: Theory, Application and Interpretation,* rev. ed. Homewood, IL: Richard D. Irwin, 1978.

Blank, L. and D. Smith, "A Comparative Evaluation of the Accelerated Cost Recovery System as Enacted by the 1981 Economic Recovery Tax Act," *The Engineering Economist,* Vol. 28, No. 1, Fall, 1982, pp. 1–30.

Crowley, J. and J. Sprague, "A New Approach to Depreciation Measurement for Public Utilities," *Engineering Economist,* Vol. 29, No. 4, Summer 1984, pp. 239–250.

Horngren, C. *Accounting for Management Control: An Introduction,* 3rd ed. Englewood Cliffs, NJ: Prentice-Hall, 1974.

8

The Impact of Taxes on Investment Decisions

Introduction

The preceding chapters have outlined some broad principles of economic modeling for the purpose of improving the decision-making process. Those models examined the various investment, revenue, and expense flows identified with the alternatives being considered. An important class of flows not yet covered are those relating to income taxes paid (or to be paid when weighing a possible future investment).

Typically, one can project the taxes relating to an investment that will have to be paid out over the decision horizon. Clearly, these must be considered an integral part of the entire array of expense flows if valid decision models are to be constructed. This chapter will examine the concepts necessary to expand the models already presented so as to take into account the effect of income taxes.

Bear in mind that not all enterprises are subject to income taxes in the United States. The corporate income tax, which will be the major focus of this chapter, is levied on all profit-seeking corporations. Although it is safe to say that most investment decision-making is conducted in a profit-seeking environment, there are clearly many exceptions. (These will be addressed in Chapter 14.) Decision-makers in organizations not subject to income taxes would omit the considerations outlined in this chapter.

Before proceeding, it is necessary to distinguish among the various types of taxes and to identify those that are most relevant for economic decision-making studies.

Property taxes are levies on real assets, are collected periodically, and are generally based on the product of their assessed value and the relevant tax rate. These are enacted by various local governmental jurisdictions, as opposed to the federal government, and there is considerable variation in their coverage and implementation. Usually land, buildings, and equipment are included, and, less typically, inventories. Although property taxes are not generally very heavy by comparison with other real expenses, they should be considered, particularly if the impact on property taxes of competing alternatives differ substantially. (For example, it is not uncommon for localities seeking new industrial investment to offer property tax concessions to prospective firms. Such concessions need to be considered as part of the overall economic attractiveness of a particular site.)

For the purpose of this chapter, taxes other than those on income may be lumped together into a single category, because they may be treated in the same way as other outflows were in previous chapters. Examples are property taxes, described above; excise taxes, which are based on the production or sale of specified products; sales taxes, which are local levies on the sale of a wide spectrum of products to end users; and user taxes, which apply to the specific use of a governmental facility or service.

Income taxes differ from property taxes. Income taxes on corporations have been in existence in the United States since 1913, but they did not become an economic factor to be reckoned with until the World War II era, when the severe requirements of the war caused a large increase in government expenditure and a concomitant need for greater revenues.

That conflict's end did not witness a reduction in activity of the federal government to prewar levels. Although the government's role has been reinterpreted through time as a result of our democratic political system, the need for taxing corporate income to finance government programs has not disappeared. In spite of some political controversy with regard to this matter, the existence of corporate income taxes is expected to continue, although actual tax rates and specific rules may very well change. Thus, corporate income taxes have continued to be significant and need to be considered when evaluating alternatives.

The largest corporate income tax is levied by the federal government. Most state and some local governments have similar provisions, though at lower rates. The aggregate effect of these taxes is the decision-maker's concern. They are based on the corporation's reported net income, which, as indicated in the preceding chapter, is based on net revenue less allowable expenses, including depreciation and interest payments.

The only practical approach for a book on the principles of economic analysis is to present the basic concepts relevant to the treatment of income taxes in decision-making, without concern for the minute levels of detail of the tax code. Because the tax rules in effect at any one time reflect the elected government's determination to influence the economy in ways consistent with its view of what course is best overall, the tax provisions have experienced frequent revisions and are sure to continue to do so.

The Income Tax Mechanism

Corporate income taxes are based on the difference between revenue and all allowable expenses, including depreciation and interest. Thus, they are generally applied to the "net income before income tax" line of the income statement of Figure 7–1. Although the relevant code is very complex, for most economy studies only the following basic features need be considered:

The Graduated Feature of the Tax

The actual application of the federal corporate income tax is graduated—that is, higher rates are applicable for successively larger marginal net income. Rates in effect in 1984 are shown in Table 8–1.

The graduated rates work as indicated in Example 8–1.

EXAMPLE 8–1

A firm reporting taxable income of $75,000 would be subject to a tax of

$$\text{Income tax} = 0.15(25,000) + 0.18(50,000 - 25,000)$$
$$+ 0.30(75,000 - 50,000) = \$15,750$$

In other words, each marginal level of income is taxed at a higher rate. Of course, at any income level one can compute the average uniform rate which when applied to the total income has an effect equivalent to the cumulative marginal applications. For Example 8–1, taxes for an income of $75,000 are equivalent to an average rate of 21 percent.

$(15,750/75,000) = 0.21$ or 21 percent

Clearly, the analyst should include the effect of state and local income taxes. In doing so, one must be mindful of the fact that they are an allowable expense

TABLE 8–1   Federal Tax Rate for Corporations, 1984

Taxable Income	Marginal Tax Rate (%)	Average Applicable Rate (at top end of range) (%)
< $25,000	15	15
25,001–50,000	18	16.5
50,001–75,000	30	21
75,001–100,000	40	21.75
> $100,000	46	approaches 46% as income increases

Source: Tax Imposed on Corporations, Stand. Fed. Tax. Rep., Commerce Clearing House, 1985, ¶444.

(or deductible from income) when computing the federal tax liability, but that the reverse is generally not permitted. The upshot of this is that the effective combined rate is not strictly additive.

Suppose that in Example 8–1 the state tax is a flat 4 percent. The total state income tax due would be computed as

State tax = 0.04($75,000) = $3,000

Since adjusted income for federal tax purposes permits deduction of state taxes,

Federal taxable income = $75,000 − $3,000 = $72,000

Then the federal tax is computed using graduated rates,

Federal tax = 0.15($25,000) + 0.18($50,000 − $25,000)

$$+ 0.30($72,000 − $50,000) = $14,850$$

The federal taxes would be higher were the state levy not deductible, but the total of federal plus state taxes is $14,850 + $3,000 = $17,850, which is greater than with the federal tax alone. The effective average aggregate rate in this case turns out to be 17,850/75,000 = 23.8 percent.

An approximation to the applicable effective rate is provided by

$$ER = SR + (1 − SR)(AFR)$$

where ER stands for the effective aggregate rate, SR the applicable average or flat state rate, and AFR the average or effective federal rate. Applying it to the illustration above,

$$ER = 0.04 + (1 − 0.04)(14,850/72,000) = 0.04 + (0.96)(0.20625) = 0.238$$

which agrees exactly with the value computed above. The formula is referred to as an approximation because its application generally involves an a priori rounded figure for AFR, which in this case would have been 21 percent. In that case,

$$ER = 0.04 + (0.96)(0.21) = 0.2416$$

which is in sufficiently close agreement to the actual value to be used for economic studies.

For the purpose of modeling investment acceptability or for comparing alternatives, two related points should be kept in mind. First, a single value for the after-tax rate is generally postulated, which includes federal and local levies. Such an effective rate, as illustrated above, simplifies the computation of after-tax flows in time value models.

The second point stems from the likelihood that most analysts conduct their studies for relatively large firms, for which taxable income ordinarily vastly exceeds $100,000 per year. This means not only that the applicable marginal graduated federal rate is at the maximum but that the equivalent average rate approaches the marginal. Thus, for all practical purposes, the effective federal portion of the tax rate for firms with relatively large profit-earning capability is the maximum applicable rate of the graduated schedule in force at any particular point in time. (For Table 8–1, this corresponds to 46 percent.)

Thus, analysts almost universally use equivalent average income-tax rates in their studies, including state and local taxes. The rates used are generally in excess of 40 percent. For simplicity, in this book, illustrations and exercises will use an effective average or flat rate of 50 percent.

Investment Tax Credits

In 1961, Congress enacted an investment tax credit as a stimulus to productive investment. The credit does not affect taxable income per se but does affect the actual taxes paid in the following way:

Suppose a firm's income taxes for the coming year would amount to $10 million, excluding consideration of the credit. Further suppose that it is considering the immediate purchase and installation of an asset to replace an outdated one which would cost $100,000. If an investment tax credit were applicable and it amounted to, say, 10 percent of the first cost, then a credit to the firm's income taxes in the amount of 0.10($100,000) = $10,000 could be realized if the investment is made. Thus, the firm's income taxes for the year, as adjusted, would be $10 million − $10,000 or $9.99 million.

Although from this perspective the credit may not seem to be of great moment, from the standpoint of considering the desirability of the investment it is quite germane. Since the credit, if realized, is worth $10,000, and if it is considered to be taken in the year of the investment, it makes the net investment cost $100,000 − $10,000 or $90,000. Thus, it makes the asset more desirable; hence the tax credit's adoption by Congress when an overall stimulus to investment was considered necessary. Clearly the effect of such a credit to reduce the real net investment cost affects the desirability of the asset and must be considered by the analyst. How this may be accomplished will become apparent in the next section.

Since its original enactment, the provisions for investment tax credits have been modified very frequently. During inflationary periods, when Congress attempted to cool down the economy, provisions for investment credits were reduced or eliminated. When an economic stimulus was thought to be needed, credits were increased. Since variability in investment tax credits seems likely to continue, the approach taken in this book is not to emphasize the provisions in effect at the time of publication but, rather, to outline the basic principles that pertain to the handling of such credits for investment decision modeling. For the remainder of this book, when investment tax credits are to be considered, the applicable percentage of stated investment cost will be given.

Capital Gains (and Losses)

Just as the investment tax credit affects the aggregate tax liability without affecting taxable income per se, so too, does a capital gain or loss. However, one significant difference between them is that capital gains or losses refer to assets that have been or will be disposed of during the tax period in question. Therefore they usually need not be considered in most decision situations. They are relevant in ex-post-facto after-tax analyses—e.g., to answer the question, "What are the after-tax flows for an asset which was just disposed of?"; in situations where one is considering the replacement of an asset with another; and in cases where the disposition terms of the asset can be anticipated.

Capital gains occur when assets are disposed of at values above their acquisition cost. Capital losses result from disposition below their book value. Capital gains have the effect of increasing the tax liability, whereas capital losses have the reverse effect. There is also a distinction between long-term and short-term gains and losses. Long-term gains or losses are those for which the asset was held for more than a certain period, say one year. (Again, legislation defining long-term vs. short-term holding periods has been changed by Congress from time to time.) Long-term gains or losses are treated differently from short-term ones in that the applicable tax rates for the former are considerably less than the rates for ordinary income. However, since the main focus of attention in this book is on fixed assets held for relatively long periods, the typical situation to be considered involves the reduced rates in effect on long-term gains or losses.

To illustrate, the sale of an asset having a book value of $3,000 at a price of $1,000 results in a capital loss (for income-tax purposes) of $2,000. Assuming that the asset had been held for longer than one year, the loss would be qualified as a long-term one. If the current rate for such losses is 25 percent, the firm would be entitled to a 0.25($2,000) = $500 income-tax credit. Just as with the investment tax credit, this is relevant to decision modeling because the real (after-tax) amount realizable from the sale of the asset is $1,000 + $500 = $1,500. Thus, it can be seen that capital gains and losses also need to be accounted for if valid after-tax analyses are to be conducted.

The exact treatment of such gains and losses has changed over time and seems likely to continue to do so. Thus, in this book, only the principles governing the treatment of gains and losses will be emphasized.

Modeling Alternatives to Include the Effect of Income Taxes

Frequently, several alternatives will differ not only in some of their pretax economic considerations but in their after-tax considerations as well. Since income taxes are real economic outflows bearing on each alternative, they should be taken into account. The time value models presented in preceding chapters can now be expanded to reflect the effects of income taxes.

It will become apparent that elements such as depreciation schedules, arrangements for acquiring physical assets (such as buying vs. leasing), or the cost of borrowing capital for the purpose of obtaining assets all have an important bearing on the relative desirability of an alternative. A variety of these will be presented and analyzed in a series of cases. What is in common to all, and a necessary step in an after-tax analysis, is the transformation of the familiar before-tax cash flows (BTCF) to a set of after-tax cash flows (ATCF).

Once that transformation has been accomplished, the analysis of alternatives proceeds according to the same time value principles as have already been outlined, the models then utilizing the after-tax flows as the relevant data. Thus, one can compute the internal rate of return, present worth, periodic worth, or incremental internal rate of return applied to the after-tax flows. This idea is the central theme of the chapter, and is symbolized by the schematic of Cash Flow Diagram D8–1.

D8–1

Transformation of Before-tax Flows into After-tax Flows

BTCF

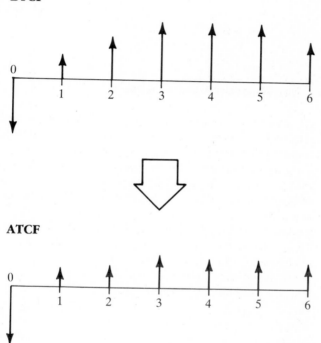

ATCF

Before-tax flows are transformed to after-tax flows by computing the income-tax effect for each time period, adjusting the before-tax amount accordingly. This can be done individually for each period, or, in some cases, where the before-tax flows and tax effects are identical for several periods, one computation will serve for all. In any event, computation of the income-tax effect for any period must take into account all of the relevant income-tax provisions.

In this chapter, attention will be limited to provisions for depreciation, interest payments, investment tax credits, and tax effects of capital gains and losses. (An example of other considerations omitted here are carryovers and carrybacks, which are income-tax effects occurring in one period that are larger than total net income for that period. When such effects are not permitted to be used in toto for one period, part of the effect may be transferred to other time periods, forward or back, to be used in those different periods.)

Although the period-by-period tax computations can be done on an ad hoc basis, it is convenient for planning (and for displaying them in the text) to resort to a tabular format such as the one below. The examples and cases that follow conform to such a format and will illustrate its use.

End of Year	Before-tax Cash Flow	Depreciation Charges	Other Tax Effects	Taxable Income	Income Taxes	After-tax Cash Flow
(1)	(2)	(3)	(4)	(5) =(2) − (3) or (2) − (4) or (2) − (3) − (4)	(6) =(5) × tax rate	(7) =(2) − (6)

After-tax Flows for Depreciable Assets

In order to examine the effect of depreciation charges on the before-tax cash flows, consider the following situation: An investment under consideration will cost $100,000 to put in place. It is expected to generate cost savings of $40,000 per year while requiring operating expenditures of $10,000 per year. It is anticipated that its useful economic life will be six years, with a salvage value of $15,000 for depreciation purposes. It is also assumed that the asset will be disposed of for $15,000 at the end of the study period, thus causing no capital gain or loss. The depreciation method used is straight-line. The desirability of the investment will be evaluated, and the relevant MARR will be presented in the course of that process.

To begin, it is necessary to identify the before-tax cash flow. The projected values will be modeled as a set of discrete time values, consistent with the treatment in Chapters 2 through 4. They are composed of the investment outlay of $100,000 at time 0; the annual cost savings of $40,000, which will be modeled as a revenue; the annual disbursement of $10,000; and the salvage value of $15,000 at time 6. Since the $40,000 savings and $10,000 expenditures have an identical effect on taxes, they will be combined, yielding a net savings of $30,000 per year. Diagram D8−2 shows the cash flows.

D8−2

BTCF ($000)

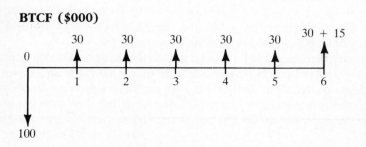

These before-tax cash flows may be analyzed using time value models as follows: Suppose the MARR to be applied here is 20 percent. The present worth of the flows is

$$PW = -100,000 + 30,000(P/A\ 20\%, 6) + 15,000(P/F\ 20\%, 6) = \$4,788.77$$

and its equivalent annual worth is

$$AW = -100,000(A/P\ 20\%, 6) + 30,000 + 15,000(A/F\ 20\%, 6) = \$1,440.01$$

The internal rate of return based on the before-tax flows is obtained by solving for i,

$$-100,000 + 30,000(P/A\ i\%, 6) + 15,000(P/F\ i\%, 6) = 0$$

The reader should verify that this yields 21.82 percent. Thus, from a before-tax point of view, the investment seems justified.

The transformation of this before-tax cash flow to one after taxes is demonstrated in Table 8–2. All figures were rounded to the nearest dollar. Note that the straight-line depreciation charge per year was computed as $(100,000 - 15,000)/6$. These depreciation charges would be "accounting" or "paper" entries, not corresponding to actual outflows. However, they are of crucial importance in that they affect the annual income taxes that would be payable, which are real outflows. In this case, annual net savings of $30,000 would increase taxable income by $15,833, after depreciation has been taken into account. This net increase in taxable income would increase taxes, a real outflow, by $7,917. Finally, combining the real net savings of $30,000 (an inflow), with the real increase in taxes attributable to it, $7,917 (an outflow), yields a net after-tax cash (in)flow of $22,083 per year. The resulting after-tax cash flow appears in Diagram D8–3.

Applying the time value models to the after-tax flows yields an internal rate of return by solving the following for i:

$$-100,000 + 22,083(P/A\ i\%, 6) + 15,000(P/F\ i\%, 6) = 0$$

Here the reader should verify that the internal rate of return is 11.47 percent. If this is compared to the before-tax internal rate of return computed above, it is seen to be approximately one-half its magnitude. (That it is less should not be a surprise since a part of the cost savings results in greater tax payments, thus

TABLE 8–2   After-tax Cash Flow   *(in dollars)*

End of Year	Before-tax Cash Flow	Depreciation Charges	Other Tax Effects	Taxable Income	Income Taxes	After-tax Cash Flow
(1)	(2)	(3)	(4)	(5)	(6)	(7)
			=(2) − (3)		=(5) × 50%	=(2) − (6)
0	−100,000					−100,000
1–6	30,000	14,167		15,833	7,917	22,083
6	15,000					15,000

D8–3

ATCF ($000)

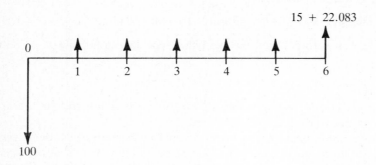

reducing the actual savings.) That it is about one-half the magnitude is related to the tax rate used, namely 50 percent. In general, the after-tax internal rate of return can be approximated from the before-tax rate by multiplying it by 1 − the effective income-tax rate used in the study. That this is only an approximation is due to a variety of factors, including the effect of the depreciation method employed on the timing of the after-tax flows.

Present, future, or annual worth measures of the after-tax flows may also be made, although the question arises as to the appropriate discount rate to use. For now, it will be assumed that the after-tax MARR is 10 percent, which is one-half the before-tax rate. (Once again, this is based on the effective tax rate of 50 percent.) Procedures to evaluate the after-tax MARR will be explored in Chapter 12.

The after-tax *PW* and *AW* are as follows:

$$PW = -100,000 + 22,083(P/A\ 10\%, 6) + 15,000(P/F\ 10\%, 6) = \$4,644.33$$

$$AW = -100,000(A/P\ 10\%, 6) + 22,083 + 15,000(A/F\ 10\%, 6) = \$1,066.37$$

and the investment is also justified from an after-tax point of view.

The Effect of Accelerated Depreciation on After-tax Flows

As mentioned in the chapter on depreciation, accelerated methods are frequently preferred because they facilitate more rapid capital recovery. The value of accelerated depreciation can be seen from Example 8–2.

EXAMPLE 8–2

Suppose sum-of-the-years'-digits depreciation is used on the asset analyzed in Table 8–2, with everything else constant. The after-tax cash flows would now be computed as in Table 8–3.

TABLE 8–3   After-tax Cash Flow   (in dollars)

End of Year	Before-tax Cash Flow	Depreciation Charges	Other Tax Effects	Taxable Income	Income Taxes	After-tax Cash Flow
(1)	(2)	(3)	(4)	(5)	(6)	(7)
				=(2) − (3)	=(5) × 50%	=(2) − (6)
0	− 100,000					− 100,000
1	30,000	24,286		5,714	2,857	27,143
2	30,000	20,238		9,762	4,881	25,119
3	30,000	16,190		13,810	6,905	23,095
4	30,000	12,143		17,857	8,929	21,072
5	30,000	8,095		21,905	10,953	19,048
6	30,000	4,048		25,952	12,976	17,024
6	15,000					15,000

These flows are depicted in Cash Flow Diagram D8–4.

D8–4

ATCF($000)

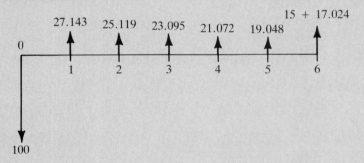

Solving the following for i yields an after-tax internal rate of return of 12.40 percent.

$$-100,000 + 27,143(P/F\ i\%, 1) + \cdots + (17,024 + 15,000)(P/F\ i\%, 6) = 0$$

This is almost a full percent more than under straight-line depreciation. Also, the annual worth at 10 percent is increased to $1,626. Thus, although the total depreciable amount ($85,000) is identical for both methods, the timing of the depreciation charges is not.

The accelerated procedure permits greater deductions at the outset, resulting in lower taxes and greater real returns toward the beginning. Since the time value models "weight" the earlier amounts correspondingly more than subsequent ones, it can now be seen why the accelerated depreciation method yields a higher annual worth or rate of return.

Another way of accelerating the depreciation process is to shorten the depreciable life. In order to grasp the significance of this approach, reconsider Example 8–2 with the depreciable life reduced to five years.

EXAMPLE 8–3

The straight-line method will once again be used. Table 8–4 shows how the after-tax cash flows can be generated.

TABLE 8–4 After-tax Cash Flow (in dollars)

End of Year	Before-tax Cash Flow	Depreciation Charges	Other Tax Effects	Taxable Income	Income Taxes	After-tax Cash Flow
(1)	(2)	(3)	(4)	(5)	(6)	(7)
				=(2) − (3)	=(5) × 50%	=(2) − (6)
0	− 100,000					− 100,000
1–5	30,000	17,000		13,000	6,500	23,500
6	30,000	0		30,000	15,000	15,000
6	15,000					15,000

Note that now the depreciation charge per year is computed as (100,000 − 15,000)/5. Apart from the changes in depreciation allowances, and the after-tax flows in years 1 through 5, the only other alteration occurs in the treatment of year 6. By that time, the asset would be fully depreciated, so there would be no corresponding charge for that year. Therefore, all of the net savings, or $30,000, would be taxable at 50 percent, leaving a net flow of $15,000. The salvage value is the same as before. These flows appear in Cash Flow Diagram 8–5.

D8–5

ATCF ($000)

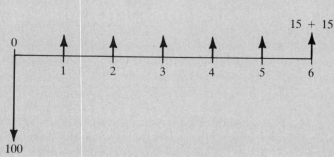

The internal rate of return is now found by solving

$$-100,000 + 23,500(P/A\ i\%, 5) + (15,000 + 15,000)(P/F\ i\%, 6) = 0$$

for i, which yields 11.96 percent. Note that this is an increase of about one-half percent ($11.96 - 11.47$), produced simply by shortening the depreciable period. Here, too, the annual worth at 10 percent is increased, to $1,382.

Finally, Example 8–4 examines the projected after-tax flows of our familiar asset under post-1985 ACRS provisions. Recall that this is a procedure combining both aspects of acceleration dealt with in Examples 8–2 and 8–3—namely, acceleration as compared with straight-line—and a shortened depreciation period. An additional characteristic of ACRS is the depreciation of the entire adjusted first cost of the asset.

EXAMPLE 8–4

Table 8–5 shows the computation of the after-tax-cash flows. The depreciation charges were obtained as the product of the allowable ACRS percentages given in Table 7–5, and the investment cost, $100,000.

Here, the treatment of the savings of year 6 is the same as in Example 8–3, since the asset would be fully depreciated by then. What is different in achieving the after-tax transformation is the way that the salvage value is handled. If the asset can be disposed of for $15,000 at the end of year 6, as expected, then those proceeds would be taxable as ordinary income. That is because the full investment cost would already have been depreciated, thus requiring the salvage value of $15,000 to be regarded as recovery of previously charged depreciation.

Also, the $-$1,000 in income taxes for year 2 should be interpreted as a

TABLE 8–5   After-tax Cash Flow *(in dollars)*

End of Year	Before-tax Cash Flow	Depreciation Charges	Other Tax Effects	Taxable Income	Income Taxes	After-tax Cash Flow
(1)	(2)	(3)	(4)	(5)	(6)	(7)
				$=(2)-(3)$	$=(5)\times 50\%$	$=(2)-(6)$
0	-100,000					-100,000
1	30,000	20,000		10,000	5,000	25,000
2	30,000	32,000		-2,000	-1,000	31,000
3	30,000	24,000		6,000	3,000	27,000
4	30,000	16,000		14,000	7,000	23,000
5	30,000	8,000		22,000	11,000	19,000
6	30,000	0		30,000	15,000	15,000
6	15,000			15,000	7,500	7,500

credit to offset other income taxes that the firm has for that year. The after-tax cash flows are shown in Cash Flow Diagram D8–6.

D8–6

ATCF ($000)

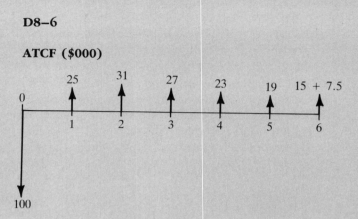

Then the internal rate of return may be computed by solving the following for i:

$$-100,000 + 25,000(P/F\ i\%, 1) + \cdots + (15,000 + 7,500)(P/F\ i\%, 6) = 0$$

which yields 13.13 percent, the highest result so far. The annual worth under ACRS increases to $2,030, also the highest result.

Clearly, this combination of accelerated depreciation (including shortened depreciable period) and full depreciation of investment cost enhances the return on capital. It is not surprising, therefore, that such a provision was enacted by the federal government in 1981, during a period of weak aggregate investment. By this act, some projects which decision-makers had thought to be marginal could then be viewed as sufficiently attractive to warrant investment. The overall effect was to provide a stimulus to investment.

The Effect of an Investment Tax Credit on After-tax Cash Flows

In order to analyze in more detail the effect of investment tax credits, suppose that in Example 8–4 a credit of 10 percent of the adjusted investment cost is permitted.

EXAMPLE 8–5

The credit would amount to $10,000 (0.10 × $100,000). Although this credit does not change the taxable income per se, it can be used to offset income

TABLE 8–6   After-tax Cash Flow: Investment Tax Credit   (*in dollars*)

End of Year	Before-tax Cash Flow	Depreciation Charges	Other Tax Effects	Taxable Income	Income Taxes	After-tax Cash Flow
(1)	(2)	(3)	(4)	(5)	(6)	(7)
				=(2) − (3)	=(5) × 50% −(4)	=(2) − (6)
0	− 100,000					− 100,000
1	30,000	20,000	10,000 cr	10,000	− 5,000	35,000
2	30,000	32,000		− 2,000	− 1,000	31,000
3	30,000	24,000		6,000	3,000	27,000
4	30,000	16,000		14,000	7,000	23,000
5	30,000	8,000		22,000	11,000	19,000
6	30,000	0		30,000	15,000	15,000
6	15,000			15,000	7,500	7,500

taxes in the year following purchase and installation of the asset, which would occur at the end of year 1. Table 8–6 shows how this may be derived. The rest of the flows are identical to those of Example 8–4.

These flows are diagrammed in Cash Flow Diagram D8–7.

D8–7

ATCF ($000)

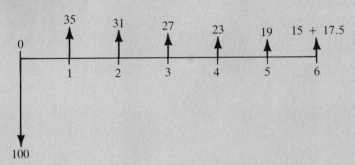

The internal rate of return for these after-tax flows, obtained from solving

$$- 100,000 + 35,000(P/F\ i\%, 1) + \cdots + (15,000 + 7,500)(P/F\ i\%, 6) = 0$$

for i, is 16.63 percent. This is a substantial increase over the rate applicable without the investment tax credit. The annual-worth measure also increases to $4,117. The value of such a credit is, therefore, considerable and does serve to stimulate capital investment.

If one reviews the entire series of changes that were introduced in Examples 8–2 through 8–5, encompassing both accelerated depreciation and investment tax credits, it is clear that the combination of these provisions, sanctioned by the federal government, makes investments more attractive. The government, by permitting such treatments, is pursuing a policy to enhance the prospects of productive investment.

Evaluating Buy-vs.-Lease on an After-tax Cash Flow Basis

Frequently, needed equipment may be obtained in more than one way—that is, it may not have to be purchased. One alternate possibility is that of a lease, or rental arrangement. The lessor may be the manufacturer of the equipment or a subsidiary, or an independent leasing organization.

The differences between buying and leasing may be summarized as follows: when an asset is purchased, the relatively large investment outlay required influences the asset structure of the organization in that one form of asset is converted to another. For example, when cash is used to acquire a fixed asset, the firm's balance sheet would show a reduction of cash and a corresponding increase in fixed assets. Once the asset is purchased, it may be depreciated according to the principles previously described. Thus, ownership of the asset has a further bearing on future income taxes in that depreciation charges reduce taxable income, and on the level of reported fixed assets in that book values are periodically reduced by the amount of the depreciation charged. Of course, it is also true that the care and maintenance of a purchased asset is the sole responsibility of the owner.

In contrast, when an asset is leased, an agreement is made permitting its use for a fee which is typically paid periodically. Thus, the user need not expend scarce capital in order to gain the benefit of use of the asset. This is frequently among the most desirable features of leasing. Leasing has no direct effect on the user firm's asset structure as expressed in its balance sheet; since the lessor maintains ownership of the asset, it is his balance sheet that is affected. Of course, the user also does not receive the benefit of depreciating the asset. However, although the user receives no reduction of taxable income and income taxes attributable to depreciation, the lease payments are a deductible expense. This latter characteristic is an important one, and is a prominent factor in the widespread practice of leasing. Finally, another aspect of leasing potentially attractive to the user is that maintenance responsibilities may be included in the lease.

Since the buy-vs.-lease alternatives affect taxes in different ways, it is necessary to perform an after-tax cash flow analysis when analyzing them. In order to get some insight into the way such alternatives may be compared, consider Example 8–6.

Example 8–6

Suppose that the equipment under study in Examples 8–2 through 8–5 may be leased with no effect on maintenance costs. Thus, the cost savings and operating expenses under a lease arrangement are the same as before. Suppose that the difference, under leasing, is that the acquisition cost (and salvage value) is replaced with an annual lease payment of $22,000 made at the beginning of each period (i.e., times 0–5). Further suppose, for the purpose of the analysis, that the after-tax MARR is 10%. Finally, assume that both the ACRS depreciation system and a 10-percent investment tax credit are applicable if the asset is purchased. In other words, the lease alternative should be compared with the analysis done in Example 8–5. The derivation of after-tax flows for the lease alternative is shown in Table 8–7.

TABLE 8–7 After-tax Cash Flow: Leasing (*in dollars*)

End of Year	Before-tax Cash Flow	Depreciation Charges	Other Tax Effects	Taxable Income	Income Taxes	After-tax Cash Flow
(1)	(2)	(3)	(4)	(5)	(6)	(7)
				=(2) − (3)	=(5) × 50%	=(2) − (6)
0	−22,000			−22,000	−11,000	−11,000
1	8,000			8,000	4,000	4,000
2	8,000			8,000	4,000	4,000
3	8,000			8,000	4,000	4,000
4	8,000			8,000	4,000	4,000
5	8,000			8,000	4,000	4,000
6	30,000			30,000	15,000	15,000

The − $22,000 before-tax cash flow in period 0 represents the first lease payment. Since it is a permissible expense for tax purposes, income taxes would be reduced by $11,000 for the period (assuming a 50-percent rate), and the net outflow would be $11,000. For periods 1–5 the before-tax flow is made up of the $30,000 savings less the $22,000 lease payment. There is no depreciation applicable here, and an $8,000 before-tax savings increases taxes by $4,000, thus leading to a $4,000 after-tax savings. Period 6 savings are unencumbered by a lease payment since it was assumed that these would be prepaid.

These flows are diagrammed in Cash Flow Diagram D8–8.

D8–8

ATCF ($000)

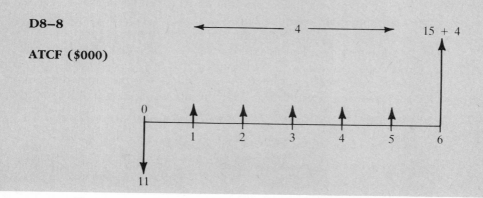

The annual worth of these flows, using an interest rate of 10 percent, is calculated as

$$AW = [-11,000 + 4,000(P/A\ 10\%, 5)$$
$$+\ 15,000(P/F\ 10\%, 6)](A/P\ 10\%, 6) = \$2,899.99$$

Since this is less than the annual worth of the after-tax flows for ownership under ACRS with an investment tax credit, leasing is less desirable than owning, given the present terms and assumptions used in the model. (One such assumption is that sufficient capital is available for purchase of the asset.) The after-tax flows for ownership developed in Example 8–6 are duplicated in Cash Flow Diagram D8–9.

D8–9

ATCF ($000) (Ex. 8–5)

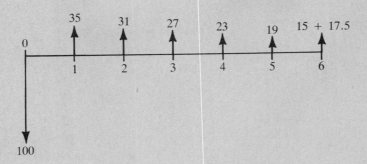

By comparison, these flows give an annual worth of

$$AW = [-100,000 + 35,000(P/F\ 10\%, 1) + 31,000(P/F\ 10\%, 2)$$
$$+\ \cdots\ +\ (15,000 + 7,500)(P/F\ 10\%, 6)](A/P\ 10\%, 6) = \$4,117.09$$

Although it is possible to arrive at a conclusion in the example above, it is clear that any comparison hinges on the terms specific to each alternative. In that light it may be of interest to examine the rental payments in a more general way. In particular, one might wish to conduct a sensitivity analysis regarding the lease payments. In doing so, the indifference rental, or the lease payment for which owning and leasing appear equally desirable may be computed. (This is an application of an idea developed in Chapter 6.) Assuming that the relevant comparison is with ownership as described by Example 8–5, one can find the indifference point by solving the following equation for x, the magnitude of lease payments.

$$-x(0.5) + (30 - x)(0.5)(P/A\ 10\%, 5) + 30(0.5)(P/F\ 10\%, 6) =$$
$$-100,000 + 35,000(P/F\ 10\%, 1) + \cdots + (15,000 + 7,500)(P/F\ 10\%, 6)$$
$$= \$17,930.99$$

Note that (0.5) is the effective income-tax rate and the "right side" is the present worth of the ownership alternative as described in Example 8–5, or $17,930.99. The reader may verify that the solution for x is $19,787 (to the nearest dollar). In other words, any value of lease payments greater than $19,787 makes leasing less desirable than owning, and vice versa. (See Figure 8–1.)

This indifference approach is valuable in that it permits the user to judge a range of lease payment values that would be acceptable in a particular situation. Thus, it might prove to be a useful tool in negotiating, both for lessors and lessees.

Evaluating After-tax Cash Flows When Capital Is Borrowed

An implicit assumption in most of the preceding examples was that the source of the investment funds was internally generated and that sufficient capital was available to purchase the assets. Frequently this is not the case. Often money is borrowed to help finance some part of the firm's capital equipment additions. When that happens, one must treat not only the flow of funds directly attributable to the loan, such as interest payments, but the indirect effect of the loan interest payments on income taxes as well. As an illustration, reconsider Example 8–5, in which the asset is acquired with an investment tax credit and is depreciated under ACRS.

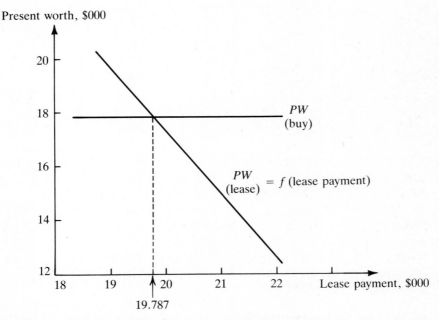

FIGURE 8–1 Present Worth vs. Lease Payment Amount, Buy vs. Lease Alternative.

EXAMPLE 8–7

Suppose that the firm assembles the $100,000 capital required by borrowing $35,000 and putting up the balance from its own funds. (In fact, it is more typical to consider the capital expenditures, borrowing, and internally generated amounts in aggregate. However, if those aggregates are in the ratio of 35 percent attributable to borrowed funds, the simplifying assumption above will work well.) Furthermore, assume that interest is paid on the $35,000 borrowed at an annual rate of 9 percent, with the principal to be repaid at the end of year 6. This is most similar to what would prevail if the company issued bonds with a 9-percent coupon rate. (Other types of loans which are amortized periodically would be handled a little differently. There one would need to calculate the value of the annual repayment amount attributable to interest, in order to compute the income tax effects.) The transformation of before-tax flows and interest payments into after-tax cash flows appears in Table 8–8.

TABLE 8–8 After-tax Cash Flow *(in dollars)*

End of Year	Before-tax Cash Flow	Depreciation Charges	Interest Expense Effects	Taxable Income	Income Taxes	Loan Cash Flow	After-tax Cash Flow
(1)	(2)	(3)	(4)	(5)	(6)	(7)	(8)
				$=(2) - (3)$ $- (4)$	$=(5) \times$ 50%		$=(2) - (6)$ $+ (7)$
0	−100,000					+35,000	−65,000
1	30,000	20,000	3,150	6,850	3,425		
					−10,000 cr	−3,150	33,425
2	30,000	32,000	3,150	−5,150	−2,575	−3,150	29,425
3	30,000	24,000	3,150	2,850	1,425	−3,150	25,425
4	30,000	16,000	3,150	10,850	5,425	−3,150	21,425
5	30,000	8,000	3,150	18,850	9,425	−3,150	17,425
6	30,000	0	3,150	26,850	13,425	−3,150	13,425
6	15,000			15,000	7,500		7,500
6						−35,000	−35,000

Note that the interest payments, although paid once per year, are listed twice for analysis: column (4) indicates the interest payments that reduce taxable income; column (7) includes all aspects of the loan flows (borrowed capital, interest payments, and repayment of the loan) in order to arrive at the after-tax cash flow. The investment tax credit appears at the end of year one. Also, period 6 entries are broken into three parts: once for the "savings" attributable to the installed equipment; the second is for the salvage value; and the third is for the loan repayment. The net after-tax flow for period 6 is − $14,075.

These flows are depicted in Cash Flow Diagram D8–10.

D8–10

ATCF ($000)

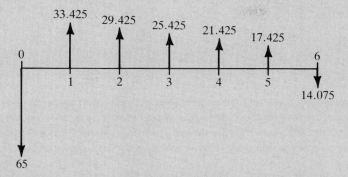

The internal rate of return for this investment, given by solving

$$-65{,}000 + 33{,}425(P/F\ i\%, 1) + \cdots + -14{,}075(P/F\ i\%, 6) = 0$$

for i is 29 percent. Thus, if the firm had only $65,000 available, but could borrow the remainder as described by the terms in the model, the equipment would be desirable, given an after-tax MARR of 10 percent.

Bear in mind that even if the firm had $100,000 available, it might still be desirable to borrow $35,000 for the machine, if the balance of the firm's capital could be made to earn at least 10 percent after tax. That this is so can be demonstrated by an incremental-rate-of-return analysis on the firm's $35,000 not needed for the machinery, combined with the 29-percent return on the machine with borrowed capital, or simply by comparing the annual worth of both alternatives. The annual worth of Example 8–7, given by

$$[-65{,}000 + 33{,}425(P/F\ 10\%, 1) + \cdots +$$
$$-14{,}075(P/F\ 10\%, 6)](A/P\ 10\%, 6) = \$6{,}042.09$$

is superior to the annual worth of Example 8–5, expressed as

$$[-100{,}000 + 35{,}000(P/F\ i\%, 1) + \cdots + (15{,}000$$
$$+ 7{,}500)(P/F\ i\%, 6)](A/P\ 10\%, 6) = \$4{,}117.09$$

Example 8–7 is an illustration of the principle of leverage, which provides that it may be advantageous to borrow funds, if they can be put to productive use at a rate of return greater than the interest rate paid for those funds.

Conclusion

The approach used through most of this chapter was to take a single alternative and to compute the after-tax cash flows under a variety of assumptions. The reader should bear in mind that the same after-tax viewpoint should be used in the comparison of actual alternatives. After-tax time value measures for each of the alternatives can be compared when ranking is performed. The reader should test his understanding of the after-tax computations by performing the comparison-of-alternatives exercises at the end of this chapter on an after-tax basis.

Exercises

Many of these exercises have appeared previously. Now the task is to perform after-tax transformations on the before-tax flows, then go on to compute measures of effectiveness. For these problems, assume an income tax rate of 50 percent. You may be interested in comparing your results with those obtained on a pretax basis.

8–1 Engineers at a plant that manufactures wood furniture are considering the installation of a system to reclaim and utilize the sawdust that is a by-product of the process. It includes a preheater for the hot water system that will burn the dust safely, ductwork throughout the plant, and a set of blower motors to propel the dust and deliver it to the preheater. The system as currently designed requires investment of $225,000. It is expected to save $40,000 per year in conventional fuel, as well as reducing dust cleaning and dispposal. The system is expected to serve for 15 years, after which it is considered to be worthless. Assume that the equipment will be depreciated on a straight-line basis over the 15 years. What after-tax internal rate of return would these savings provide?

8–2 *a.* Your firm is considering the purchase of a new computer system. The total cost, including peripherals, would be $350,000. The maintenance contract would amount to $20,000 per year. A five-year study period is being used, at the end of which time the system is thought to be worth $100,000. Assume that the equipment is depreciated on a straight-line basis over five years, with a salvage value of $100,000. What would the equivalent annual after-tax cost of purchase and maintenance be? Assume that the after-tax MARR is 8 percent.
b. Repeat the problem, assuming sum-of-the-years'-digits depreciation, and then ACRS depreciation. Compare your results.
c. A lease alternative is presented by the vendor in which maintenance is included. For the five-year period the annual lease payment (prepaid) would be $110,000. What would the equivalent after-tax annual cost be for this alternative? (Remember the lease is prepaid, which means the payments occur at times 0 through 4.)

8–3 The owner of an existing office building is investigating the prospects of conserving electricity. The building is heated electrically with thermostats in each lease area. A computer-controlled system is being examined which will permit the building manager to set and control the heat for a given temperature in each area, as opposed to giving the tenants the ability to set the heat levels themselves. It is felt that this will reduce waste associated with uncontrolled access to the thermostats and would save $16,000 per year. The system will cost $75,000 to install and will last for 10 years, at which time it will have a salvage value of $10,000. Assume that the system is depreciated under ACRS with a 5-year recovery period. In addition, a 3-percent investment tax credit is applicable. If the after-tax MARR is 5 percent, what after-tax external rate of return would be provided?

8–4 Your firm developed a new product, investing $2 million. These assets are depreciated using the declining-balance method with $k = 0.178$. Its sales, less out-of-pocket expenses, for 5 years are: $300,000, 400,000, 500,000, 300,000, 250,000. At the end of the fifth year, the productive assets associated with the product are disposed of for $750,000. Compute the after-tax internal rate of return of the firm's investment.

8–5 *a.* Five design configurations for a metal processing by-product recovery system are being evaluated for their economic characteristics. The alternative designs are listed in order of increasing investment requirements. If at least $300,000 is available, and the after-tax MARR is 6 percent, which design is most attractive? Assume straight-line depreciation over 5 years with no salvage value.

	Yr.	A	B	C	D	E
Investment ($000)	0	−75	−100	−180	−250	−280
Annual savings generated (revenue) ($000)	1–5	+25	+27	+43	+75	+82

b. If the capital available were only $200,000, which design would now be most attractive?

8–6 *a.* Four proposals for a machining process are being considered. Each proposal will satisfy the demands for quantity and quality of product. The differing cost structures of the proposals are given below:

	Yr.	A	B	C	D
Investment ($000)	0	−50	−60	−100	−150
Annual costs ($000)	1–4	−130	−125	−113	−90

If the after-tax MARR is 8 percent, which alternative is cost-minimizing? Assume sum-of-the-years'-digits depreciation over 4 years.
b. If only $75,000 is available, what is the recommended course of action?

8–7 *a.* Four mutually exclusive, cost-saving (or revenue-enhancing) projects are being studied. Data follow:

	Yr.	A	B	C	D
Investment ($000)	0	−150	−250	−300	−450
Annual cost savings (revenue) ($000)	1–4	+32	+70	+90	+125
Salvage value ($000)	4	+40	+60	+70	+95

If the after-tax MARR is 6 percent, which is superior, assuming $450,000 is available? Assume ACRS depreciation with 3-year recovery periods.
b. What if only $300,000 is available?

8–8 One out of four cost-minimizing proposals, shown below, is to be selected.

	Yr.	A	B	C	D
Investment ($000)	0	−125	−160	−190	−230
Annual costs ($000)	1–4	−140	−132	−110	−100
Salvage value ($000)	4	+35	+50	+60	+75

Which is best, if the after-tax MARR is 8 percent? Assume ACRS depreciation with 3-year recovery periods and a 4-percent investment tax credit.

8–9 Four cost-minimizing configurations need to be analyzed. The data follow:

	Yr.	A	B	C	D
Investment ($000)	0	−125	−145	−170	−200
Annual costs ($000)	1	−80	−80	−80	−65
	2	−80	−75	−70	−65
	3	−80	−70	−60	−65
	4	−80	−65	−50	−65
Salvage value ($000)	4	+20	+30	+40	+50

Which is best if the after-tax MARR is 5 percent? Assume straight-line depreciation over 4 years with salvage values as indicated.

8–10 The data for four cost-minimizing alternatives are as follows:

	Yr.	A	B	C	D
Investment ($000)	0	−95	−130	−150	−175
Annual costs ($000)	1	−60	−50	−40	−38
	2	−60	−50	−40	−38
	3	−60	−50	−40	−38
Salvage value ($000)	3	+50			
Annual costs	4		−50	−40	−38
Salvage value	4		+50	+60	
Annual costs	5				−38
Salvage value	5				+60

Analyze the data, assuming the after-tax MARR is 8 percent. Assume ACRS depreciation with 3-year recovery periods and a 3-percent investment tax credit.

8–11 Two processes are being considered to fulfill the same function. The first requires an immediate investment of $100,000 and will provide revenues of $300,000 and $360,000 in each of the first two years, $510,000 for the third year, $505,000 for each of the next two years, $400,000 in the sixth year, and $300,000 in the seventh. Costs will be $310,000 in the first year, $350,000 in the second, $490,000 in the third, $475,000 in each of the next two years, $360,000 in the sixth, and $250,000 in the seventh. In addition, the salvage value of the process at the end of the seventh year is expected to be $30,000. The other alternative requires a $125,000 initial investment. Revenues would be $400,000 per year for five years and costs would be $350,000 per year. The salvage value at the end of the fifth year would be $40,000.

If funds are worth 6 percent after taxes, which investment is superior? Assume ACRS depreciation with 5-year recovery periods. Use annual worth as a yardstick.

8–12 Repeat Exercise 8–5 using incremental internal rate of return as a yardstick.

8–13 Two minicomputers are being compared for use by an engineering consulting firm. Although other characteristics of the machines will be considered, an economic analysis yielded the following data: System *A* would cost $10,000 to acquire and $3,000 per year to operate, including power, materials, and a maintenance contract. It is expected to be used for 6 years, after which it would have a salvage value of $2,000.

System *B* would require a $12,000 investment but would cost only $2,000 per year to operate, including maintenance. Its economic life is also felt to be realistically estimated at 6 years, after which its salvage value would be $3,000. If funds are valued at 8 percent after taxes, and these are viewed as mutually exclusive alternatives, which is superior? Assume ACRS depreciation with 5-year recovery periods and a 4-percent investment tax credit.

8–14 A machine may be purchased for $20,000. Operating costs are expected to be $3,500 per year with maintenance being $500 per year. Its anticipated economic life is 7 years, after which it is projected that its market value would be $5,000. As an alternative to the purchase arrangement, the manufacturer offers to lease the machine for 7 years at an annual cost of $3,900, the lease payments to be prepaid for each year. Operating expenses are expected to be the same for the leased machine as for the purchased one; however, maintenance under the lease will be the responsibility of the manufacturer. If capital is worth 8 percent after income taxes, which alternative is superior? Assume a 5-year recovery period under ACRS for the purchased machine, with an investment tax credit of 3 percent. Also assume that one-fourth of the funds needed for purchase of the machine would be borrowed at an interest cost of 12 percent per year.

8–15 Energy costs for a certain process are now $12,000 per year. They are expected to increase at an average of 5 percent per year over the next decade, due to increasing activity. A microprocessor-based control system designed to cut down on the energy required by 10 percent can be obtained and installed for $8,000, its expected economic life being 10 years. If the market value of the system after 10 years is $2,000, and funds are valued at 8 percent after taxes, can the system be justified? Assume that the system is depreciated under ACRS with a 5-year recovery period.

8–16 Your company is considering one of two forklift trucks. *A* costs $11,000, requires $2,000 per year to operate, and is expected to have a 3-year economic life. Its salvage value at that time is projected to be $4,000. *B* costs $15,000 but requires $1,500 per year to operate. It is expected that its economic life will be 5 years, with a salvage value of $4,000. If funds are worth 6 percent after taxes, which is superior? Assume ACRS depreciation with 3-year recovery periods.

8–17 Three cost-saving machines are being considered for an existing function. Data follow:

	A	B	C
Investment($)	60,000	80,000	90,000
Annual costs($)	142,000	136,000	128,000
Economic life(yrs.)	5	5	5

If funds are worth 10 percent after income taxes, which alternative is superior? Assume that these machines are depreciated under ACRS with 5-year recovery periods, that there is a 4-percent investment tax credit, and that one-third of the capital is borrowed at an interest rate of 12 percent per year.

8–18 *a.* A machine is purchased for $22,000 and is depreciated under SYD for three years, with an expected salvage value of $4,000. It generates net income per year of $11,000 for four years. After the fourth year it is disposed of for $5,000. Assuming a 50-percent income-tax rate, what are the after-tax flows?
b. What is the annual worth of these flows if the after-tax MARR is 10 percent?

8–19 Your firm is convinced of the desirability of a new piece of equipment which would help cut down on annual costs. It requires an immediate investment of $150,000 and would provide annual net savings of $35,000 in the first year, rising by $5,000 per year. It is expected that the equipment would be usable for 6 years (even though its recovery period under ACRS would be 5 years). It is anticipated that after the sixth year the equipment would be worth $15,000.

The firm's decision involves comparing the purchase of the equipment, under the terms as indicated, with the option of renting it for 6 years at $30,000 per year. Determine which option has the higher after-tax annual worth if the MARR is 10 percent. The lease payments would be made at the beginning of each annual period.

8–20 Three alternatives are to be evaluated on an after-tax basis. Data for them follow:

	A	B	C
Investment($)	75,000	105,000	120,000
Annual operating costs($)	45,000	40,000	30,000
Economic life(yrs.)	5	7	6
Salvage value($)	10,000	13,000	18,000

All three would be depreciated under ACRS with 5-year recovery periods.

Which alternative has the highest measurable economic worth? The after-tax MARR is 12 percent.

Selected References

Blank, L. and D. Smith, "A Comparative Evaluation of the Accelerated Cost Recovery System as Enacted by the 1981 Economic Recovery Tax Act," *The Engineering Economist*, Vol. 28, No. 1, Fall, 1982, pp. 1–30.

Buck, J. and T. Hill, "Generalized Depreciation Methods and After-Tax Project Evaluation," *The Engineering Economist,* Vol. 22, No. 2, Winter, 1977, pp. 79–96.

Commerce Clearing House, *Standard Federal Tax Reports.* Chicago: Commerce Clearing House, periodically.

Lohmann, J., E. Foster and D. Layman, "A Comparative Analysis of the Effect of ACRS on Replacement Economy Decisions," *The Engineering Economist,* Vol. 27, No. 4, Summer, 1982, pp. 247–260.

Long, M. "Using a Before- or After-Tax Discount Rate in the Lease-Buy Decision," *The Engineering Economist,* Vol. 26, No. 4, Summer, 1981, pp. 263–274.

9

Replacement Analysis

Introduction

Preceding chapters have dealt with a variety of principles for evaluating the desirability of investments. When a single investment was under consideration (independent alternative), or where choices existed between various candidates (mutually exclusive alternatives), it was implicitly assumed that a new situation was being faced, in that no existing asset was available to continue performing services similar to the alternatives in question. Frequently this is not the case. Often it is necessary to analyze the relative desirability of an existing asset (referred to as the defender) as compared with one or more new assets or designs (called, not surprisingly, challengers).

This class of problems is treated in a separate chapter because, first, it is a very commonly faced situation. Furthermore, several considerations set it apart from the problems previously discussed. Among them are the probable existence of sunk costs associated with the defender and the probable disparity in lives when the expected remaining term of the defender is compared with that of the challenger.

Reasons for Replacement

There are several possible reasons for seeking to replace an existing asset. Often, more than one of the following may pertain simultaneously:

173

Deterioration

The asset, having had extensive use, no longer operates as reliably as originally, or may need frequent or extensive repair. Other symptoms of excess deterioration may include declining product quality, lower productivity, difficulty in performing its function at previously set standards, or higher operating costs.

Inadequacy

In this case the asset still performs as it was intended, but changing needs make it inadequate for present and/or future projected requirements. This typically occurs as activity changes in character or increases in volume. Generally, two types of remedies present themselves. One is to augment the existing asset with another or similar one. The other is to replace the existing one with an asset having sufficient capacity under the revised conditions facing the firm. In any case, the market value of the existing asset will tend to be relatively high, in that it still retains capacity for productive use.

Obsolescence

Continuous, if irregular, improvement in the means of production has been observable since the dawn of the industrial revolution. The development of new methods, procedures, equipment, and technology often makes existing assets less desirable even though they have useful life remaining. This may be due to the greater productivity, lower unit costs, or increased quality offered by the innovation. Thus, although the premature replacement of an otherwise satisfactory asset does not provide for its full capital recovery, it may be warranted if it results in greater productivity or improved performance.

Whatever the reason(s) for considering a replacement of an existing asset, the analyst should include the following factors in conducting a before-tax analysis:

For the challenger:
Its acquisition or capital cost;
Its periodic benefits, costs, or cost savings;
Its economic life; and
Its salvage value.

For the defender:
Its periodic benefits, costs, or cost savings;
Its remaining economic life;
A current estimate of its salvage value at the end of its remaining useful life; and
Its market or trade-in value at the time the replacement is being considered, not
its book value.

The last item for the defender is important, in that potential confusion exists on how to treat the investment that has already been made in the existing asset. A full discussion of this issue follows, under the heading of "Sunk Costs." That, in turn, is followed by an illustrative replacement example. Replacement analyses should be conducted on an after-tax basis in organizations subject to income taxes. However, a before-tax analysis will be presented first, in order to outline the basic principles involved. The necessary modification for an after-tax analysis will be presented later on in the chapter.

Sunk Costs

When an asset is being considered for replacement, a logical pitfall which traps many decision-makers concerns the treatment of sunk costs. This may be defined as invested capital already committed to a certain use which may not or will not be recovered under a given set of circumstances. However, the sunk costs are related to past decisions and actions and should not be considered when the defender is currently being re-evaluated.

For instance, a machine purchased one year ago for $20,000 with a five-year depreciable life under ACRS may now be felt to be inadequate for current purposes and also obsolescent because of rapid technological progress in its field. As a result of obsolescence, the market value of the machine, were it to be sold, is estimated to be only $5,000. Recall that under ACRS, the machine would have a book value of $20,000 - 0.20($20,000) = $16,000$. In this instance, the unrecoverable capital, or sunk cost, would be considered to be $11,000, or the difference between the book value of $16,000 and the realizable market value of $5,000.

The pitfall concerning such sunk costs is that they should not be taken directly into account when evaluating the desirability of retaining the existing asset. Thus, it would be an error, and one all too commonly made, to retain the asset under the rationale that one could not afford to part with it because of the large investment not yet recovered. To take such a position would be to compound an error already made, or to "cry over spilled milk." One must face the fact that errors of judgment, forecasting, and the like cannot be totally avoided, and that future technological developments cannot always be accurately predicted. In short, past decisions which led to the present circumstances are not relevant to the current choices which must be faced: that is, whether or not to replace the defender with the best available challenger. What is relevant and must be considered, therefore, is the present usefulness and market value of the existing asset as compared with any challenger.

Replacement Analysis

The concepts useful in replacement analysis are illustrated in the following situation. It will first be examined on a pretax basis, in order to focus on the most basic issues. Suppose that the machine mentioned previously in the section on sunk costs shows signs of obsolescence. To repeat its history, it was purchased one year ago for $20,000. Operating costs were $12,000 in the first year and are expected to continue at that level. Now it is desired to evaluate a challenger which can be obtained for $22,000. That it is even being considered at this point is due to the dramatic annual cost savings projected for it. Estimates are that the challenger would cut annual costs in half, to $6,000. Sale of the existing machine (or trade-in value, which is conceptually equivalent) would net $5,000 now, after all expenses of removal and transport are paid. For simplicity, it will be assumed here that the salvage value of the challenger is zero. (In later examples positive salvage values for the challenger will be used, which is a more realistic expectation.) Also assume that the economic life of the challenger is 5 years and that the defender can be used for 5 more years. What is the comparative desirability of this challenger?

The answer to this question revolves about the relative worth of a net investment of $17,000 now ($22,000 for the new machine, less $5,000 realizable from the old) in order to obtain savings of $6,000 per year for 5 years. Thus, one may consider an incremental-rate-of-return analysis as describing the relevant issues here. In other words, the incremental investment of $17,000 would earn a return of 22.5 percent, found by solving

$$-17,000 + 6,000(P/A\ i\%, 5) = 0 \quad \text{for } i$$

If the pretax MARR is 16 percent, the incremental investment appears desirable and the challenger may, indeed, be favored.

Note that the sunk costs associated with the defender are not relevant. The $20,000 investment associated with it has already been made, and nothing can change that. The $11,000 unamortized capital value ($16,000 book value less $5,000 trade-in or sales value) needs to be recognized and accounted for. However, it should not enter into the replacement decision to be made, for it will not be recovered more quickly if such an attractive challenger is ignored. In short, the cost savings to be provided by the challenger are seemingly worth investing in now. Past errors, if they are judged so, associated with the purchase of the defender should not be avoided or forgotten. Instead, they should be reviewed for the purpose of improving present and future studies.

At this point, another look at the situation above is in order. It is more common to examine each alternative, defender vs. challenger, individually, rather than incrementally, as above.

EXAMPLE 9–1

Assume that the before-tax MARR is 16 percent. For the defender,

$$AW = -5,000(A/P\ 16\%, 5) - 12,000 = -\$13,527$$

For the challenger,

$$AW = -22,000(A/P\ 16\%, 5) - 6,000 = -\$12,719$$

Here, the capital cost attributed to the defender is $5,000, even though no outlay would be made if it were retained. The idea is that it implicitly costs $5,000 to continue to use the defender, because it is that amount that would be forgone if the defender were retained. Thus, the $5,000 market value may be viewed as an implicit investment. Of course, the conclusion based on this view is the same as before, under the incremental-rate-of-return analysis. Since the annual worth is greater for the challenger (its costs are less), it is more desirable.

Lastly, note that the study period for both alternatives was identical. There is justification for this treatment in this case because replacement of the defender is being considered so early in its economic life and because the projected life of the challenger is about the same as that of the remaining useful life of the defender. Thus, this illustrates the case of equal study periods between defender and challenger. Later examples will employ different scenarios.

After-tax Considerations

Strictly speaking, there is at least one exception to the principle of avoiding any consideration of sunk costs when evaluating the defender's attractiveness. If an existing asset is disposed of at an amount different from its book value, there are income-tax ramifications which should be included in the replacement analysis. This is why replacement analyses should be done on an after-tax basis.

Thus, if an asset is sold for less than its book value, as would occur in the illustration above, the firm would be entitled to an income-tax credit. The credit would be based on the unrecovered capital or sunk cost and would be referred to as a capital loss. Income-tax credits for corporations based on long-term capital losses (for assets held more than one year) are calculated at a lower rate than that for ordinary income. Since that rate has often been changed by Congress, in this book the applicable rate used for computing capital-loss credits is a round number, 25 percent. When conducting studies, the reader should check on the current rate.

Therefore, in Example 9–1, a decision to sell the machine for $5,000 would result in a declared capital loss of $11,000. This would entitle the firm to an income-tax credit of .25(11,000), or $2,750. Now when one weighs the implicit investment in the defender, one must include the $2,750 tax credit, for retaining the defender is tantamount to forgoing both its $5,000 sales value and the capital-loss credit attendant to that sale. Since the tax credit, unlike the sunk cost, is a real, present effect, it should be considered when comparing the defender to a challenger. Including the capital-loss effect this way is part of the adaptation required to conduct an after-tax analysis. Since there may be other income-tax consequences, it is advisable to conduct all replacement analyses on an after-tax basis.

The procedure for an after-tax replacement analysis follows the form already developed for Example 9–1, modified by the after-tax transformation introduced in the preceding chapter. To illustrate this process, Example 9–1 will be repeated on an after-tax basis. For simplicity, no consideration will be given to the effect of investment tax credits.

EXAMPLE 9–2

Table 9–1a shows the transformation of before-tax flows for the defender.

The tax credit that would be forgone if the defender is retained is combined with its market value at time 0. This example differs from the after-tax trans-

TABLE 9–1a   After-tax Cash Flow—Defender *(in dollars)*

End of Year	Before-tax Cash Flow	Depreciation Charges	Other Tax Effects	Taxable Income	Income Taxes	After-tax Cash Flow
(1)	(2)	(3)	(4)	(5)	(6)	(7)
				=(2) – (3)	=(5) × 50% +2,750 cr	=(2) – (6)
0	–5,000				+2,750 cr	–7,750
1	–12,000	6,400		–18,400	–9,200	–2,800
2	–12,000	4,800		–16,800	–8,400	–3,600
3	–12,000	3,200		–15,200	–7,600	–4,400
4	–12,000	1,600		–13,600	–6,800	–5,200
5	–12,000	0		–12,000	–6,000	–6,000

formations shown in Chapter 8 in that here only costs are involved. Thus, the before-tax cash flows in years 1–5 are treated as negative numbers. Note, also, that depreciation charges in years 1–4 correspond to the last 4 annual depreciation allowances permitted under ACRS for a 5-year asset. (The first year's allowance for the defender would already have been accrued or taken and is therefore not germane to the current decision.) By year 5 the defender would be fully depreciated, so there would be no corresponding charge for that year.

Column 5 may be interpreted as the total of chargeable expenses which would serve to reduce income taxes. Column 6 then represents the reduction in income taxes attributable to the indicated expenses. Finally, column 7 shows the net after-tax cash flow, which is the difference between the anticipated cash expenses and the tax effects expected for the period. The resulting after-tax cash flows are shown in Cash Flow Diagram D9–1 and yield an annual worth (cost) of – $6,218, assuming an after-tax MARR of 8 percent.

D9–1

ATCF ($000)

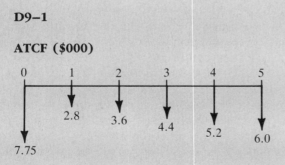

$$AW = [-7{,}750 - 2{,}800(P/F\ 8\%, 1) - \cdots$$
$$- 6{,}000(P/F\ 8\%, 5)](A/P\ 8\%, 5) = -\$6{,}218.21$$

Table 9–1b shows the transformation of the before-tax flows for the challenger, and Cash Flow Diagram D9–2 depicts the flows.

TABLE 9–1b   After-tax Cash Flow—Challenger *(in dollars)*

End of Year	Before-tax Cash Flow	Depreciation Charges	Other Tax Effects	Taxable Income	Income Taxes	After-tax Cash Flow
(1)	(2)	(3)	(4)	(5)	(6)	(7)
				$=(2)-(3)$	$=(5)\times 50\%$	$=(2)-(6)$
0	−22,000					−22,000
1	−6,000	4,400		−10,400	−5,200	−800
2	−6,000	7,040		−13,040	−6,520	520
3	−6,000	5,280		−11,280	−5,640	−360
4	−6,000	3,520		−9,520	−4,760	−1,240
5	−6,000	1,760		−7,760	−3,880	−2,120

D9–2

ATCF ($000)

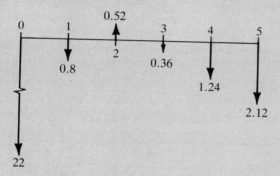

Since they give an annual worth (cost) of − $6,245, the challenger is deemed less desirable on an after-tax basis.

$$AW = [-22,000 - 800(P/F\,8\%, 1) - \cdots$$
$$- 2,120(P/F\,8\%, 5)](A/P\,8\%, 5) = -\$6,245.13$$

Note the effect of income taxes on replacement analysis illustrated by this example. On a pretax basis, the challenger was favored; on an after-tax basis, the defender was found more desirable (although the difference was small). This result reinforces the principle that it is preferable to conduct replacement studies on an after-tax basis.

Unequal Study Periods

In Examples 9–1 and 9–2, the defender and challenger could sensibly be compared over the same time horizon, namely 5 years. Only one year had elapsed since the defender had been acquired, making it a reasonable expectation that the machine could be used for 5 more years, and the challenger had a projected 5-year life. Thus, although annual worth was used as a comparative measure, present worth would have served equally well. However, it frequently occurs that comparisons present themselves with differing time horizons for the challenger and defender. This section deals with that situation.

EXAMPLE 9–3

Suppose that the same defender as in Example 9–2 is being evaluated. Now, however, the challenger is a machine with an anticipated economic life of 10 years. Its investment cost is $30,000, with annual disbursements expected to be $7,000 per year. It is forecast that the challenger will have a salvage value of $3,000 after 10 years. Thus, these alternatives present themselves over differing time spans (i.e., 5 more years for the defender, as opposed to 10 years for the challenger). They can be compared by computing the equivalent annual cost of each. These are given below, on before- and after-tax bases.

Pretax Basis, Defender. The annual worth of the before-tax cash flow associated with the defender was found in Example 9–1 to be $-\$13,527$.

Pretax, Basis, Challenger. The equivalent annual worth of this challenger can be obtained as follows:

$$AW = -\$30,000(A/P\ 16\%,\ 10) - 7,000 + 3,000(A/F\ 16\%,\ 10) = -\$13,066$$

Thus, on a pretax basis, the challenger promises lower annual costs and is therefore superior.

After-tax Basis, Defender. The annual worth of the defender's after-tax flows, computed in Example 9–2, was $-\$6,218$.

TABLE 9–2 After-tax Cash Flow, 10-yr.—Challenger *(in dollars)*

End of Year	Before-tax Cash Flow	Depreciation Charges	Other Tax Effects	Taxable Income	Income Taxes	After-tax Cash Flow
(1)	(2)	(3)	(4)	(5)	(6)	(7)
				=(2) − (3)	=(5) × 50%	=(2) − (6)
0	− 30,000					− 30,000
1	− 7,000	3,000		− 10,000	− 5,000	− 2,000
2	− 7,000	5,400		− 12,400	− 6,200	− 800
3	− 7,000	4,800		− 11,800	− 5,900	− 1,100
4	− 7,000	4,200		− 11,200	− 5,600	− 1,400
5	− 7,000	3,600		− 10,600	− 5,300	− 1,700
6	− 7,000	3,000		− 10,000	− 5,000	− 2,000
7	− 7,000	2,400		− 9,400	− 4,700	− 2,300
8	− 7,000	1,800		− 8,800	− 4,400	− 2,600
9	− 7,000	1,200		− 8,200	− 4,100	− 2,900
10	− 7,000	600		− 7,600	− 3,800	− 3,200
10	+ 3,000				1,500	1,500

After-tax Basis, Challenger. The transformation of before-tax to after-tax flows is given for the challenger in Table 9–2. ACRS depreciation percentages for a 10-year asset were used. Again, consideration of investment tax credits was omitted for simplicity.

The first of the two entries for year 10 corresponds to the disbursement and depreciation charge for that year. The second entry for year 10 reduces the gross $3,000 salvage value by $1,500, since the asset would already be fully depreciated, and the sales value realized would thus be taxed as recovery of previously charged depreciation. These after-tax flows give an annual worth of

$$AW = [-30,000 - 2,000(P/F\,8\%, 1) - 800(P/F\,8\%, 2)$$
$$\cdots - (3,200 - 1,500)(P/F\,8\%, 10)](A/P\,8\%, 10) = -\$6,236$$

On an after-tax basis, the defender would be preferable, since its equivalent annual costs are lower. Recall that this is a reversal of preferences from the pretax analysis. Thus, this example not only illustrates a technique for handling unequal initial study periods but once again points out that the after-tax analysis, being more comprehensive, is a better vehicle for replacement analyses.

It should be recalled that the use of annual worth for unequal (initial) study periods involves an implicit assumption—namely, that the choice of the shorter-lived asset results in the computed annual worth (cost) for years 6–10. When this implicit assumption is not valid, an adjustment needs to be made in order to compare the defender and challenger. For instance in Example 9–3, if the defender cannot be replaced with a similar asset and the only other way the defender can be utilized over a 10-year span is with an extensive overhaul, then that cost should be included when calculating the defender's equivalent annual worth.

Should the Defender Be Used for One More Year?

Examples 9–1 through 9–3 have considered a defender with much of its useful life remaining. Most often a challenger must be compared to a defender which is near the end of its useful life. In such cases it is not appropriate to make the computation of annual worth for the defender over a long study period, such as the useful life of the challenger, since the services of the defender cannot be had (or cannot be had economically) for very much longer.

One way to analyze the situation is to compare the expected cost or benefit of using the defender for one more year to the equivalent uniform annual cost or benefit of the challenger. If the annual cost or benefit of the defender for the next year is superior to that of the challenger, it would be preferable to retain the defender for at least one more year. If this comparison is favorable to the challenger, a further step should be taken before it can safely be chosen. That is, the equivalent annual cost or benefit of the defender for its remaining life (including the next year) should be computed, if it can realistically be expected to serve for more than one year. This should then be compared to the equivalent annual cost or benefit of the challenger. If that comparison also favors the challenger, then the defender should be replaced. Finally, if the defender is retained after such an analysis, the process may be repeated in the subsequent year with more current data.

EXAMPLE 9—4

This procedure is illustrated by Example 9—4. Suppose, now, that the defender of Examples 9—1 through 9—3 had been used for 4 years. Thus, it would have one more year of use for which a depreciation charge can be made under its ACRS schedule. Furthermore, imagine that the challenger is the one of Example 9—3—that is, it has an anticipated equivalent annual cost after taxes of $6,236. Since the defender has already been used for 4 years and is unlikely to last very much longer, should it be held for one more year or replaced by the challenger now?

An after-tax analysis of the defender will be made with the MARR at 8 percent as before. It is now estimated that the defender can be sold for $2,000 immediately, $1,500 one year hence, and $500 in 2 years. Annual disbursements are still $12,000. The task is to compute the cost of retaining the defender for one more year, and to compare it to $6,236, the equivalent annual cost of the challenger.

First, one must compute the after-tax implied capital investment in the defender. On a before-tax basis, the retention of the defender is tantamount to forgoing its $2,000 sales value. However, at the end of year 4, the book value is only $1,600. (ACRS provides for 92 percent of depreciation of investment cost after 4 years for a 5-year asset. Here the book value would be 0.08(20,000) = $1,600.) If the firm sells the asset for more than its book value (but less than its investment cost) the difference, here $400, is treated as income and taxed accordingly. (Assume a 50-percent rate.) Thus, sale of the asset for $2,000 would increase income taxes by $200. Sale of the asset would then entail forgoing only $1,800.

Next one needs to find the after-tax amount realizable if the asset is disposed of after one year. The pretax amount is projected to be $1,500. However, if $1,500 is obtained at that time, the whole amount would be treated as income, since the asset would have already been fully depreciated. The amount realizable after taxes would thus be only 0.5(1,500) or $750.

Finally, the $12,000 in disbursements for year 5 can be transformed to $5,200 after taxes, since income taxes would be reduced by $6,800 as a result of those expenses.

These after-tax flows for holding the defender for one more year are depicted in Cash Flow Diagram D9—3.

Therefore, if we let DC_j be the equivalent annual defender cost for j more years,

$$DC_1 = -[2,000 - 400(0.5)](A/P\ 8\%, 1)$$

$$+ [1,500(0.5)](A/F\ 8\%, 1) - 5,200 = -\$6,394$$

D9–3

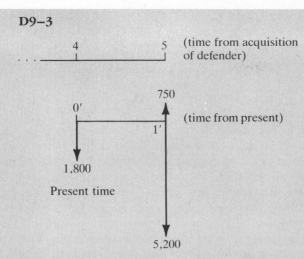

Since this compares unfavorably with the equivalent annual cost of $6,236 for the challenger, the defender warrants further consideration.

The next step is to compare the equivalent annual cost of the challenger to that of the defender for its anticipated remaining life. If we assume that the defender can be used for only two more years, with the salvage value at the end of that time being $500, the flows are as shown in Cash Flow Diagram D9–4.

D9–4

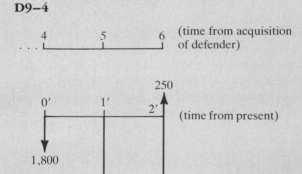

Thus,

$$DC_2 = -[2,000 - 400(0.5)](A/P\,8\%,\,2) + [500(0.5)](A/F\,8\%,\,2)$$

$$- [5,200(P/F\,8\%,\,1) + 6,000(P/F\,8\%,\,2)](A/P\,8\%,\,2) = -\$6,474$$

The rationale used for generating the after-tax values in DC_2 is the same as for DC_1. Since the equivalent annual cost of retaining the defender over its remaining life exceeds that of the challenger, the defender should, indeed, be replaced.

Asset Life that Minimizes Annual Cost

Frequently an analyst faces a decision in which consideration needs to be given to how long an asset should be held. For instance, in the case of a fleet of identical trucks being purchased, one might wish to arrive at a policy decision on how long to keep each vehicle before replacement. Having come to a conclusion on that life span, one might then stagger the schedule of purchase and replacement of trucks in order to smooth out the annual disbursements for truck purchases. Or, a single defender may need to be examined to determine how much longer it should be held before replacement by an identical or similar asset. In any event, one useful approach is to compute the optimal life of the asset over its anticipated serviceable future.

When only costs are at issue, as is most frequently the case, this optimal life is the time period that minimizes equivalent annual costs. It may be computed by calculating the equivalent annual cost of acquiring, operating, and disposing of the asset for every practically conceivable time period and choosing the period that minimizes such cost. This may be symbolized by

$$OL = \min_{k} (EUAC_k) = -P(A/P\ i\%, k) + MV_k(A/F\ i\%, k)$$

$$- [\text{SUM}_{J} Disb_J (P/F\ i\%, j)](A/P\ i\%, k) \qquad (9\text{–}1)$$

where OL is the optimal or minimum cost life in years, $EUAC$ is the equivalent uniform annual cost of owning and operating the asset over the specified interval, MV_k is the projected sales, trade-in, or market value of the asset at the end of k periods, and $Disb_J$ is the anticipated cost associated with operating the asset during period j. As before, it is preferable to conduct such analyses on an after-tax basis, in spite of the fact that doing so involves more computation.

EXAMPLE 9–5a

A machine used in large numbers costs $4,000 to purchase. It would be depreciated over 3 years under ACRS (which provides for annual depreciation allowances amounting to 33, 45, and 22 percent of investment cost respectively). The projected operating costs and market value of the machine are given in Table 9–3.

Pretax Analysis   Although an after-tax analysis was recommended, it will be more convenient for expository purposes to begin without consideration of income taxes.

TABLE 9—3 Projected Market Value and Disbursements

End of Year	Projected Market Value	Projected Disbursements
0	$4,000	
1	2,500	$1,500
2	1,500	1,700
3	700	2,000
4	200	2,400
5	0	3,000
6	0	3,600

Table 9—4 contains the equivalent annual costs for the machine on a pretax basis, using a 16-percent MARR.

Consider the equivalent annual cost if the machine is held for 3 years. The anticipated flows are depicted in Cash Flow Diagram D9—5.

TABLE 9—4 Equivalent Uniform Annual Costs

End of Year	Projected Market Value	Projected Disbursements	$EUAC_k$
0	$4,000		
1	2,500	$1,500	$3,640
2	1,500	1,700	3,390
3	700	2,000	3,290
4	200	2,400	3,235.3
5	0	3,000	3,234.8
6	0	3,600	3,275

D9—5

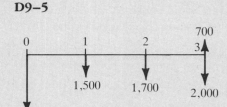

These flows may be converted to equivalent annual costs using

$$EUAC_3 = -4,000(A/P\ 16\%, 3) + 700(A/F\ 16\%, 3) - [1,500(P/F\ 16\%, 1)$$

$$+ 1,700\ (P/F\ 16\%, 2) + 2,000(P/F\ 16\%, 3)](A/P\ 16\%, 3) = -\$3,290.16$$

which appears transformed as Cash Flow Diagram D9—6.

D9–6

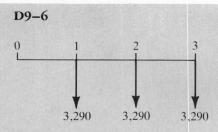

The pretax analysis indicates that annual costs may be minimized by retaining the asset for 5 years. Of course, this assumes that it is replaced with an asset with the same equivalent annual costs.

Analyses of this sort often yield a U-shaped pattern of equivalent annual costs having a unique minimum. This can occur when the "capital consumption" or decline in value of the asset decreases, while periodic disbursements increase as the asset is held longer. Thus, the increasing disbursements cause equivalent annual costs to rise as longer retention horizons are considered.

In order to conduct an after-tax analysis one must take into consideration all factors that bear on the taxes to be paid. To simplify the exposition, only the effect of depreciation and disposing of the asset at other than book value will be treated here. Other effects, such as investment tax credits, will be omitted. The reader interested in including such effects can extend the coverage of this section by following procedures analogous to those actually presented.

EXAMPLE 9-5b

After-tax Analysis Since a single table showing all the computations in an after-tax analysis would be too large, it is presented in two parts. Table 9–5a shows the expected decline in market value of the machine as time progresses,

TABLE 9–5a Tax Effects of Sale or Disposition of Machine

End of Year	Market Value	Depreciation Charge	Book Value	Income-tax Effect of Sale	After-tax Sale Value
0	$4,000				
1	2,500	$1,320	$2,680	$-180(0.25) = -45	$2,545
2	1,500	1,800	880	620(0.5) = 310	1,190
3	700	880	0	700(0.5) = 350	350
4	200	0	0	200(0.5) = 100	100
5	0	0	0	0	0
6	0	0	0	0	0

TABLE 9–5b   Tax Effects of Expenses and Depreciation

End of Year	Expenses (BTCF)	Depreciation Charge	Taxable Income	Income-tax Effect	After-tax Cash Flow
1	$– 1,500	$1,320	$– 2,820	$– 1,410	$ – 90
2	– 1,700	1,800	– 3,500	– 1,750	50
3	– 2,000	880	– 2,880	– 1,440	– 560
4	– 2,400	0	– 2,400	– 1,200	– 1,200
5	– 3,000	0	– 3,000	– 1,500	– 1,500
6	– 3,600	0	– 3,600	– 1,800	– 1,800

annual depreciation charges, book value, and the tax effects of disposition of the machine if it is made at the time indicated. Table 9–5b shows the effect on taxes of depreciation charges and expenses, and the after-tax flows attributable to those quantities.

The negative values for expenses indicate outflows; for taxable income and income taxes, negative values indicate that they are thereby reduced by the indicated quantity; for after-tax cash flows they indicate the actual net outflows. (Note that period 2 shows an inflow.)

When the after-tax values are used in Equation 9–1 with the after-tax MARR at 8 percent, the results shown in Table 9–6 are obtained. To illustrate the computation of one of the entries, consider holding the machine for 3 years. The expected uniform annual costs on an after-tax basis would be

$$EUAC_3 = -4,000(A/P\ 8\%, 3) + 350(A/F\ 8\%, 3) - [90(P/F\ 8\%, 1)$$

$$- 50\ (P/F\ 8\%, 2) + 560(P/F\ 8\%, 3)](A/P\ 8\%, 3) = -\$1,633$$

TABLE 9–6   Expected Uniform Annual Costs

Hold Asset k Periods	EUAC$_k$
1	$– 1,865
2	– 1,694
3	– 1,633
4	– 1,598
5	– 1,600
6	– 1,627

As with the pretax computation, this can be portrayed by Cash Flow Diagram D9–7.

D9–7 Anticipated After-tax Flows

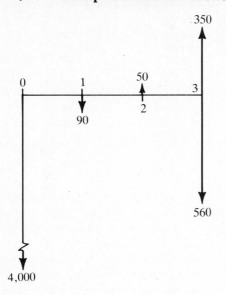

Exercises

9–1 A town's public works department is analyzing a potential challenger to an existing sweeper vehicle. Given the information below, compute the most economic life of the new machine (i.e., its minimum equivalent uniform annual cost, EUAC). Since the organization is not subject to income taxes, a pretax analysis is appropriate. Assume a 10-percent MARR.

Projected Market Value and Disbursements

End of Year	Projected Market Value	Projected Disbursements
0	$35,000	
1	28,000	$ 5,000
2	22,000	6,000
3	17,000	7,000
4	13,000	8,000
5	10,000	9,000
6	9,000	10,000

9–2 An engineer at a voluntary hospital is analyzing the economic characteristics of a new centrifuge. Its acquisition cost, projected market values, and operating costs are given below. Indicate how the minimum EUAC may be computed by writing an equation expressing EUAC as a function of how long the machine is retained. Then compute the EUAC assuming the machine is held for 5 years. Since the hospital is exempt from income taxes, do a pretax analysis. Use a 12-percent MARR.

Projected Market Value and Disbursements

End of Year	Projected Market Value	Projected Disbursements
0	$16,000	
1	13,000	$1,200
2	11,000	1,400
3	10,000	1,600
4	8,000	2,000
5	7,000	2,400
6	6,000	2,600
7	5,000	2,600
8	4,000	2,800
9	3,000	2,800
10	2,000	2,800

9–3 *a.* A firm will purchase a large number of vehicles over the next few years. Use the following information to compute the minimum equivalent uniform annual cost associated with retaining each vehicle for a given time:

Projected Market Value and Disbursements

End of Year	Projected Market Value	Projected Disbursements
0	$20,000	
1	15,000	$18,000
2	11,000	19,000
3	8,000	21,000
4	6,000	24,000
5	5,000	28,000
6	4,000	33,000

Do the analysis on a pretax basis, using a 20-percent MARR.
b. Repeat the analysis on an after-tax basis. Assume an income-tax rate of 50 percent and an after-tax MARR of 10 percent. Assume that the vehicles are depreciated under ACRS with a 3-year recovery period.

9–4 **Projected Market Value and Disbursements**

End of Year	Projected Market Value	Projected Disbursements
0	$10,000	
1	8,000	$3,500
2	6,500	4,000
3	5,000	5,000
4	4,000	5,500
5	3,500	6,000
6	3,000	6,000

The marketing department of your company requires a small fleet of automobiles for its systems engineers to visit clients. Compute the minimum EUAC for the vehicle described above. Do an after-tax analysis, assuming an income-tax rate of 50 percent. Use straight-line depreciation, a depreciable period of 3 years, and a salvage value, for depreciation purposes, of $5,800. The after-tax MARR is 10 percent. The capital gain or loss rate is 25%.

9–5 A furniture manufacturer is considering the purchase of a new lathe. Using the following data, write an expression for the EUAC as a function of how long the machine is retained. Then compute the EUAC, assuming the machine is held for 6 years. Do it on an after-tax basis, assuming a tax rate of 50 percent. Use an after-tax MARR of 12 percent. Assume an investment tax credit of 3 percent. Depreciation is under ACRS with a 5-year recovery period.

Projected Market Value and Disbursements

End of Year	Projected Market Value	Projected Disbursements
0	$26,000	
1	23,000	$3,000
2	20,000	3,200
3	18,000	3,400
4	16,000	3,800
5	14,000	4,200
6	12,000	4,600
7	10,000	5,000
8	8,000	5,400
9	7,000	5,800
10	6,000	6,200

9–6 A press was purchased 3 years ago at a cost of $32,000. It was depreciated under ACRS with a 5-year recovery period. Operating expenses and market values for the coming years are given in the following table. On the basis of current performance, it is expected that the existing press can be used for 5 more years. A challenger has been offered, requiring investment of $40,000. It, too, would be depreciated under ACRS with a 5-year recovery period. Its operating expenses and market values are also given below. Assuming that the challenger has a 5-year economic life, should the defender be replaced now? Assume a 50-percent income-tax rate and an after-tax MARR of 10 percent.

Projected Market Value and Disbursements

End of Year	Defender Projected Market Value	Defender Projected Disbursements	Challenger Projected Market Value	Challenger Projected Disbursements
0	$15,000		$40,000	
1	13,000	$7,000	35,000	$5,000
2	11,000	7,500	30,000	5,250
3	10,000	8,000	26,000	5,500
4	9,000	8,500	22,000	5,750
5	8,000	9,000	19,000	6,000

9–7 Your firm purchased a milling machine exactly 3 years ago at a cost of $110,000. It has been used to produce parts whose revenue equivalent attributable to it is $85,000 per year. Actual operating expenses have been $45,000 per year. The machine has been depreciated under ACRS with a 5-year recovery period.

Currently the firm has been offered a new machine with superior productivity. It will cost $225,000 to acquire, although a $35,000 trade-in allowance would be given for the defender. Assume that it would be depreciated over 5 years, using ACRS, and that its useful life is also 5 years. Include the factor that an investment tax credit of 4 percent would be available for the replacement (although it was not in effect for the defender). The challenger is expected to produce twice the equivalent revenue of the defender, $170,000, while out-of-pocket expenses would increase to only $65,000. In considering the economic worth of the challenger, assume that its market value after 5 years would be $40,000. The anticipated market values for the defender, should it be retained now, would be $25,000 in one year and $18,000 in 2 years.

a. Do a pretax analysis to see whether the defender or challenger has greater economic worth. Assume a pretax MARR of 20 percent.

b. Repeat the analysis on an after-tax basis, assuming an after-tax MARR of 10 percent and a tax rate of 50 percent.

9–8 A piece of road-grading equipment was purchased 4 years ago for $42,000. It was depreciated using the straight-line procedure over 6 years, with a salvage value, for depreciation purposes, of $6,000. It is expected that it can be utilized for 3 more years, with operating and maintenance expenses and projected market values as given in the following table. A new model is now available for $50,000. It can be used for 8 years, as the table indicates. It would be depreciated under ACRS with a 5-year recovery period. Compare the defender and challenger on an after-tax basis, assuming an income-tax rate of 50 percent. Assume an after-tax MARR of 10 percent.

Projected Market Value and Disbursements

End of Year	Defender Projected Market Value	Defender Projected Disbursements	Challenger Projected Market Value	Challenger Projected Disbursements
0	$16,000		$50,000	
1	12,000	$12,000	45,000	$ 8,000
2	8,000	13,000	35,000	8,500
3	4,000	15,000	31,000	9,000
4			27,000	9,500
5			23,000	10,000
6			20,000	11,000
7			17,000	12,000
8			14,000	13,000

9–9 A grinding machine was purchased 5 years ago at a cost of $28,000. It was depreciated under ACRS with a 5-year recovery period. Although it is fully depreciated, its service is still satisfactory, albeit at a higher level of maintenance expense than originally. A new model is available, with maintenance and operating costs and market values as indicated in the following table. Should the grinder be replaced now or used for one more year? The after-tax MARR is 12 percent. If purchased, the new machine would be depreciated under ACRS with a 5-year recovery period, and there would be an investment tax credit of 3 percent.

Projected Market Value and Disbursements

End of Year	Defender		Challenger	
	Projected Market Value	Projected Disbursements	Projected Market Value	Projected Disbursements
0	$5,000		$40,000	
1	4,000	$8,000	30,000	$3,000
2	3,000	9,000	25,000	3,250
3			21,000	3,500
4			17,000	4,000
5			13,000	4,500

9–10 A clothing manufacturer needs to increase its capacity to produce garment labels. Currently these are produced on a single machine which has been fully depreciated. It is worth $1,500 now and has operating costs of $2,500 per year. It is felt that the existing machine can be used for 5 more years, at which time it would be worthless. After investigating the various possibilities, the methods and standards department has narrowed down the options to two:

(1) To purchase another machine of similar capacity, labeled B. This will provide the additional needed service. Such a machine will cost $5,000, can serve for 10 years, and will have operating and maintenance costs of $2,000 per year with no salvage value. It would be depreciated under ACRS with a 5-year recovery period. (To evaluate this option over 10 years, assume that in 5 years the original label-maker is replaced with another one like B, which would be used for the second 5-year period. It would have a value of $1,500 at the end of the tenth year.)

(2) To acquire a larger label-maker, C, at a cost of $8,500. It would provide all the needed capacity, would last for 10 years with no salvage value at that time, being depreciated under ACRS with a 5-year recovery period, and would have operating costs of $4,000 per year. Use an after-tax MARR of 10 percent to evaluate the two options.

9–11 A group of forklift trucks was purchased 3 years ago for $8,000 each. They were depreciated under the straight-line method over 5 years with no salvage value (for depreciation purposes). It is felt that these trucks can be used for another 2 years at most. Operating expenses would be $4,000 per year; the market value of the trucks would be zero if they are not disposed of now. A forklift truck dealer has offered new models costing $11,000. In addition, the salesman indicated that a trade-in of the existing trucks can be negotiated. Purchase of the new trucks will result in an investment tax credit of 4 percent.

They would be depreciated under ACRS with a 3-year recovery period. The new trucks would have operating expenses of $2,500 per year. They would be usable for 5 years, at which time the market value would be nil. The firm uses a 50 percent income-tax rate for study purposes, and an after-tax MARR of 10 percent. What value of trade-in allowance for the old trucks would make the after-tax economic worth of the defender and challenger the same? Write an equation in terms of TIA, the trade-in allowance. Then proceed with the solution.

9–12 A bottling machine used by a food processor has been in use for 5 years. It is fully depreciated but can be used for 3 more years, after which time it will be worthless. Operating expenses are $10,000 per year. A new machine can be bought now for $60,000. An investment tax credit of 3 percent is applicable. It would be depreciated under ACRS with a 5-year recovery period and would have operating costs of $7,000 per year. If it is acquired now, the machine vendor would grant an $8,000 trade-in value for the existing machine. Find the usable life of the challenger that will make the after-tax economic worth of both alternatives equal. Use an income-tax rate of 50 percent and an after-tax MARR of 9 percent. (For simplicity, assume that the market value of the challenger at the end of its useful life is zero.)

9–13 A cement mixer is fully depreciated and can no longer be used as is. However, a $5,000 overhaul would permit it to be used for 3 more years, after which it would be worthless. Currently it is worth $1,000 to a used-parts dealer. Operating costs for the overhauled mixer would be $7,000 in the first year, rising by $500 per year. A new mixer would cost $75,000 with a 4-percent investment tax credit applicable. It would be depreciated under ACRS with a 5-year recovery period and could be used for 8 years, after which it, too, would be worth $1,000. Operating costs for the new mixer would be $4,000 in the first year, rising by $500 per year. Should the defender be overhauled or should the new mixer be purchased? The income-tax rate is 50 percent and the after-tax MARR is 10 percent.

Selected References

Lohmann, J., E. Foster and D. Layman, "A Comparative Analysis of the Effect of ACRS on Replacement Economy Decisions," *The Engineering Economist,* Vol. 27, No. 4, Summer, 1982, pp. 247–260.

Swalm, R., "Economics of Machine Selection and Replacement—A Bibliography," *The Engineering Economist,* Vol. 6, No. 3, Spring, 1961, pp. 51–57.

10

The Impact
of Inflation

Introduction

Chapters 8 and 9 included material which permits the analyst to take
into account the effect of income taxes on the desirability of a project. This was a vital elaboration of the time value modeling process
described earlier. This chapter presents a further enhancement of the
modeling process, one that helps take into account the potential effects that future price changes will have on a project's attractiveness. Although the need for accounting for future price increases is
not as imperative as the need to take into account income taxes,
there are many situations in which potential errors of judgment would
be made if future price changes were not taken into consideration.
Hence, some guidelines will also be given to indicate when the impact
of future price changes should not be ignored.

There are only two directions in which the general level of prices for
a specific set of goods or services may move over time: an upward
movement of such prices, referred to as inflation, is associated with
a decline in the real value of currency; a downward movement of
prices, called deflation, is associated with an increase in the real value
of currency. It is not the purpose of this book to devote attention to
a detailed examination of the causes or possible remedies for either
of these conditions. However, a fundamental understanding of these
phenomena, their measurement, and their ramifications for economic
investment decision-making at the establishment level is important
and will be covered in this chapter.

195

Measuring Price Changes through Time

Although it is necessary to characterize the behavior of price changes for specifically defined categories of goods or services over time, any method of doing so is, by the nature of the phenomenon, imperfect. Not all elements in the defined category experience the same price changes over time; consumers or industrial users adjust their purchases in response to price changes, tending always to substitute less costly alternatives; and in a dynamic free market, improvements in quality due to technological innovation continually occur, thus making temporal comparisons difficult.

Nonetheless, since quantifiable characterization of the overall rate of price changes is necessary, it can be estimated by the use of time series index measures. These use a weighted average of the prices for a specifically defined class of products. The weights correspond to the relative importance of each product or product subclass in the collection, or "market basket," and the prices are the observed market prices for those products at the time in question. The ratio of this weighted average and a similar one using prices for equivalent products in the base period give an index value comparing the general level of prices in the period in question to that in the base period.

A formula for such an index measure, called a Laspeyres Index, is given by

$$I_n = \frac{\Sigma \, q_0 p_n}{\Sigma \, q_0 p_0}$$

where I_n is the index measure for the general price level in period n as compared with the base period 0, q_0 are the relative weights for each product or product subclass, which in a Laspeyres index are based on the relative importance as experienced in the base period, and p_n and p_0 are the observed prices in periods n and the base period, respectively.

For instance, if the base period for a certain collection of goods were, say, 1976, the period in question were 1983, and the index value for 1983, using the above formula, were 1.5, then the interpretation would be that the general price level for the goods and services described by the index had risen by fifty percent during the period. Of course, this represents an average, with some prices rising more than others and some prices possibly even declining over that period. (Recall, also, that the index is, at best, a good approximation to the overall price-change over the period, based on the reasons given previously.)

Index measures such as these may be combined with the exponential formulas developed for time value analysis. For the example above, a fifty-percent overall price increase over seven years is equivalent to an average annual increase given by

$$1.5 = 1(1 + i)^7$$

or $i = 5.96$ percent.

Because price indexes are computed and compiled for various purposes, they are based on different collections of goods and services, or market baskets. The best-known is the Consumer Price Index, compiled by the Bureau of Labor Statistics of the U.S. Department of Labor. It is designed to measure retail price changes affecting urban wage earners. Thus, its market basket includes such commodities as food, housing, and clothing and such services as health care and transportation. As measured by this index, the period since World War II has been one of general inflation. Figure 10–1 indicates that in the late 1970s the inflation rate was significantly higher than in the late 1950s and early 1960s. It should not be surprising, then, that attention to consideration of inflation in engineering economic analysis did not become significant until then.

Although the Consumer Price Index provides a general indication of the movement of prices, taken alone it is inadequate for engineering economic analysis, because industrial analysts need to consider price changes for industrial goods and commodities. An alternate set of indexes, also prepared by the Bureau of Labor Statistics, is designed for that purpose. They are referred to as Producer Prices and Price Indexes and are compiled and reported monthly. These may be described briefly as follows:

> Producer price indexes measure average changes in prices received in primary markets of the United States by producers of commodities in all stages of processing. These data were previously presented as the Wholesale Price Index. The name "Producer Price Indexes" is now being used to reflect more accurately the coverage of the data. . . .

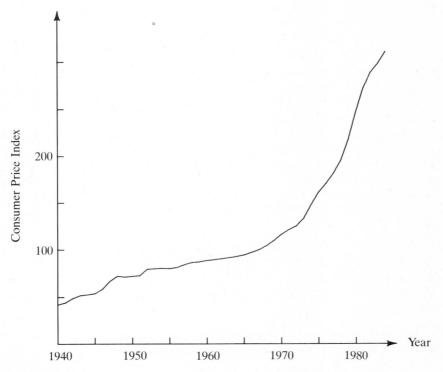

FIGURE 10–1

Sources: U.S. Dept. of Commerce, Statistical Abstract of the United States, and Historical Statistics of the United States.

Producer price indexes can be organized by stage of processing or by commodity. The stage-of-processing structure organizes products by degree of fabrication (i.e., finished goods, intermediate or semifinished goods, and crude materials). The commodity structure organizes products by similarity of end-use or material composition.

Finished goods are commodities that will not undergo further processing and are ready for sale to the ultimate user, either an individual consumer or a business firm. Capital equipment (formerly called producer finished goods) includes commodities such as motor trucks, farm equipment and machine tools. Finished consumer goods include foods and other types of goods eventually purchased by retailers and used by consumers. Consumer foods include unprocessed foods such as eggs and fresh vegetables, as well as processed foods such as bakery products and meats. Other finished consumer goods include durables such as automobiles, household furniture, and jewelry, and nondurables such as apparel and gasoline.

Intermediate materials, supplies, and components are commodities that have been processed but require further processing before they become finished goods. Examples of such semifinished goods include flour, cotton yarns, steel mill products, belts and belting, lumber, liquefied petroleum gas, paper boxes, and motor vehicle parts.

Crude materials for further processing include products entering the market for the first time which have not been manufactured or fabricated but will be processed before becoming finished goods. Scrap materials are also included. Crude foodstuffs and feedstuffs include items such as grains and livestock. Examples of crude non-food materials include raw cotton, crude petroleum, natural gas, hides and skins, and iron and steel scrap.

For analysis of general price trends, stage of processing indexes are more useful than commodity grouping indexes. This is because commodity grouping indexes sometimes produce exaggerated or misleading signals of price change by reflecting the same price movement through various stages of processing. . . .

In calculating producer price indexes, price changes for the various commodities are averaged together with weights representing their importance in the total net selling value of all commodities as of 1972. The detailed data are aggregated to obtain indexes for stage-of-processing groupings, commodity groupings, durability of product groupings, and a number of special composite groupings. Each index measures price changes from a reference period which equals 100.0 (usually 1967. . . .*

Each month, the Bureau of Labor Statistics issues its Producer Price report in the form of 13 tables, listed below. It should be noted that these include price indexes for the output of selected industries. It is expected that by 1986 Table 4 will cover all 493 SIC (Standard Industrial Classification) mining and manufacturing industries.

1. Producer price indexes and percent changes by type of processing;
2. Producer price indexes and percent changes for selected commodity groupings by stage of processing;

*U.S. Department of Labor, Bureau of Labor Statistics, *Producer Prices and Price Indexes* (monthly).

3. Producer price indexes for selected state-of-processing groupings, seasonally adjusted;
4. Producer price indexes for the net output of selected industries and their products;
5. Producer price indexes by durability of product;
6. Producer prices and price indexes for commodity groupings and individual items;
7. Producer prices and price indexes for refined petroleum products by region;
8. Producer price indexes for special commodity groupings;
9. Producer price indexes for the output of selected SIC industries;
10. Producer price indexes for the output of selected census product classes;
11. Producer price indexes and percent changes for total railroad freight and selected STCC (Standard Transportation Commodity Code) groups;
12. Producer price indexes and percent changes for selected telephone services; and
13. Producer price indexes and percent changes for postal services.

The analyst will find the producer price indexes useful from two perspectives. They will provide both a relevant overall measure for the degree of price change experienced for a period and estimates for components that may be of special interest. For example, an equipment investment decision for the food processing industry may require estimates of future price changes of the raw materials used in the process. Time trend data for such a raw material may be developed from the successive entries from one or more of the applicable tables of the thirteen listed above.

There are other price indexes that may be relevant to the analytical needs of the decision-maker. For example, construction cost indexes are published by a number of nongovernmental organizations. One such is the Boeckh index.* *Engineering News Record* also publishes a building cost index, which is reported in *Construction Review* and the *Statistical Abstract of the U.S.*

Incorporating Anticipated Future Price Changes

The Rationale for Trying to Anticipate Price Changes

If all factors in an economic feasibility study were affected identically by future price changes, there would be no need to make adjustments to the analytical procedures outlined in the preceding chapters. They would give valid results as they stand. However, such a condition almost never holds in practice. The various factors being considered, such as labor, energy, raw materials, equipment, manufactured components, and the like are expected to be affected differently by future price changes. Furthermore, the disparities among anticipated future effects may

---

*The Boeckh index is maintained by the American Appraisal Co., 525 E. Michigan St., Milwaukee, WI 53202. It is reported in various publications, including *Construction Review* and *Statistical Abstract of the U.S.*

be quite dramatic. For example, in a given study, energy costs might be expected to rise dramatically while the cost of certain types of electronic equipment might be expected to decline significantly. Other factors of an after-tax analysis, such as depreciation charges or contractual interest payments on long-term debt, will not change at all.

The upshot of all this is that different long-term price changes for the various components of a study can have a measurable effect on the economic worth of the project. Thus, an attempt, however imperfect, should be made to take into account the effects of future price changes on the desirability of an alternative when they are thought to be quantitatively significant. Not to do so would be to give a distorted picture of the alternative in question at the time decisions are made.

As a simple illustration, during a period of rapidly rising energy prices consideration is likely to be given to investments in new equipment or processes to reduce energy consumption. The true economic worth of these alternatives is likely to be seriously underestimated if future energy price increases are ignored.

Of course, if the anticipated magnitude of future price changes is expected to be insignificant, or if all factors are expected to be affected similarly, then the added measures described in this chapter may not be necessary. This was essentially the case prior to the 1960s, when inflation was not as severe a problem as it subsequently became.

Procedures To Be Employed

There are essentially two ways in which future price changes may be taken into account. The procedures are referred to as:

1. **Constant- (or Real-) Dollar Method.** Here one must express all present and future flows in constant dollars (sometimes referred to as real dollars). By constant dollars is meant the monetary value of the flow of factors expressed in terms of the currency value at a particular point in time, usually today's currency. Thus, the flows at any point in time are expressed in terms that represent constant purchasing power. (This should not be confused with discounting the future or expressing the equivalent worth of future flows based on the opportunity costs of capital.) When this method is employed, the relevant MARR is the same as was used in the preceding chapters. In other words, the unadjusted MARR represents an expectation regarding a return on capital assuming no variation in the purchasing power of currency.
2. **Current- (or Actual-) Dollar Method.** In this method one expresses all present and future flows in current dollars (also referred to as actual or future dollars). By current dollars is meant the anticipated future flow expressed in the currency amounts of the period in question, given whatever price changes are expected to occur by that time. This differs from the preceding method in that the flows stated over the future time periods are not given in units that express constant purchasing power.

When this method is employed, the relevant MARR is *not* the same as was used in the preceding chapters. Now, since the future flows are not expressed in units of constant purchasing power, the MARR must be adjusted appropriately. Thus, if the overall effect for the future is expected to be an inflation, the MARR must be increased. In other words, since the MARR represents an expectation regarding a return on capital without consideration of the possibility of a variation in the purchasing power of currency, when that purchasing power is expected to decrease, the MARR must be increased in order to compensate. (Clearly the MARR would be decreased to compensate for an expected deflation.)

If the unadjusted MARR is designated by i, and the overall rate of change for prices is given by p, then the adjusted MARR would be given by

$$(1 + i)(1 + p) - 1$$

Thus, for instance, if the unadjusted MARR is 10 percent and an overall inflation rate of 6 percent is expected, then the combined MARR to be used for this method would be $(1.1)(1.06) - 1 = 0.166$, or 16.6 percent.

It should be emphasized that each method must be implemented with its appropriate MARR, else an invalid recommendation may result. For example, the use of the first method, constant dollars, in connection with the evaluation of a revenue-enhancing alternative, when applied with an inflation-adjusted MARR, would unfairly penalize the investment in that the future flows would be discounted too heavily.

Although both methods are conceptually and computationally equivalent, the current- (actual-) dollar method is generally favored because it is direct in application and understandable when communicating the elements of the study to others.

Example 10–1

Example 10–1 will serve to demonstrate the fundamental equivalence of the two methods described above. For this purpose, let us return to the simple cash flow situation previously encountered in Example 3–1. That cash flow is repeated in Cash Flow Diagram D10–1.

D10–1

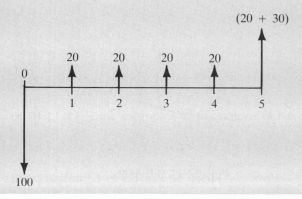

The reader should verify that the present worth of these flows is − $10.882 (thousand) if the unadjusted MARR is 12 percent. Now, suppose that there is in effect an average long-term inflation rate of 6 percent. If one begins with method 2, the current- or actual-dollar method, it is necessary to adjust all future flows by the 6 percent inflation rate. This can be done as in Table 10–1.

TABLE 10–1 Current- or Actual-Dollar Flows

Year	Real- or Constant-Dollar Flow	Current- or Actual-Dollar Flow
0	− 100	= − 100
1	20	$20(1.06) = 21.2$
2	20	$20(1.06)^2 = 22.472$
3	20	$20(1.06)^3 = 23.820$
4	20	$20(1.06)^4 = 25.250$
5	50	$50(1.06)^5 = 66.911$

The cash flow diagram for actual or current dollars would then be represented in Diagram D10–2.

D10–2

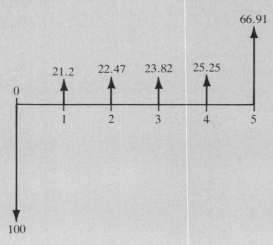

Given the current or actual cash flows, it is necessary to discount them at an adjusted MARR that takes into account the declining purchasing power of the dollar. Since the inflation rate is 6 percent, the adjusted MARR would be $(1.12)(1.06) − 1 = 0.1872$ or 18.72 percent. Then, if one follows method 2 and discounts the current or actual flows at the adjusted MARR, one obtains,

$$PW = -100 + 21.2\,(P/F\ 18.72\%, 1) + 22.472\,(P/F\ 18.72\%, 2) + \cdots$$

$$+\ 66.911\,(P/F\ 18.72\%, 5) = -10.882$$

By comparison, method 1, using constant or real dollars, requires one to discount by the unadjusted MARR of 12 percent the flows of column 2, Table 10–1, which are expressed in terms of constant purchasing power. This was done previously in Chapter 3 and agrees with the present worth of − 10.882 obtained under the current-dollar method.

Practical Considerations in Accounting for Price Changes

Projecting Price Changes   Both methods require estimates of future price changes. This is a difficult problem, and one that will not be dealt with in depth here. The reader should consult the references for this chapter for more detailed guidance.

Most attempts at projecting future price increases begin with some form of extrapolation of past price changes. That is why the price indexes mentioned previously are so important. Thus, if a certain index showed price changes of 3, 5, and 4 percent in each of the last 3 years, that would be equivalent to a 3-year change of $(1.03)(1.05)(1.04) - 1 = 0.12476$, or 12.476 percent. That is, in turn, equivalent to an average annual increase of $(1.12476)^{1/3} - 1 = 0.03997$ or 3.997 percent.

The simplicity of these computations, however, belies the degree of difficulty of correctly anticipating future changes. For instance, will the next year's change be more or less than 4 percent? Was the choice of three years of historical data appropriate to the economic situation being forecast? What economic events currently under way will have the greatest influence on future changes? What are the economic pundits' views? And so it goes.

The problem of projecting price changes cannot help but precipitate a very large number of questions—so many, in fact, that the estimation problem may frequently be separated from the rest of the economic analysis. Some large companies use economic consultants for that purpose, either internal and/or external. Of course, in smaller companies, or where such help is not available, the economic decision-maker is on his own and must do the best he can. Suffice it to say that where a price-adjusted analysis is warranted, a reasonable attempt needs to be made to project future price changes. Such an attempt is better than none and can be augmented by a sensitivity analysis.

After-tax Analysis   In taking into consideration the effects of future anticipated price changes, it does not make sense to ignore other tangible factors that have a potential effect on the economic worth of the alternative. Such a factor is income taxes. Chapters 8 and 9 clearly demonstrated the importance of taking income taxes into account, because they are real economic flows to be reckoned with. When considering future price changes, this factor takes on added significance because some of the elements of the problem, such as depreciation charges and interest payments on long-term debt, will be stable over the study period, while others, such as labor, energy costs, or revenues, may be subject to future price changes. In order to see how the inclusion of both price changes and income taxes may affect the outcome of an analysis, as well as how to conduct one, consider Example 10–2.

Example 10–2

The basic situation will be that described in Chapter 8, in which a cost-saving investment is being considered. The amount of the required investment is $100,000, the annual gross saving is $40,000, and the annual expense associated with the process is $10,000. The equipment is expected to provide such annual savings throughout its depreciable period, 6 years, after which it will have a salvage value, for depreciation purposes, of $15,000. These flows are recapitulated in Cash Flow Diagram D10–3.

D10–3

BTCF ($000)

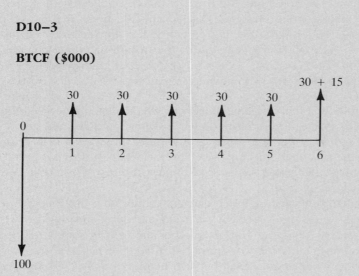

Suppose that the after-tax flows are to be generated under the same assumption employed in Table 8–2—namely, that straight-line depreciation is used. (An investment tax credit is not considered here.) Further suppose that all the future flows subject to price changes will experience an 8-percent inflation rate. Depreciation charges, of course, will remain unaffected by inflation. The revised (current-dollar) before-tax cash flows and their transformation to after-tax cash flows are depicted in Table 10–2.

Note that the current-dollar before-tax flows were computed as in Table 10–1. For example, the net current-dollar saving in year 3 was estimated at $30,000 (1.08)^3 = 37,791$. Also, the treatment of salvage value here differs from that in Example 8–2. There, it was implicitly assumed that the equipment would be worth its book value at the end of year 6. Under those conditions, then, there would be no income-tax effect on the before-tax flow of $15,000. (The reader should refer to Table 8–2 to review the process, if necessary.)

TABLE 10–2   After-tax Cash Flow, Considering Inflation   (*Current-dollar Method*)

End of Year	Before-tax Cash Flow (Constant Dollars, Given)	Before-tax Cash Flow (Current Dollars, Computed)	Depreciation Charges	Taxable Income	Income Taxes	After-tax Cash Flow
(1)	(2)	(3)	(4)	(5)	(6)	(7)
				$=(3)-(4)$	$=(5)\times50\%$	$=(3)-(6)$
0	−100,000	−100,000				−100,000
1	30,000	32,400	14,167	18,233	9,117	23,284
2	30,000	34,992	14,167	20,825	10,413	24,580
3	30,000	37,791	14,167	23,624	11,812	25,979
4	30,000	40,815	14,167	26,648	13,324	27,491
5	30,000	44,080	14,167	29,913	14,957	29,124
6	30,000	47,606	14,167	33,439	16,720	30,887
6	15,000	23,803		8,803	4,402	19,402

In this example, given an 8-percent annual inflation rate, the current-dollar market value of the equipment would amount to $23,803 at the end of year 6. Since the book value of $15,000 at time 6 is unaffected by inflation, there would be an income-tax effect to consider if one treats the asset as though it were being disposed of for $23,803 at that time. The difference between the expected market value, $23,803, and the book value, $15,000, would be treated as taxable income, being recovery of previously charged depreciation. Thus, at a tax rate of 50 percent, income taxes would be increased by $4,402 as a result of sale of the asset for $23,803, leaving a net after-tax benefit of $19,402.

The resulting after-tax cash flows are depicted in Cash Flow Diagram D10–4.

D10–4

ATCF ($000)

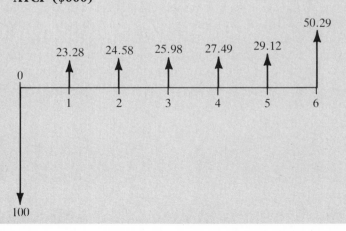

The annual worth of these flows must be obtained by using an inflation-adjusted MARR, since the before-tax flows were expressed in current dollars. That MARR would be 18.8 percent, based on the unadjusted MARR of 10 percent used previously in Chapter 8 and the 8-percent inflation rate:

$$(1.10)(1.08) - 1 = 0.188$$

Then the annual worth would be given by

$$AW = -100 + 23.284\,(P/F\ 18.8\%,\ 1) + 24.58\,(P/F\ 18.8\%,\ 2) + \cdots$$

$$+ 50.288\,(P/F\ 18.8\%,\ 6)]\,(A/P\ 18.8\%,\ 6) = -\$1.02\ \text{(thousand)}$$

The reader should note that this would not be considered a desirable proposal. Interestingly, the after-tax analysis done in Chapter 8, varying only in that inflation was not considered, differed in the substantive result. There the annual worth was positive, indicating a desirable alternative. Once again, an example has been provided to point out that an analysis, to have maximal validity, should take into account as many of the factors that can be expected to bear on the outcome as can practically be evaluated.

Of course, the reader should also not forget that the undesirability of the alternative in question here depends to some degree on the assumptions made concerning the inflation rate to be expected. As with other situations, a sensitivity analysis may be undertaken to see the extent to which the recommendation would be modified under different inflation assumptions.

Differential Effects of Future Price Changes on Individual Factors

Heretofore, Example 10–2 was treated so that all fixed factors, such as depreciation, would be constant in the analysis, while all other factors would be subject to the same future degree of price changes. The latter condition is very frequently unrealistic, as was pointed out earlier in this chapter. In fact, it is often one of the salient rationalizations for undertaking a "price-adjusted" analysis to begin with.

Now assume that in Example 10–2, a number of the factors are expected to change in price at markedly different rates. What effect will this have on the measurable economic worth of the proposal? This is an important question. In general, if the future prices of certain factors of a given proposal are expected to behave differently, even extraordinarily, as compared to other factors, then an attempt should be made to take such changes into account. The modification in treatment of Example 10–2 based on this added consideration can be seen in Example 10–3.

Example 10–3

Suppose that the basic situation of Example 10–2 is the same as before except that the annual cost savings to be derived from the investment is being based on energy savings. Further suppose that the cost of energy is anticipated to rise disproportionately with respect to other factors during the course of the study period. Therefore an attempt will be made to account for this factor separately. Suppose that energy prices are expected to rise by 10 percent per year over the study period. Then the computation of the current-dollar savings attributable to the investment is shown in Table 10–3.

TABLE 10–3   Current-dollar Energy Savings

End of Year	Constant-dollar Energy Savings	Current-dollar Energy Savings
1	40	$40(1.1) = 44$
2	40	$40(1.1)^2 = 48.4$
3	40	$40(1.1)^3 = 53.24$
4	40	$40(1.1)^4 = 58.564$
5	40	$40(1.1)^5 = 64.420$
6	40	$40(1.1)^6 = 70.862$

Further suppose that the previously stated annual costs associated with this investment, $10,000, are primarily attributable to labor expenses, which are expected to rise at an annual rate of 5 percent over the course of the study period. The current-dollar amounts for this factor can then be computed by the method used for energy savings. They are shown in Table 10–4.

TABLE 10–4   Current-dollar Labor Costs

End of Year	Constant-dollar Labor Costs	Current-dollar Labor Costs
1	10	$10(1.05) = 10.5$
2	10	$10(1.05)^2 = 11.025$
3	10	$10(1.05)^3 = 11.576$
4	10	$10(1.05)^4 = 12.155$
5	10	$10(1.05)^5 = 12.763$
6	10	$10(1.05)^6 = 13.401$

The differences between the current-dollar savings and costs yield the current-dollar net savings which will be used as the basis for the rest of the analysis (see Table 10–5).

TABLE 10–5   Current-dollar Net Savings

End of Year	Current-dollar Energy Savings	Current-dollar Labor Costs	Current-dollar Net Savings
1	44.	10.5	33.5
2	48.4	11.025	37.375
3	53.24	11.576	41.664
4	58.564	12.155	46.409
5	64.420	12.763	51.657
6	70.862	13.401	57.461

If these revised estimates are followed through so as to generate after-tax cash flows under the additional assumption that the salvage value of the equipment is expected to rise by only 4 percent per year, to $18,980, then the results would appear as in Table 10–6.

TABLE 10–6 After-tax Cash Flow, Considering Inflation, Component Factors Treated Separately (Current-dollar Method)

End of Year	Before-tax Cash Flow (Constant Dollars, Given)	Before-tax Cash Flow (Current Dollars, Computed)	Depreciation Charges	Taxable Income	Income Taxes	After-tax Cash Flow
(1)	(2)	(3)	(4)	(5)	(6)	(7)
				$=(3)-(4)$	$=(5) \times 50\%$	$=(3)-(6)$
0	−100,000	−100,000				−100,000
1	30,000	33,500	14,167	19,333	9,667	23,834
2	30,000	37,375	14,167	23,208	11,604	25,771
3	30,000	41,664	14,167	27,497	13,749	27,916
4	30,000	46,409	14,167	32,242	16,121	30,288
5	30,000	51,657	14,167	37,490	18,745	32,912
6	30,000	57,461	14,167	43,294	21,647	35,814
6	15,000	18,890		3,980	1,990	16,990

When the annual worth of these after-tax flows is computed, once again based on the inflation-adjusted MARR of 18.8 percent, one obtains

$$AW = [-100 + 23.885\,(P/F\ 18.8\%,\ 1) + \cdots$$

$$+\ (35.814 + 16.99)\,(P/F\ 18.8\%,\ 6)]\,(A/P\ 18.8\%,\ 6)$$

$$=\ +\$0.867\,(\text{thousand})$$

Thus, the alternative is now economically justified, whereas previously, when no distinction was made in the reaction of the individual factors to future price changes, it was not.

Once again, it can be seen that a valid analysis should include all the relevant variables that can practically be incorporated. In this example, the importance of taking into account the separate price effects of the savings and cost factors is due to the relatively large price increases expected for the commodity that the investment is designed to reduce—namely, energy. Thus, when this aspect of the problem is properly taken into account, the asset is seen to be desirable.

Exercises

Many of these exercises have appeared previously. Now the task is to adjust the cash flows to take price changes into account before computing measures of effectiveness. For these problems, assume an income-tax rate of 50 percent. You may be interested in comparing your results with those obtained previously.

10–1 Use your library to examine consumer and producer price index values for the last few years. For producer prices, choose several categories that are of interest to you. Try to find other than government-published indexes also.

10–2 The owner of an existing office building is investigating the prospects of conserving electricity. The building is heated electrically with thermostats in each lease area. A computer-controlled system is being examined, which will permit the building manager to set and control the heat for a given temperature in each area, as opposed to giving the tenants the ability to set the heat levels themselves. It is felt that this will reduce waste associated with uncontrolled access to the thermostats and would save $12,000 in the first year (at today's prices). Moreover, the cost of electric power is expected to increase in the future; thus, the annual savings are expected to increase at an average rate of 8 percent per year. The system will cost $75,000 to install and will last for 10 years, at which time it will have a salvage value of $10,000 at today's prices. The inflation rate applicable to used machinery is believed to be 5 percent. Assume that the system is depreciated under ACRS with a 5-year recovery period. In addition, a 3-percent investment tax credit is applicable. If the after-tax MARR is 10 percent and the overall inflation rate for the period is assumed to be 6 percent, what after-tax net present worth would be provided?

10–3 Engineers at a plant that manufactures wood furniture are considering the installation of a system to reclaim and utilize the sawdust which is a by-product of the process. It includes a preheater for the hot water system that will burn the dust safely, ductwork throughout the plant, and a set of blower motors to propel the dust and deliver it to the preheater. The system as currently designed requires investment of $225,000. It is expected to save $40,000 per year in conventional fuel at today's prices, as well as reducing dust cleaning and disposal. As a result of inflation in the price of fuel, savings are expected to increase at a rate of 8 percent per year. The system is expected to serve for 15 years, after which it is considered to be worthless. Assume that the equipment will be depreciated on a straight-line basis over the 15 years. What after-tax present worth would these savings provide? The unadjusted after-tax MARR is 10 percent, and the overall rate of inflation expected is 5 percent.

10–4 Your firm developed a new product, investing $2 million in equipment. The required assets are depreciated using the declining-balance method with $k = 0.178$. Product sales, less expenses, for 5 years are: $300,000, 400,000, 500,000, 300,000, 250,000 at today's prices. Net sales are expected to be affected by an 8-percent inflation rate. At the end of the fifth year, the productive assets associated with the product will be disposed of for $750,000 at today's prices. Such used equipment is believed to be subject to a 4-percent inflation rate. Compute the after-tax present worth of the firm's investment. The unadjusted after-tax MARR is 12 percent, and the overall inflation rate is expected to be 5 percent.

10-5 Five design configurations for a metal processing by-product recovery system are being evaluated for their economic characteristics. The alternative designs are listed in order of increasing investment requirements.

    a. With one design to be chosen, at least $300,000 available, and an unadjusted after-tax MARR of 6 percent, which is most attractive? Assume straight-line depreciation over 5 years with no salvage value and a 4-percent annual inflation rate on all future flows.

	Yr.	A	B	C	D	E
Investment ($000)	0	−75	−100	−180	−250	−280
Annual savings generated (revenue) ($000)	1–5	+25	+27	+43	+75	+82

    b. If the capital available were only $200,000, which would now be most attractive?

10-6 Four proposals for a machining process are being considered, each of which will satisfy the demands for quantity and quality of product. The difference between the proposals is their cost structure, given below:

Investment, Annual Costs ($000)

	Yr.	A	B	C	D
Investment ($000)	0	−50	−60	−100	−150
Annual costs ($000)	1–4	−130	−125	−113	−90

    a. If the unadjusted after-tax MARR is 8 percent, which alternative is cost-minimizing? Assume sum-of-the-years'-digits depreciation over 4 years with all future flows subject to a 5-percent annual inflation rate.

    b. If only $75,000 is available, what is the recommended course of action?

10-7 Four mutually exclusive, cost-saving (or revenue-enhancing) projects are being studied. Data are as follows:

	Yr.	A	B	C	D
Investment ($000)	0	−150	−250	−300	−450
Annual cost savings ($000)	1–4	+32	+70	+90	+125
Salvage value ($000)	4	+40	+60	+70	+95

    a. If the unadjusted after-tax MARR is 6 percent, which is superior, assuming $450,000 is available? Assume ACRS depreciation with 3-year recovery periods and a 4-percent annual inflation rate.

    b. What if only $300,000 is available?

10–8 One of the four cost-minimizing proposals shown below is to be selected:

	Yr.	A	B	C	D
Investment ($000)	0	−125	−160	−190	−230
Annual costs ($000)	1–4	−140	−132	−110	−100
Salvage value ($000)	4	+35	+50	+60	+75

Which is best, if the unadjusted after-tax MARR is 8 percent? Assume ACRS depreciation with 3-year recovery periods, and a 4-percent tax credit. Also assume that all future flows are affected by a 6-percent inflation rate.

10–9 Four cost-minimizing configurations need to be analyzed. The data follow:

	Yr.	A	B	C	D
Investment ($000)	0	−125	−145	−170	−200
Annual costs ($000)	1	−80	−80	−80	−65
	2	−80	−75	−70	−65
	3	−80	−70	−60	−65
	4	−80	−65	−50	−65
Salvage value ($000)	4	+20	+30	+40	+50

Which is best if the unadjusted after-tax MARR is 5 percent? Assume straight-line depreciation over 4 years with salvage values as indicated. All future flows are subject to a 3-percent annual inflation rate.

10–10 The data for four cost-minimizing alternatives are as follows:

	Yr.	A	B	C	D
Investment ($000)	0	−95	−130	−150	−175
Annual costs ($000)	1	−60	−50	−40	−38
	2	−60	−50	−40	−38
	3	−60	−50	−40	−38
Salvage value ($000)	3	+50			
Annual costs	4		−50	−40	−38
Salvage value	4		+50	+60	
Annual costs	5				−38
Salvage value	5				+60

Analyze the data, assuming the unadjusted after-tax MARR is 8 percent. Assume ACRS depreciation with 3-year recovery periods, and a 3-percent investment tax credit. All future flows are subject to a 6-percent inflation rate.

10–11 Two processes are being considered to fulfill the same function. The first requires an immediate investment of $100,000 and will provide revenues, at today's prices, of $300,000 and $360,000 in each of the first two years, $510,000 in the third year, $505,000 in each of the next two years, $400,000 in the sixth year, and $300,000 in the seventh. Costs in today's prices will be $310,000 in the first year, $350,000 in the second, $490,000 in the third, $475,000 in each of the next two years, $360,000 in the sixth, and $250,000 in the seventh. In addition, the salvage value of the process at the end of the seventh year is expected to be $30,000 at today's prices.

The other alternative requires a $125,000 initial investment. Revenues and costs at today's prices would be $400,000 and $350,000 per year respectively, for 5 years. The salvage value at the end of the fifth year would be $40,000 at today's prices.

If funds are worth 6 percent after taxes but unadjusted for inflation, which investment is superior? Assume ACRS depreciation with 5-year recovery periods. Assume that the price basis for revenue can increase by only 4 percent per year because of stiff international competition. The price basis for costs and salvage values will be subject to a 6-percent inflation rate. The overall inflation rate is thought to be 6 percent. Use annual worth as a yardstick.

10–12 Two minicomputers are being compared for use by an engineering consulting firm. Although other characteristics of the machines will be considered, an economic analysis yielded the following data: System *A* would cost $10,000 to acquire and would require $3,000 per year in today's prices to operate, including power, materials, and a maintenance contract. It is expected to be used for 6 years, after which it would have a salvage value of $2,000 at today's prices.

System *B* would require a $12,000 investment but only $2,000 per year to operate and maintain at today's prices. Its economic life is also felt to be realistically estimated at 6 years, after which its salvage value would be $3,000 at today's prices. If funds are valued at 8 percent after taxes with no adjustment for inflation, and these are viewed as mutually exclusive alternatives, which is superior? Assume ACRS depreciation with 5-year recovery periods, and a 4-percent investment tax credit. Assume a 6-percent inflation rate for operating and maintenance costs, with a zero rate for salvage values.

10–13 A machine may be purchased for $20,000. Operating costs are expected to be $3,500 per year at today's prices with maintenance being $500 per year. Its anticipated economic life is 7 years, after which it is projected that its market value would be $5,000, also at today's prices. The manufacturer offers a leasing alternative to the purchase arrangement, specifying that the machine may be leased for 7 years at an annual cost of $3,900, the lease payments to be prepaid for each year. Operating expenses are expected to be the same for the leased machine as for the purchased one; however, maintenance under the lease will be the responsibility of the manufacturer. If capital is worth 8 percent after income taxes before adjusting for inflation, which alternative is superior?

Assume a 5-year recovery period under ACRS for the purchased machine, with an investment tax credit of 3 percent. Also assume that one-fourth of the funds needed for purchase of the machine would be borrowed at an interest cost of 12 percent per year. In addition, operating and maintenance costs and market values are expected to experience a 6-percent inflation rate. (This is also the overall inflation rate.) Lease and interest costs are not subject to future price changes.

10–14 Energy costs for a certain process are now $12,000 per year at current prices. They are expected to increase at an average of 5 percent per year over the next decade, because of increasing activity. A microprocessor-based control system designed to reduce the energy required by 10 percent can be obtained and installed for $8,000, its expected economic life being 10 years. If the market value of the system after 10 years is $2,000 at today's prices and funds are valued at 8 percent after taxes without an inflation adjustment, can the system be justified? Assume that the system is depreciated under ACRS with a 5-year recovery period. Also assume that future savings are subject to a 6-percent inflation rate and future market value to a 2-percent inflation rate; and that the overall inflation rate will be 4 percent per year.

10–15 Your company is considering one of two forklift trucks. A costs $11,000, requires $2,000 per year in operating expense at today's prices, and is expected to have a 3-year economic life. Its salvage value at today's prices is projected to be $4,000. B costs $15,000 but only $1,500 per year to operate at today's prices. It is expected that its economic life will be 5 years, with a salvage value at today's prices of $4,000. If funds are worth 6 percent after taxes prior to an inflation adjustment, which is superior? Assume ACRS depreciation with 3-year recovery periods. Also assume a 4-percent inflation rate for all future flows.

10–16 Three cost-saving machines are being considered for an existing function. Data are as follows:

	A	B	C
Investment ($)	60,000	80,000	90,000
Annual costs ($)	142,000	136,000	128,000
Economic life (yrs.)	5	5	5

If funds are worth 10 percent after income taxes prior to an inflation adjustment, which alternative is superior? Assume that these machines are depreciated under ACRS with 5-year recovery periods, there is a 4-percent investment tax credit, and one-third of the capital is borrowed at an interest rate of 12 percent per year. Assume that all future flows are subject to a 6-percent inflation rate.

10–17 A machine is purchased for $22,000, is depreciated under SYD for 3 years, with an expected salvage value of $4,000 at today's prices. It generates net income per year of $11,000 for 4 years at today's prices. After the fourth year it is disposed of for $5,000, also at today's prices. What is the annual worth of these flows if the unadjusted after-tax MARR is 10 percent? All future flows are subject to a 6-percent inflation rate.

10–18 Your firm is convinced of the desirability of a new piece of equipment which would help cut down on annual costs. It requires an immediate investment of $150,000 and would provide annual net savings of $35,000 in the first year, rising by $5,000 per year, at today's prices. It is expected that the equipment would be usable for 6 years (even though its recovery period under ACRS would be 5 years). It is anticipated that after the sixth year the equipment would be worth $15,000 at today's prices.

The firm's decision involves comparing the purchase of this equipment, with the terms as indicated, with the option of renting it for 6 years at $30,000 per year. Determine which option has the higher after-tax annual worth if the unadjusted MARR is 10 percent. The lease payments would be made at the beginning of each annual period and are not subject to inflationary changes. All other real flows would be subject to a 6-percent inflation rate.

10–19 Three alternatives are to be evaluated on an after-tax basis. Data are shown below:

	A	B	C
Investment ($)	75,000	105,000	120,000
Annual operating costs ($)	45,000	40,000	30,000
Economic life (yrs.)	5	7	6
Salvage value ($)	10,000	13,000	18,000

All three would be depreciated under ACRS with 5-year recovery periods.

Which alternative has the highest measurable economic worth? The unadjusted after-tax MARR is 12 percent. All future flows are subject to a 6-percent inflation rate.

10–20 Data for a deferred-investment alternative vs. a full-investment one are depicted in Cash Flow Diagrams D10–5 and D10–6. All the data are expressed in terms of today's prices.

D10–5

Full ($000):

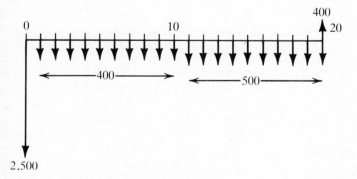

D10–5 may be interpreted to mean that investment is $2.5 million, operating and maintenance expenses are $400,000 per year for the first 10 years, $500,000 per year for the last 10 years, with a salvage value of $400,000 at year 20 (all in terms of today's prices).

D10–6

Deferred ($000):

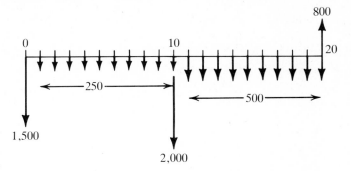

In D10–6 there is a subsequent investment of $2 million in year 10. The other parameters of initial investment, operating expense, and salvage value also differ from the previous configuration. Here, too, the data are in today's prices. If the unadjusted MARR is 12 percent, which has a more desirable present worth? Assume that the overall inflation rate is 6 percent, the rate for construction costs (deferred investment) is 10 percent, and the rate for operating costs and salvage values is 5 percent. Do not consider any income-tax effects.

10–21 A bridge project is being considered under the following two configurations: Under the first, a four-lane bridge can be constructed for $8 million. It is expected to serve the anticipated needs for the next 20 years. Maintenance would be $80,000 per year for the first 5 years, going up by $25,000 for the next 5 years and similarly for the remaining 10 years. These expenses are in today's prices. The salvage value at year 20 is $2 million, also in today's prices.

The other alternative is to construct a two-lane bridge now at a cost of $5 million. Maintenance would be $55,000 per year for the first 5 years, going up to $70,000 for the next 5, stated at today's prices. At the tenth year two more lanes would need to be added to accommodate the anticipated growth in traffic. This would cost an additional $6 million at today's prices. Maintenance of the combined four-lane facility would then be $130,000 in years 11–15, and $155,000 in years 16–20. Salvage value for the second alternative would be $2.5 million at year 20. Maintenance and salvage values are also stated in terms of today's prices.

If funds are worth 10 percent before adjustment for inflation, which alternative is superior? Assume that maintenance and salvage values are subject to a 5-percent inflation rate, while construction costs (deferred investment) are subject to a 10-percent inflation rate. The overall inflation adjustment is 8 percent. Treat this as a public sector project for which income taxes are not relevant.

10–22 An engineer at a voluntary hospital is analyzing the economic characteristics of a new centrifuge. Its acquisition cost, projected market values, and operating costs, expressed in today's prices, are shown in the following table. Indicate how the minimum EUAC may be computed by writing an equation expressing EUAC as a function of how long the machine is retained. Then compute the EUAC assuming the machine is held for 5 years. Since the hospital is exempt from income taxes, do a pretax analysis. Use an unadjusted MARR of 12 percent and assume that all future flows are subject to an inflation rate of 6 percent.

Projected Market Value and Disbursements

End of Year	Projected Market Value	Projected Disbursements
0	$16,000	
1	13,000	$1,200
2	11,000	1,400
3	10,000	1,600
4	8,000	2,000
5	7,000	2,400
6	6,000	2,600
7	5,000	2,600
8	4,000	2,800
9	3,000	2,800
10	2,000	2,800

10–23 Your firm purchased a milling machine exactly 3 years ago at a cost of $110,000. It has been used to produce parts whose revenue equivalent attributable to it is $85,000 per year at today's prices. Operating and maintenance expenses have been $45,000 per year at today's prices. The machine has been depreciated under ACRS with a 5-year recovery period. Currently a new machine with superior productivity has been offered to the firm. It will cost $225,000 to acquire, although a $35,000 trade-in allowance would be given for the defender. Assume that it would be depreciated over 5 years using ACRS and that its useful life is also 5 years. Include the factor that an investment tax credit of 4 percent would be available for the replacement (although it was not in effect for the defender). The challenger is expected to produce twice the equivalent revenue of the defender, $170,000 at today's prices, while expenses would increase to only $65,000 at today's prices. In considering the economic worth of the challenger, assume that its market value after 5 years would be $40,000. The anticipated market values for the defender, should it be retained now, would be $25,000 in one year and $18,000 in two years. (These market values are also expressed in terms of today's prices.)

Do an analysis on an inflation-adjusted after-tax basis to determine whether the defender or challenger has greater economic worth. Assume an unadjusted after-tax MARR of 10 percent and a tax rate of 50 percent. Assume also that all future flows are subject to an inflation rate of 8 percent.

10–24 A piece of road-grading equipment was purchased 4 years ago for $42,000. It has been depreciated using the straight-line procedure over 6 years, with a salvage value, for depreciation purposes, of $6,000. It is expected that it can be utilized for 3 more years, with operating and maintenance expenses and projected market values expressed in today's prices as shown in the table following.

A new model is now available for $50,000. It can be used for 8 years, as the table indicates. It would be depreciated under ACRS with a 5-year recovery period. Compare the defender with the challenger on an after-tax basis, assuming an income-tax rate of 50 percent. Assume an unadjusted after-tax MARR of 10 percent. Treat projected disbursements as though a 6-percent inflation rate is applicable, and projected market value as though a 3-percent rate applies. Assume an overall inflation rate of 5 percent.

Projected Market Value and Disbursements (*in today's dollars*)

End of Year	Defender		Challenger	
	Projected Market Value	Projected Disbursements	Projected Market Value	Projected Disbursements
0	16,000		50,000	
1	12,000	12,000	45,000	8,000
2	8,000	13,000	35,000	8,500
3	4,000	15,000	31,000	9,000
4			27,000	9,500
5			23,000	10,000
6			20,000	11,000
7			17,000	12,000
8			14,000	13,000

10–25 A grinding machine was purchased 5 years ago at a cost of $28,000. It was depreciated under ACRS with a 5-year recovery period. Although it is fully depreciated its service is still satisfactory, albeit at a higher level of maintenance expense than originally. A new model is available, with maintenance and operating costs and market values as indicated in the table. Should the grinder be replaced now or used for one more year? The unadjusted after-tax MARR is 12 percent. If purchased, the new machine would be depreciated under ACRS with a 5-year recovery period and there would be an investment tax credit of 3 percent. Assume that projected disbursements are subject to a 6-percent inflation rate and projected market values to a 4-percent rate, with an overall inflation rate of 5 percent.

Projected Market Value and Disbursements (*in today's dollars*)

End of Year	Defender		Challenger	
	Projected Market Value	Projected Disbursements	Projected Market Value	Projected Disbursements
0	5,000		40,000	
1	4,000	8,000	30,000	3,000
2	3,000	9,000	25,000	3,250
3			21,000	3,500
4			17,000	4,000
5			13,000	4,500

Selected References

Allen, R. G. D., *Index Numbers in Theory and Practice,* Chicago: Aldine, 1975.

Davidson, S., C. Stickney and R. Weil, *Inflation Accounting,* New York: McGraw-Hill, 1976.

Freidenfelds, J. and M. Kennedy, "Price Inflation and Long Term Present Worth Studies," *The Engineering Economist,* Vol. 24, No. 3, Spring, 1979, pp. 143–160.

Jones, B., *Inflation in Engineering Economic Analysis,* New York: John Wiley & Sons, 1982.

Oakford, R. and A. Salazar, "The Arithmetic of Inflation Corrections in Evaluating 'Real' Present Worths," *The Engineering Economist,* Vol. 27, No. 2, Winter, 1982, pp. 127–146.

Remer, D. and S. Ganiy, "The Role of Interest and Inflation Rates in Present Worth Analysis in the United States," *The Engineering Economist,* Vol. 28, No. 3, Spring, 1983, pp. 173–190.

11

Measuring Costs in Industry

Introduction

Classical break-even analysis involves the study of the operations of an establishment (or one of its subunits). The basic ideas explored in Chapter 6 required definitions regarding only a relatively small set of cost categories. In order to get insights into related classes of decision-making problems, it is necessary to extend those basic ideas to a wider understanding of the meaning and measurement of costs in their many forms and applications.

This chapter, therefore, is devoted to an explanation of the measurement of costs in industrial practice (job and process costing), the rationale behind standard cost systems, incremental costs, learning curves, and life cycle costing. (The concepts of opportunity costs and sunk costs were presented elsewhere.)

Job Order Costing

This section deals with the practical need to identify costs in the typical industrial milieu. The context chosen for the following explanation is that of a manufacturing plant. However, the principles outlined are equally applicable to other kinds of establishments. (See the references for this chapter.)

First it will be necessary to describe the sort of manufacturing situation that lends itself to job order costing. The type of plant for which such a system is applicable is the so-called job shop, an establishment in which many different products are made and where at any given time a number of different products or jobs may be in process. Such a plant is designed to be flexible in its capabilities. An example could be an electrical parts plant capable of producing any of the thousands of different elements of the firm's product line. Clearly not all of the products would be processed simultaneously. Products or jobs would be scheduled, taking into account the firm's output requirements as well as the desire to operate the plant as efficiently as possible.

The aim of the costing procedure is to allocate the total costs incurred in producing the various parts, products, or jobs in a way that reflects the real burden imposed by each. The chief difficulty in carrying this out stems from the fact that, although some of the costs are directly traceable to the product or job, others are not. The costs that are not directly traceable must be allocated to the products or jobs in some arbitrary (although reasonable) fashion. The following sections describe how this may be done.

Some definitions are required.

Direct Costs

Direct costs are those that are directly traceable. Usually, the only direct costs of the manufactured product, part, or job are those for labor and materials. These are accounted for as the product makes its way through the plant. All but the very smallest of plants are divided into departments based on functional responsibilities, and accounting systems are generally designed to permit costs to be traced at least to the department level. For instance, a plant may have a receiving department, a machine shop, a plating or paint shop, an assembly department, and a shipping and storage department.

Where it is deemed necessary to keep track of costs in a more detailed fashion, departments may be divided into cost centers. For instance, the machining department may be divided into three cost centers, one for manually operated machines, one for numerically controlled machines, and one for computer numerically controlled machines. (Note that although some costs may be direct from the point of view of operating a department, they may not be direct from the point of view of a particular job or product. The cost of lighting the receiving department may be traceable, for instance, given appropriate metering, and thus is direct from the point of view of operating the department. However, such a cost is not traceable to the production of individual products, parts, or jobs.)

Example 11—1a

The basic principles can be illustrated by a plant producing eyeglasses. The product in question is a gold-plated, metal frame, Model #12345. Flat wire is used for the lens frames and nose bridges; round wire is used for the earpieces. These components are gold-plated. In addition, other hardware is needed, including the nose rests, the connectors between the glasses and earpieces (also gold plated), and the plastic ends for the earpieces. Suppose that 4,000 units of Model #12345 are needed. Since historical records indicate that some scrap in the production process is likely to occur, 4,100 units are put into production.

By the time the job has made its way through the plant, the job cost sheet might show direct costs as in Table 11–1.

TABLE 11–1 Direct Costs

Identification: Eyeglass Frame
Job: Part #12345
Quantity ordered: 4,100
Quantity completed: 4,019

Date ordered: June 13, 1985
Date completed: June 30, 1985

DIRECT MATERIALS

Date	Item	Amount	Price	Cost	Total DM
6/13	Flat wire	5,125 ft	$0.36/ft	$ 1,845	
6/13	Round wire	4,780 ft	$0.24/ft	$ 1,147	
6/15	Gold	117 t oz	$350/t oz	$40,950	
6/20	Earpiece ends	8,200	$0.03 ea	$ 246	
6/20	Connectors	8,200	$0.09 ea	$ 738	
6/25	Nose rests	8,200	$0.10 ea	$ 820	$45,746

DIRECT LABOR

Date	Department or Cost Center	Hours	Rate	Cost	Total DL
6/13–25	Fabricating	160	$12	$1,920	
6/20–25	Plating	14	$10	$ 140	
6/25–30	Assembly	80	$ 6	$ 480	$ 2,540

It then remains to determine the indirect costs.

Indirect Costs

The indirect costs associated with a job or product are those that are not directly traceable. They are allocated so as to apportion a reasonable share of the total untraceable costs to the job or product in question. This is often known as overhead. Included in overhead are supervision, employee fringe benefits, depreciation of plant and equipment, indirect materials and supplies, utilities, property taxes, and insurance.

Indirect costs are allocated by the use of overhead rates. The rate is multiplied by an appropriate measure of activity to arrive at the amount of overhead assigned to the product or job. This can be done in a variety of ways. For instance, the plant may use a single overhead rate, or it may develop overhead rates for various departments or cost centers. Also, since the job shop is, by definition, an entity that processes many different jobs, activity cannot be gauged by the units of output, because the various products absorb overhead differently. The alternative is to use a measure of input as a barometer of activity. In doing so, it is important to choose one that realistically indicates the absorption of overhead. A number of input measures are used, the most common of them being direct labor hours, direct labor cost (dollars), and machine or process time.

Example 11–1b

Returning to Example 11–1a, hypothetical overhead costs have been entered in the cost sheet of Table 11–2.

TABLE 11–2 Direct Costs and Overhead

IDENTIFICATION: Eyeglass Frame
Job: Part #12345
Quantity ordered: 4,100
Quantity completed: 4,019

Date ordered: June 13, 1985
Date completed: June 30, 1985

DIRECT MATERIALS

Date	Item	Amount	Price	Cost	Total DM
6/13	Flat wire	5,125 ft	$0.36/ft	$ 1,845	
6/13	Round wire	4,780 ft	$0.24/ft	$ 1,147	
6/15	Gold	117 t oz	$350/t oz	$40,950	
6/20	Earpiece ends	8,200	$0.03 ea	$ 246	
6/20	Connectors	8,200	$0.09 ea	$ 738	
6/25	Nose rests	8,200	$0.10 ea	$ 820	$45,746

DIRECT LABOR

Date	Dept. or Cost Center	Hours	Rate	Cost	Total DL
6/13–25	Fabricating	160	$12	$ 1,920	
6/20–25	Plating	14	$10	$ 140	
6/25–30	Assembly	80	$ 6	$ 480	$ 2,540

OVERHEAD COST

Date	Dept. or Cost Center	Hours	Overhead Rate Machine Hours	Overhead Rate Direct Labor Hours	Cost	Total OHD
6/13–25	Fabricating	160	$17		$ 2,720	
6/20–25	Plating	14		$19	$ 266	
6/25–30	Assembly	80		$11	$ 880	$ 3,866

Table 11–2 indicates how overhead was computed for each cost center. The rates were based on machine hours in one department and direct labor hours in others. The sum for the departments gives the total overhead allocated to the product in question. Total product or job cost is then the sum of all direct and overhead costs.

This method of costing has been, and will continue to be, the prevalent one in industry because the job shop, in its varied forms, has been the dominant form of establishment. It will remain so as it evolves into the "factory of the future."

Process Costing

Process costing is applicable where the volume of output is relatively large and consequently requires the exclusive use of a given facility over the long term. Examples of such situations include chemical processing, high-volume assembly operations, and certain types of food processing. Process costing is simpler than job order costing because each department's costs can readily be associated with production. In other words, all the department's costs are direct and coincide with product cost. In process costing, then, there is no need to develop overhead rates because there are no indirect costs to contend with.

The basic procedure in process costing calls for the identification of production centers; the recording of each production center's costs for the period in question; the measurement of each production center's output for that period; the computation of the ratio of cost and output to obtain average unit cost for each production center; and the summing of the unit costs of the production centers that are required for a particular product to obtain that product's total unit cost. (See the references for this chapter.)

Standard Costs

The preceding sections dealt with the question of how to measure historical costs. This is an industrial task of first-level importance, since rational decisions cannot be made without such information. However, the function of control requires that ongoing performance be compared to some set of realistic expectations. Although it is sometimes considered sufficient to use historical performance as a yardstick with which to compare ongoing activities, such a viewpoint can hardly be regarded as ideal, since historical effectiveness can vary anywhere from excellent to unacceptable. A more sophisticated perspective is one in which performance is compared to a standard that is based on careful study of the capabilities of the productive system under normal operating conditions. This is what a standard cost system is intended to provide. Thus, a standard cost system may be viewed as an enhancement of a firm's cost reporting system.

A standard cost may be defined as the expected cost of a product or service, based on the anticipated quantities and prices of the needed inputs. A standard cost system therefore requires data of four types for each product, part, or service:

Standard materials quantities,
Standard materials prices,
Standard labor quantities; and
Standard labor rates.

Standard Materials Quantities

Standard materials quantities indicate how much of each constituent is expected to be required, usually expressed as units per production lot size. They are based primarily on the bill of materials, which is a list of the nominal components of the product and their quantities. (A pump assembly, for example, may require, among other things, seven screws of a particular type, one impeller, and one valve. Thus, a lot of 1,000 pumps would nominally require 7,000 screws, 1,000 impellers, and 1,000 valve components.) Clearly, the bill of materials is a final statement of the design of the product or component.

However, standard materials quantities are generally greater than the product of the production lot size and the unit bill of material requirements. Some shrinkage is inevitable in the production process because of a variety of errors leading to spoiled work in process. Thus, the standard materials quantities supplement the bill of materials quantities by an amount corresponding to the normal waste anticipated in the production process. Using the pump example, suppose that careful analysis indicated that it would be reasonable to expect that 14 impellers per 1,000 pumps assembled would be found defective or would be broken during assembly. The standard materials quantity of impellers per lot of one thousand pumps would then be 1,014.

Standard Materials Prices

Standard materials prices represent realistic estimates of the actual net prices of the materials used. Since prices tend to vary more than quantities in the standard cost system, a good deal of vigilance is required to maintain a valid set of prices when such a cost system is implemented.

Standard Labor Quantities

Standard labor quantities indicate the number of standard hours required for each operation specified in the production process. They are derived from production specifications and include allowances for such things as machine downtime and worker fatigue and personal time. Of the four types of data required for a standard cost system, the standard labor quantities are generally the most difficult and costly to estimate and maintain. Most firms that employ such systems assign the responsibility for accumulating and revising data on standard labor quantities to a specialized staff department. (A common name for such a group is the Methods and Standards Department.)

Under that framework the data may be obtained in a variety of ways. One way is to undertake time studies of the various functions. Another is to construct estimates of the aggregate time required for an operation from references that provide recognized standards for each of the individual work elements. Finally, historical experience may be used to assign standard times if the experience is considered to represent a realistic future standard. In practice, any combination of these may be used.

As an example, suppose that impeller fabrication required two steps, casting and trim. If they required direct labor time of 0.25 minute and 0.15 minute, respectively, including allowances, then total time required would be 0.4 minute per unit. Based on lots of 1,000 units, the standard labor quantity would then be 400 minutes or 6.67 standard hours per lot.

Standard Labor Rates

Standard labor rates are estimates of the wages (and possibly fringe benefits) attributable to the labor operations specified by the standard labor quantities. As with materials prices, wages tend to increase; thus, standard labor rates need to be adjusted periodically.

For the standard cost system to be valid, the standard labor rates should be compiled for labor classes that are as nearly homogeneous as possible. Most firms have a number of labor grades to which standard labor rates need to be assigned. The operation of the casting machine for the impeller, for instance, would be categorized in a higher labor grade than that for custodial duties or simple materials handling. When computing the standard cost of an item, then, as realistic a unit labor cost as possible should be made.

Example 11—1c

Table 11—3 illustrates the computation of standard costs for Example 11—1c.

TABLE 11—3 Standard Costs

IDENTIFICATION: Eyeglass Frame
Part: #12345
Batch Quantity: 4,050*

Standard Cost/Batch: $55,142
Standard Cost/Unit: $13.62

DIRECT MATERIALS

Dept.	Item	Quantity	Price	Cost	Total DM
F	Flat wire	5,200 ft	$0.35/ft	$ 1,820	
F	Round wire	4,800 ft	$0.22/ft	$ 1,056	
P	Gold	120 t oz	$360/t oz	$43,200	
F	Earpiece ends	8,200	$0.03 ea	$ 246	
F	Connectors	8,200	$0.10 ea	$ 820	
F	Nose rests	8,200	$0.11 ea	$ 902	$48,044

DIRECT LABOR

Dept. or Cost Center	Hours	Rate	Cost	Total DL
Fabricating	175	$12.50	$2,188	
Plating	15	$11	$ 165	
Assembly	90	$ 5.50	$ 495	$ 2,848

OVERHEAD COST		Overhead Rate			
Dept. or Cost Center	Hours	Machine Hours	Direct Labor Hours	Cost	Total OHD
Fabricating	175	$17		$2,975	
Plating	15		$19	$ 285	
Assembly	90		$11	$ 990	$ 4,250

*Standards indicate 4,100 units must be placed into production in order to obtain 4,050 units of output, because of production spoilage.

Comparing Performance to Standard

In order to compare actual performance to standard, detailed records must be kept at each step of the production process. For example, the material actually used in a process would be computed from the materials requisitions and entered onto an appropriate form. The number of labor hours actually required would also need to be recorded for each process. These records then permit the computation of the applicable quantity variances experienced in practice. (One could hardly expect performance to correspond exactly to standard. The question is not the existence of variances but their magnitudes. Small variances do not indicate that a process is out of control, but large ones may indicate such a condition. The function of the standard cost system is to alert decision-makers when a problem does exist.)

Incremental Costs

Often decisions need to be made in situations in which it would be inappropriate to consider only fixed or variable costs. Such is the case when incremental costs may be relevant. Incremental costs are the cost differences that may be identified with or attributable to some change in operation or design. For example, incremental costs may be associated with a change in the rate of output of a manufacturing plant. Perhaps a needed increase in output can be obtained by operating with one hour of overtime for the existing single shift. Or, suppose the additional needed output can be subcontracted. In that case a "make or buy" choice may exist. It would be important to be able to identify the incremental costs associated with each alternative (i.e., purchase from an outside vendor vs. the use of overtime with internal production).

Although the basic concept of incremental cost is relatively simple, determining it in practice may be another matter. The true cost categories may not be readily apparent, and frequently it is difficult to estimate their magnitudes. Example 11–2 will illustrate the point.

Example 11—2

Suppose a firm is faced with the choice of the "make" option (using overtime), or a "buy" option to be described. How should the costs be evaluated so as to help make a rational selection? An examination of the incremental costs under each option would be an important component of the decision. (Clearly other issues would be relevant, such as the effect of an intermediate-term reliance on overtime on the physical plant and on worker morale, the quality of self-produced goods compared with a purchased product, the ability of an outside vendor to deliver the goods on time and so on.)

Table 11—4 presents the economic data provided by the firm's cost accounting department.

TABLE 11—4   Product Costs

Direct labor	$1.90 per unit
Direct materials	$3.40 per unit
Nondirect costs (overhead)	$2.95 per unit

Note that the overhead has already been calculated in the following way: Assume that it was observed that the "consumption" of overhead was most closely related to labor utilization. Therefore, the decision was made to allocate overhead based on the various products' direct-labor requirements per unit. This was accomplished by computing the ratio of the total observed overhead burden for the economic unit and the sum of the direct-labor dollars for all the products processed in that unit. Assume that this turned out to be $1.55 of overhead per dollar of direct labor. For the product currently under examination, this would in turn be equivalent to

$$(\$1.55/\$ \text{ of direct labor}) \times (\$1.90 \text{ of direct labor/unit of production})$$

$$= \$2.95 \text{ per unit}$$

Since the total standard cost of this item is $8.25, it might seem that this would also be the unit incremental cost of producing the product. However, that might not be the case. The soundness of the $8.25 estimate depends on whether the unit direct-labor costs are different under the assumption that overtime is being employed and the degree to which the overhead figure, $2.95 per unit, is valid under those conditions.

Recall that the overhead figure was based on an allocation of observed nondirect costs at full production. It does not necessarily follow that the "absorption" of overhead would exhibit the same pattern under overtime. In fact, in most industrial situations, one would expect the unit overhead requirements to be less under overtime. This is due to the fact that at least some of the nondirect costs would be fixed and, thus, less per unit as output is increased.

It would also be true that the ratio of overhead cost to direct-labor dollars would be reduced by the use of overtime, since wage rates under overtime are generally higher than for straight time. Taking the latter point into account will influence the estimate for direct-labor dollars per unit. Suppose that it is reasonable to assume that a fifty-percent premium for overtime wages is applicable, thus increasing the dollar direct-labor requirement by fifty percent. What will the true incremental cost be? Table 11–5 provides part of the answer.

TABLE 11–5 Incremental Costs

Direct labor	$1.90(1.5) = $2.85 per unit
Direct materials	$3.40 per unit

Overhead is omitted from Table 11–5 to emphasize the fact that the unit overhead figure given in Table 11–4 is likely to be invalid for this purpose. It would be necessary to re-examine the real additional overhead requirements expected if overtime were used. Such an amount would need to be identified if a proper incremental analysis is to be done. The likelihood is that such a study would suggest a lower overhead figure for the incremental output.

Suppose that such an analysis indicates that the incremental overhead per unit would be $0.67. What would that tell the analyst? Under those assumptions, the total incremental cost would be $6.92 per unit. Thus, if the price needed to be paid to an outside vendor for the product was $7.50, it would be more economic to use overtime.

Note the potential pitfall if the true nature of incremental costs is ignored. If the "full production" overhead figure, $2.95 per unit, were combined with the contents of Table 11–5 for direct labor and materials, the incremental cost would appear to be $9.20, far in excess of the $7.50 quoted by the outside vendor. In these circumstances it is quite possible that an error would result from sending the production out to subcontract when it could be made at lower cost "in house."

Learning Curves

The "learning curve," or "progress function," as it is sometimes called, is an empirically measured phenomenon. It was first reported in 1936 by T. P. Wright, an aerospace industry engineer, who observed that with each doubling of cumulative production, the total man-hours needed per plane was reduced to 80 percent of the former level. The improvement, attributable to individual and group learning applied to assembly of the planes, was described as a learning curve. It has since found application in many manufacturing industries, and is particularly suitable for the assembly aspects of manufacturing.

In most applications, total direct labor hours are plotted on the vertical axis, although sometimes cost is shown. On the horizontal axis, cumulative production is plotted. As an illustration, consider the data in Table 11–6.

TABLE 11−6 Learning Curve

Cumulative production	1	2	4	8	16	32	64	128
Direct labor hours (as % of time required for first item)	1	.8	.64	.512	.410	.328	.262	.210

Notice the plots of these eight points on the graphs in Figures 11−1a and 11−1b, the latter of which is on double logarithmic paper. They indicate a decreasing exponential relationship between direct labor hours and cumulative production. The importance of the learning curve stems from its ability to project time or cost required when improvement is known to occur. It is thus particularly useful for cost estimating, budgeting, production planning and scheduling, and pricing.

Example 11−3 applies the concept of the learning curve to the cumulative average of direct labor hours required.

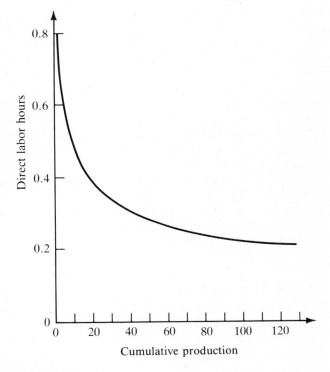

FIGURE 11−1a

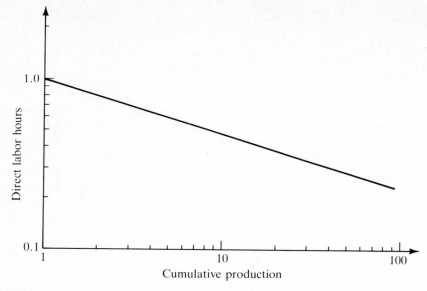

FIGURE 11—1b

Example 11—3a*

A firm undertook a contract to supply 1,000 units of a newly designed product. It was estimated that direct labor would average 10 hours per unit. One thousand units were therefore expected to require about 10,000 hours of direct labor. After 6 months of production, a report was requested, asking for: (1) the average unit labor requirements to date; (2) the labor hours required for the latest units produced; (3) the cumulative production volume at which average unit labor hours could be expected to reach the target of 10 hours per unit; and (4) the nature of the learning curve in force. Production records indicated the following:

Month	Units Produced	Direct-labor Man-hrs. Used
3	14	410
4	9	191
5	14	244
6	18	284
7	20	238
8	38	401

*Adapted from F. Moore, *Manufacturing Management,* Irwin, 1965, p. 88.

From this it was possible to compute:

Month	Cumulative Production	Cumulative Direct-labor Man-hrs.	Average Unit Man-hrs.
3	14	410	29.3
4	23	601	26.1
5	37	845	22.8
6	55	1,129	20.5
7	75	1,367	18.2
8	113	1,708	15.1

Note that, in this example, data for the direct labor man-hours for each unit are not available. The data consist of monthly averages. However, the principles for handling individual units and averages are similar. A plot of the average unit direct labor used vs. cumulative production, given in Figures 11–2a and b, clearly indicates that improvement had been taking place. Furthermore, these computations directly answer some of the questions that were asked. For example, the average unit labor hours used to date was 1,708/113 = 15.1 man-hours. The time required for the latest (month's) units, 401/38 = 10.6 man-hours, indicates that the average unit direct-labor target had not yet been reached. However, the remaining questions cannot be accurately addressed by simply plotting the progress function and "eyeballing" the results. To answer those

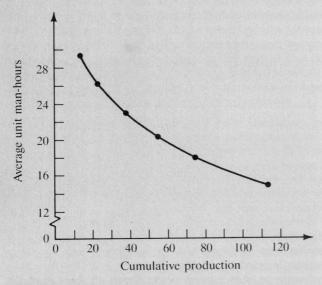

FIGURE 11–2a

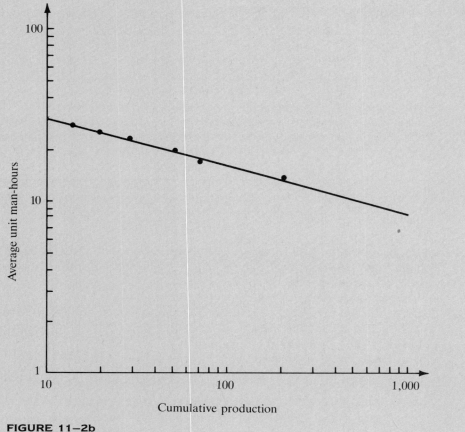

FIGURE 11–2b

questions accurately, it is necessary to utilize the mathematical description of the learning curve and to estimate the best fit to that function, using statistical estimation techniques.

Mathematics of the Learning Curve

An equation is needed to describe the phenomenon of decreasing assembly time or cost as production accumulates. The simple model suggested by Wright is an exponential function, which is still the most widely used form in industry:

$$Y_i = Y_1 i^{-b}$$

where

Y_i = the direct labor hours or cost for the ith production unit (thus, Y_{10} might represent the labor hours required for the tenth unit, while Y_1 would represent that for the first unit produced);

i = the cumulative production count; and

b = a measure of the rate of reduction of labor or cost required as production goes on.

Figure 11–3 represents such a curve. Note that because the rate of change remains the same, the absolute improvement per increase in production decreases over time. Or, putting it another way, a very much larger future addition to production would be needed to obtain the kind of incremental improvement that was previously achieved. Note, also, its correspondence in shape to Figure 11–1a, containing the eight data points for which there was a 20-percent improvement for each doubling of cumulative production. Wright referred to this as an 80-percent learning curve.

Examining this 80-percent function for any two points that represent a doubling of production will yield the corresponding value of b. For example, the fortieth unit should require only 80 percent of the direct labor hours of the twentieth unit, or

$$\frac{Y_{40}}{Y_{20}} = \frac{Y_1(40)^{-b}}{Y_1(20)^{-b}} = \frac{40^{-b}}{20^{-b}} = 0.80$$

This can easily be solved for b.

Since

$$\frac{40^{-b}}{20^{-b}} = 0.80$$

$$-b(1n40) - (-b1n20) = 1n.8, \text{ and}$$

$$b = \frac{1n.8}{1n20 - 1n40} = 0.322$$

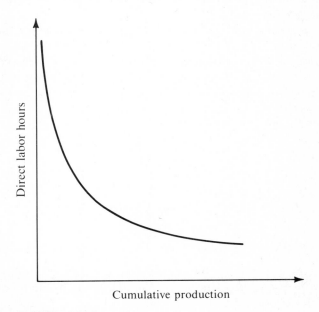

Cumulative production

FIGURE 11–3

Thus, an 80-percent learning curve may be described by the equation

$$Y_i = Y_1 i^{-0.322}$$

Learning curves of between 70 and 90 percent have been commonly reported in industry. The appropriate value of b in each case may be found by a procedure similar to the above. If the rate of improvement is denoted by r, reference thus being made to a $1 - r =$ percent learning curve, and noting that for any doubling of production, x,

$$1nx - 1n2x = -0.6931$$

then

$$b = \frac{1n(1 - r)}{-0.6931}$$

The reader is now ready for a brief return to Example 11–3.

Example 11–3b

Recall that two questions remained: the cumulative production volume at which average unit labor hours could be expected to reach the target of 10 hours per unit; and the nature of the learning curve in force. Both of them require that the learning curve (or rate of improvement) be specified mathematically. This can be done in a variety of ways. For example, one could choose any two values of cumulative production that represent a doubling and observe the percentage improvement between them. (This can't be done here because the data for individual units were not reported.)

Or, better yet, improvement over the entire range of production can be compared to the total production experienced. Thus, Y_{last} can be compared to Y_1 in order to compute b. (This also cannot be done in Example 11–3b, for the reason given above.) But just to illustrate the procedure, suppose that by the time 100 items have been produced, a savings of 91.65 percent has been observed. (In other words, the ratio of the time taken for the one hundredth, and the time taken for the first, is 0.0935.) What has been the degree of learning? Since

$$Y_i = Y_1 i^{-b}$$

$100^{-b} = 0.0935$, which gives

$$b = 0.5146$$

The reader should verify that this exponent corresponds to a 70-percent learning curve.

Although the two previously mentioned approaches to characterizing the learning curve cannot be used in this example, the following observation can be made: the cumulative production for month 7, 75, is just about double the cumulative production of month 5, 37. Also note that the improvement can be estimated by the ratio of the average unit man-hours for those months, which is $1 - (18.2/22.8)$, or 0.2018. Thus, a 79.82-percent learning curve can be hypothesized for this situation, and this answers the fourth of the original questions posed at the beginning of Example 11–3a. With the learning curve specified, attention can be directed to the remaining unanswered question.

The target of 10 hours per unit (for the average) may be estimated as follows: First the exponent b is required.

$$b = \frac{ln(0.7982)}{-0.6931} = 0.3252$$

Then one must estimate Y_1, since it is a basic parameter of the learning curve and was not specified in this problem. To do so, assume that $Y_{113} = 15.1$. (In practice, it is likely to be very close, if not equal, to this average.) Then one must solve,

$$15.1 = Y_1(113)^{-0.3252}$$

for Y_1. This yields $Y_1 = 70.249$ units. Now, using that result, the volume of production for which average labor hours per unit would reach 10 hours can be estimated.

$$10 = 70.249(Y_i)^{-0.3252}$$

or $Y_i = 401.28$

Thus, one would expect that if the learning curve described by the production through 113 units continues to have the same effect in the future, the average of 10 hours per unit will be reached by about the 402nd unit. This is an encouraging result, since the company hoped to achieve such an average over 1,000 units.

At this point it should be pointed out that there are two commonly used versions of the dependent variable in learning curves: direct labor hours or cost per unit; and average labor hours or cost per unit. Clearly, as production increases, the average more closely approximates the unit amount. This may be depicted in Figure 11–4. For values of production less than 100, a good approximation to the average, based on the unit amount, can be made from the following:

$$\overline{Y}_I = \frac{Y_1(i)^{-b}}{1 - b}$$

where $\overline{Y}_I$ = the average labor hours or cost per unit for cumulative production of I units.

The data of Example 11–3 were in the form of averages, which is one reason why the computed values were said to be approximations.

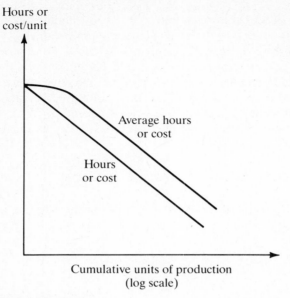

FIGURE 11–4

Statistical Estimation of the Learning Curve from Observations

The procedure employed to estimate the learning curve in the preceding section is weak in the sense that only two points were used. This sacrifices too much available information. It is better, in statistical estimation procedures, to approximate the learning curve by using all available data. In this section it is assumed that the reader has familiarity with simple linear regression analysis, in which a linear function is fitted to existing data. (Readers not familiar with such techniques can consult the references for this chapter.)

Recall that learning-curve data will appear exponential when plotted on linear axes. Furthermore, the form of the exponential function commonly used for the learning curve,

$$Y_i = Y_1 i^{-b}$$

when plotted on double logarithmic axes would appear exactly linear. (See Figure 11–1b.) In other words, the curve-fitting task can be transformed to one in which the data plotted on double log paper is estimated by a linear function. The way to do this is to transform the original exponential observations to logarithmic scale and then perform the linear regression procedure on the transformed data, illustrated with Example 11–3, with data plotted in Figure 11–2a. A plot of the same data on double logarithmic paper appeared in Figure 11–2b. Such plots would typically indicate a scatter diagram that would be well approximated by a linear function. The graphing of the learning curve data on double logarithmic paper is equivalent to performing a logarithmic transformation of the data. Thus, a description of the learning curve on linear axes such as

$$Y_i = Y_1 i^{-b}$$

is equivalent to

$$\log Y_i = \log Y_1 - b\log(i)$$

That this is linear in the logarithmic scale can be seen if $\log(i)$ is labeled X' and $\log Y_i$ is labeled Y'. Then

$$Y' = \log Y_1 - bX'$$

$\log Y_1$ is simply a constant and is the intercept of the linear function on the double logarithmic scale.

Since the estimators for the parameters of a linear fit are given by

$$B = \frac{\Sigma XY - (\Sigma X)(\Sigma Y)/n}{\Sigma X^2 - (\Sigma X)^2/n}$$

$$A = \overline{Y} - B\overline{X}$$

where B is the slope and A is the y-intercept, it is then necessary to make the computations shown in Table 11–7 to complete the curve-fitting exercise.

Example 11–3c

TABLE 11–7   Linear Regression Computations, Logarithmic Transformed Data, Base 10

X Cumulative Production	Y Average Unit Man-hrs.	X' logX	Y' logY	(X' Y')	(X')²
14	29.3	1.146	1.467	1.681	1.313
23	26.1	1.362	1.417	1.930	1.855
37	22.8	1.568	1.358	2.129	2.459
55	20.5	1.740	1.312	2.283	3.028
75	18.2	1.875	1.260	2.363	3.516
113	15.1	2.053	1.179	2.420	4.215
		9.744	7.993	12.806	16.386

$$B = \frac{12.806 - \dfrac{(9.744)(7.993)}{6}}{16.386 - \dfrac{(9.744)^2}{6}} = -0.3109$$

$$A = \frac{7.993}{6} - \frac{(-0.3109)(9.744)}{6} = 1.837$$

In other words, the fitted learning curve could be plotted as a straight line on double logarithmic paper with a slope of -0.3109 and a y-intercept of 1.837. Since the intercept of the double logarithmic curve is $\log Y_1$, a revised estimate of Y_1 for Example 11–3c can now also be obtained.

$$\log Y_1 = 1.837, \quad Y_1 = 68.707$$

Note that the estimate above also permits the learning curve to be expressed in terms of linear axes.

$$Y_i = 68.707(i)^{-0.3109}$$

These estimates yield a result somewhat different from the preceding one based on only two points. If the last two questions of Example 11–3a are addressed once again, this time using the learning curve derived from the regression analysis, the following results are obtained:

To see the nature of the learning curve that corresponds to an exponent, b, of 0.3109, form the following ratio:

$$\frac{Y_{40}}{Y_{20}} = \frac{Y_1(40)^{-0.3109}}{Y_1(20)^{-0.3109}} = \frac{40^{-0.3109}}{20^{-0.3109}} = \frac{0.3176}{0.3940} = 0.806$$

or an 80.6 percent learning curve. (Recall that the earlier approximation, based on only two points, was a 79.82 percent learning curve.)

An 80.6 percent learning curve will indicate a larger cumulative production until an average of 10 hours per unit is reached.

$$10 = 68.707(Y_i)^{-0.3109}$$

or $Y_i = 492$ units.

Other statistical measures indicating the goodness of fit of the curve to the data may also be obtained. (See the references for details.)

Life Cycle Costing

Life cycle costing is based on the observation that many types of engineering projects can be expected to incur costs in phases, each of which needs to be characterized separately. For the true costs of the project to be identified at the outset, care must be taken not to overlook any phase or to seriously misstate (typically underestimate) the costs associated with it. For example, the development of a new model of a helicopter may require:

Market research;
Prototype-related research;
Engineering product development and design;
Production method development—equipment, process design; and

Production and delivery—plant operating costs, product warranty, field costs, replacement parts production costs.

The variation of life cycle costs through time may be more clearly understood when depicted graphically, as in Figure 11–5. It is common in industrial situations for engineers and managers to portray and communicate their life cycle costing estimates this way.

Attention to life cycle costing has increased in recent years in response to the greater complexity of engineered products and processes. For many types of government procurement, life cycle costing is required of the vendor. Many industrial firms active in the private marketplace have adopted life cycle costing also, particularly for durable goods manufacture, where product or model design and development is an important component of the total organizational activity.

Life cycle costing may incorporate any cost concept or measurement technique. For example, production costs may be estimated using the existing job order costing or standard cost system, and learning curves may be employed. Also, a measure

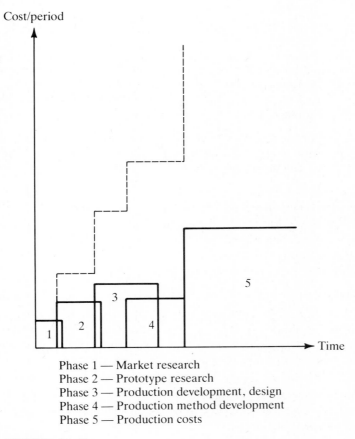

Phase 1 — Market research
Phase 2 — Prototype research
Phase 3 — Production development, design
Phase 4 — Production method development
Phase 5 — Production costs

FIGURE 11–5

of the total life cycle cost is likely to be expressed using time value techniques such as present worth or equivalent annual worth.

Exercises

11-1 Inquire at a local manufacturing company to determine what sort of costing system is used. See if you can obtain a sample cost report for a particular job or product. How was overhead attributed to the job or product?

11-2 Universal Manufacturing uses a job order costing system. The overhead rate for the department where Job 555 is processed is based on direct-labor hours. The September report shows the following information:

Cost, Job 555	
Direct labor	$ 4,000
Direct materials	$12,000

Also, 400 direct-labor hours were recorded for Job 555 in September. Total direct-labor hours budgeted for the year (at normal volume) is 700,000. Total plant overhead budgeted for that year is $10,000,000. Compute the cost of Job 555, including direct labor, materials, and overhead.

11-3 A plant using a job order costing system receives a reorder of a plastic handle produced in one of its departments on injection molding machinery. After the job, Number 123, is completed, the following information is available:

Materials issued: 4,000 pounds at $0.20 per pound
Unused material returned to stores: 300 pounds
Direct-labor hours used: 500 at $12 per hour
Overhead: $8 per direct-labor hour
Number of parts begun in lot: 20,000
Number of acceptable parts completed: 19,950

Compute the unit cost of the part.

11-4 A company using a job order costing system budgets for $40,000 of overhead per month (normal volume). Normal volume is 8,000 hours of direct labor per month. During a given month, Job 59 accounts for 100 hours of direct labor.

    a. How much overhead would be charged to the job?

    b. If the actual volume for the month was 7,500 direct-labor hours, how much total overhead would be charged?

    c. If the actual overhead costs recorded for that month were $39,000, what overhead variance occurred?

11-5 *a.* Given the data below for a given month, compute the variance of actual labor cost from standard. Use the standard wage rate to measure the quantity variance component of the total variance.

Standard direct-labor rate per hour	$8
Standard direct-labor costs	$64,000
Actual direct-labor hours	8,200
Actual direct-labor costs	$65,000

b. Now use the actual wage rate to compute the labor quantity variance. If the two quantity variances differ, indicate why this may be so and whether one may be preferable to the other.

11–6 A manufacturing company has a standard cost system in place. In one department consecutive monthly data are as follows:

	Sept.	Oct.
Standard direct-labor rate per hour	$10	$10
Standard direct-labor costs	$344,000	$319,000
Actual direct-labor hours	35,000	32,000
Actual direct-labor costs	$346,500	$323,200

What happened to the labor quantity variance from September to October?

11–7 Standard direct-labor costs are $12 per unit for a certain product. The standard wage rate is $10 per hour and 6,000 direct-labor hours were used for the product during the month. The actual direct-labor costs for that month were $62,000, and 5,300 units were actually produced. Compute the standard direct-labor cost of the output, the direct-labor quantity variance in dollars, and the direct-labor rate variance.

11–8 A certain product is manufactured primarily from one material which has a standard price of $0.75 per pound. The bill of materials specifying the standards calls for 2 pounds of material per unit. In a given period, 2,000 pounds of material were purchased at a cost of $1,400 with an allowance of 1 percent for prompt remittance of the bill. Shipping charges were $40. A work order calling for 1,500 pounds of material was issued, and 735 units were completed. Compute the materials cost variance for the period.

11–9 Incandescent light bulbs may be thought of as having three components: base, filament, and glass bulb. Suppose that the filament and base are assembled first, then the glass bulb is mounted and the vacuum formed before the unit is sealed. Data for the individual processes are as follows: For every 100 bases started, 99 acceptable units are completed. For every 100 filaments started, 98 acceptable units are completed. For every 100 bulbs started, 96 acceptable units are completed. Furthermore, 2 percent of the filament–base assemblies are scrapped and 5 percent of the final assemblies (filament–base–bulb) must be scrapped.

a. How many of each subcomponent must be started into production for every 1,000 bulbs of output desired? These are the standard quantities per 1,000 units of output.

b. If material for bases costs $0.05 per unit started, material for filaments costs $0.02 per unit started, and material for bulbs costs $0.01 per unit started, compute the standard materials cost for the finished bulb.

11–10 Suppose an engineering consulting firm has purchased a computer system with greater capacity than it currently needs, anticipating future growth. It considers selling computer time temporarily until its own needs increase. Suppose that the firm computes the incremental operating cost at $250 per hour, with a standard overhead cost of $250 per hour. If it can sell time for from $400 to $450 per hour on a short-term agreement, would this be justified? Explain.

11–11 *a.* A firm specializing in heat-treating parts for clients faced an economic downturn with trepidation. The operating volume was below the break-even point. A potential new customer inquired about heat-treating a particular part used in its guns. The heat-treating company analyzed the requirements and concluded that the marginal direct cost per unit was $0.75. Standard overhead charges for an item of this sort based on normal operating volume was about $0.50 per unit. If the customer offered $1.15 per unit on a short-term basis, would this be desirable?

b. What pitfall exists if the customer attempts to bargain for a long-term contract at this price? Explain.

11–12 A manufacturing company is considering farming out the production of a subcomponent it uses in large quantities. Unit costs at current normal volume are as follows:

Direct labor	$ 9
Direct materials	10
Variable overhead	5
Allocated fixed overhead	3
Total	$27

An independent supplier offers the part for $23 per unit. However, if it is purchased from an outside vendor, storage costs would increase. These are estimated at $1 per unit, at normal volume. Also, ordering costs must then be included, estimated at $0.50 per unit at normal volume. Assume that if the part continues to be produced, no new equipment is needed and there would be no change in the plant's fixed overhead. Should the part be made or purchased? Explain.

11–13 An established manufacturer of electric tools has received an unusual offer from an overseas customer to supply a slightly different version of an existing model, because of a fire in the foreign company's plant. Capacity of the domestic plant is estimated at 1.2 million standard units per year. The forecast for domestic sales is 900,000 standard units. The foreign order is equivalent to 400,000 standard units. The average price for units sold in the domestic market would be $75; the price offered by the foreign customer is $65. Cost data are as follows:

Direct labor	$15
Direct materials	25
Variable overhead	5
Allocated fixed overhead	10
Total	$55

Should the foreign order be accepted? Assume that the foreign order must either be supplied in full or rejected.

11–14 A recently formed company has two related products that it produces and sells. Information for each product is given below (in thousands). Fixed overhead was allocated to each product on the basis of direct labor hours.

	Product		
	X	**Y**	**Total**
Revenues	$500	$375	$875
Cost of Sales			
Direct labor	100	120	220
Direct material	150	150	300
Manuf. overhead	125	100	225
Sales Expenses			
Variable	25	20	45
Fixed	10	15	25
Net Income Before Taxes	90	−30	60

A suggestion was made to drop product Y, since it appears to be unprofitable. Given the fact that no other replacement is imminent and that the products do complement one another, what should be done? What would be the effect on short-term profitability if product Y is dropped?

11–15 Suppose a new model of your product is being assembled. After testing and production design has been completed, the first unit is assembled, requiring 148 direct-labor man-hours. Assuming that a 72-percent learning curve is in effect, derive the applicable learning curve, using the exponential form specified by Wright. Using this curve, how many labor hours would the 25th unit require?

11–16 In estimating the learning curve in effect for a certain assembly, data were recorded for the number of direct-labor hours required for the 50th and 100th item of cumulative production. The 50th unit required 18 direct-labor hours; the 100th required 15.8 hours. What degree of improvement was occurring? What would the equation of the learning curve be? How much time would the 500th item require, assuming the same rate of improvement?

11–17 A manufacturing firm monitors the learning or progress in direct-labor hours for each of its models. A new model just introduced was expected to require an average of 70 hours of direct labor per unit over the production run. The 150th unit has just been completed, requiring 80 hours of direct labor; the first unit took 300 hours.

a. What learning curve is in effect (for individual units of production)?

b. How much direct labor will be needed for the final unit of the production run, the 750th?

c. Use the approximation given in the chapter to determine the cumulative average unit labor requirements after the end of the production run.

11–18 A department tracking improvement in assembling a new model collected the following data:

Cumulative Production	1	2	3	4	5	6	7	8	9
Direct-labor Hours	200	160	145	135	120	118	112	105	102

Use the regression technique outlined in the chapter to estimate the best learning curve fit to the data. With that learning curve assumed to be in effect, how many hours will the 20th unit require?

11–19 Data on direct-labor hours for a product made under contract to another company revealed the following:

Cumulative Production	10	20	30	40	50	60
Direct-labor Hours	48	40	36	34	32	30

Use the regression technique outlined in the chapter to estimate the best learning curve fit to the data. With that learning curve assumed to be in effect, how many hours will the 100th unit require? What will the average unit labor requirement have been after the 100-unit order is completed?

Selected References

Blanchard, B. *Life Cycle Cost,* Portland, OR: M/A Press, 1978.

Brigham, E. and J. Pappas, *Managerial Economics,* 2nd ed., Hinsdale, IL: Holt, Rinehart and Winston, 1976.

Collier, C. and W. Ledbetter, *Engineering Cost Analysis,* New York: Harper & Row, 1982.

Devore, J. *Probability and Statistics for Engineering and the Sciences,* Monterey, CA: Brooks/Cole, 1982.

Draper, N. and H. Smith, *Applied Regression Analysis,* New York: John Wiley & Sons, 1966.

Engineering News-Record, *Construction Costs* (This issue published annually).

Ostwald, P. *Cost Estimating for Engineering and Management,* Englewood Cliffs, NJ: Prentice-Hall, 1974.

Shillinglaw, G. *Managerial Cost Accounting,* 4th ed., Homewood, IL: Richard D. Irwin, 1977.

Stuart, R. *Cost Estimating,* New York: John Wiley & Sons, 1982.

12

Aggregate Investment Decision-Making— Capital Budgeting

Introduction

This chapter will provide a brief overview of the current approaches to setting the MARR for project evaluation and to the theories and procedures in use for capital budgeting. Capital budgeting is a process by which an organization determines the amount of investment capital available for a given period of time and determines which of several alternative projects will be chosen for funding and implementation. In this chapter it is assumed that the projects or alternatives under consideration have been studied and analyzed with respect to one of the yardsticks already covered: for example, after-tax *PW* or *AW*.

It is beyond the scope of this book to provide a very detailed development of the theory and practice involved. However, the working engineer or manager will need at least a rudimentary understanding of the relevant concepts and techniques when analyzing projects to be considered by the organization. For a more detailed coverage of this topic, the reader may consult some of the references for this chapter.

Sources of Capital

There are two major sources of capital available to the firm: equity capital, associated with that provided by ownership; and debt capital, provided by lenders. Each will be considered in turn.

245

Equity

The corporate form of organization is the most widespread because it provides for limited liability of the owners and it facilitates capital formation. Its equity capital includes funds directly invested by owners as well as profits not paid out as dividends but retained by the firm for future investment. Limited liability is based on the fact that the corporation is the legal entity from which financial liability springs. Thus, owners risk only the amount of their investment, no more. They cannot ordinarily be held personally responsible for the corporation's debts.

Publicly held corporations are distinguished by the fact that the owners' shares may be easily traded in the free market, typically at an exchange such as the New York or American Stock Exchanges. Although many corporations are privately held, the publicly held corporation is the dominant organizational form in which engineers and managers find themselves conducting economic analyses as detailed in this book. They are dominant not because they are more numerous than privately held corporations but because they are much larger, as measured by assets, sales, production, employment, yearly investment, and just about any other objective measure. Equity ownership in the publicly held corporation is represented by shares, of which the major types are common and preferred.

Common shares are the most fundamental units of ownership. If a publicly held corporation has 1,000,000 such shares outstanding, then each share represents 1/1,000,000 of ownership. As a group, the holders of common shares expect to be the main beneficiaries of whatever profits the corporation may earn: they expect that as profits are earned, dividends will be paid to them and, better yet, increased as profits increase; or that the market price of the shares will rise; or some combination of the above. This may be thought of as an unlimited horizon.

Along with these rosy expectations, however, comes risk. If profits decline or disappear, dividends may be decreased or even suspended, and the market price for the shares may decline. Thus, while employees and creditors would continue to receive what is owed to them, common shareholders' returns may be reduced. In the event of a financial catastrophe, the shareholders' rights have the lowest priority. The claims of employees and creditors must be met before theirs can be considered.

The common shareholders have legal authority over the management of the enterprise. They exercise their rights by electing a board of directors whose function is to represent the entire class of owners and by voting on various important issues that the board puts before them from time to time. The board usually cannot involve itself in the day-to-day operations of the enterprise but does exercise judgment on key issues which the management proposes.

Sometimes corporations find it desirable to attract equity capital having less risk than that embodied in common shares; this may be accomplished by issuing preferred shares. Preferred shares, found in numerous forms, will not be discussed in detail here. It is sufficient to say that the preferred share is something of a compromise. It is less risky for owners than common shares because its dividends have priority over common shareholders. The dividend is either guaranteed or contingent on profits being earned. Often, the amount of dividends on preferred shares is constant over time, irrespective of the profit performance of the firm.

Because dividends are usually fixed, prices of preferred shares are not nearly as likely to fluctuate as much as those of common shares. Also, preferred shareholders' claims may have priority over those of common shareholders in the event of bankruptcy. Thus, in return for less downside risk, this form of ownership has a more limited upward horizon.

The corporation can raise equity capital directly by issuing common or preferred shares at any time, either when the corporation is initially chartered, when it "goes public" (changes from private to public ownership), or after it has been in operation as a public corporation for some time. It can also raise equity capital indirectly by not paying out all its profits as dividends. This is called retained earnings, and the cumulative value of this capital source is accounted for in the corporation's balance sheet. This is a major source of capital for many firms. In some cases growth companies maintain a "no dividend" or "low dividend" policy in order to maximize retained earnings as a capital source.

Debt

Another source of capital funds is long-term borrowing. Prevailing interest rates vary over time, as government policy and overall business conditions change. However, once contracted, such debt carries fixed interest charges. These are considered a cost of doing business from an income-tax point of view. The payment of interest and repayment of principal are legal commitments, unlike dividends, which are contingent on profits. The creditors' claims also have priority over those of shareholders in the event of bankruptcy.

As with preferred shares, long-term debt can take many forms. The corporation's obligations may be secured or unsecured by particular assets; the length of term of the debt can be quite variable, as can certain conditions regarding prepayment of the debt; and the debt obligation can sometimes be convertible under a given formula to other forms, such as common shares. However, from the point of view of this chapter, the salient difference between debt and equity capital is the stability of the terms, once they have been agreed upon by the borrower and lenders.

There is generally thought to be a short-run limit to the amount of long-term debt a firm can undertake in practice, irrespective of how many worthy projects it may have available. This contradicts pure theory, which indicates that capital funds can be obtained without limit if the users to which they are put return more than their cost. The practical limit stems from the fact that lenders, as a group, look carefully at dramatic changes that firms make. Large short-run increases in long-term debt may negatively affect investor confidence in the firm and may therefore be construed as very undesirable.

Setting the MARR

Up to this point in the book, analyses of alternatives have been undertaken with the MARR given. Now attention will be directed to the problem of assigning the appropriate value for the MARR. In theory, the MARR should represent the opportunity cost of capital for the project being evaluated. That is, if the project with its attendant risk is not undertaken, what yield would be provided by the next best investment at similar risk? Or, in that event, what would be forgone?

A variety of conceptual approaches to estimating the opportunity cost of capital for the firm as a whole will be explored in the pages that follow. There are also differences in industrial practice as it relates to applying these ideas to the problem of setting the MARR.

Weighted-cost-of-capital Approach

In this approach, the cost of capital is estimated, giving recognition to the composition of capitalization as described above. In other words, it is based on the observation that capital is attributable to both equity and debt of various types. In addition, this approach is based on the fact that the MARR must be at least enough to provide for future expected returns to these sources of capital.

Although capital sources can be classified in greater detail, for present purposes assume that four categories are relevant: common shares, preferred shares, retained earnings for equity, and bonds for long-term debt. Before discussing the weighted average, it is necessary to outline what is meant by the cost of capital associated with each. It is easiest to begin with debt.

Suppose that the firm's long-term debt is composed of a single issue of bonds, bearing an interest rate of 10 percent per year. This 10 percent per year may be thought of as a rental for the use of the funds; thus, it would be the debt cost of capital. If the firm has more than one long-term debt obligation, it may arrive at an average cost by a weighted average for all the debt instruments. For example, if two bond issues are outstanding, one in the amount of $10 million, paying 8 percent, and another amounting to $3.33 million, paying 12 percent, the average cost for debt is itself a weighted average, 9 percent.

$$D = \frac{(0.08)(10) + (0.12)(3.33)}{10 + 3.33} = 0.09$$

Attributing a cost of capital to equity is not as direct, since owners do not receive guaranteed or contractual payments for the use of their capital. Rather, owners invest their capital in the expectation of future returns. However, in principle, there is a certain similarity to debt for estimating the cost of equity capital in that the firm must expect to be able to earn enough in the future to reward shareholders for their investment and financial risk. This expectation of future payout is a cost of capital. It is most directly estimatable for preferred shares, since the payout is typically in the form of a set dividend. Thus, if the corporation has one issue of preferred shares paying $6 per share, and the issue price received by the firm from the investors was $50 per share, then the cost of capital associated with those preferred shares would be 6/50, or 12 percent.

For common shares, the estimation is still less direct, although the principle is the same. The question to be answered is "What is the payout needed for the common-shares capital raised"? Again one needs to consider the dividend as a real payout. But, as has already been indicated, some firms follow a policy of minimizing the dividend in order to have more capital available for reinvestment and faster growth. Thus, the dividend is not enough in estimating the cost of capital attributable to common shares.

One can attack this problem by taking the viewpoint of the common shareholder. What return would be sufficient to convince him to invest his funds? Clearly he expects a combination of ongoing dividend payments and future price appreciation of the common shares he holds. In other words he expects,

$$C = \frac{\text{dividend}}{\text{share price}} + \text{growth rate (of share price)}$$

For example, if a corporation's common shares are selling for $25 per share, if it pays $1 per share per year in dividends, and if it is felt that investors expect, on the average, a 12-percent per year appreciation in the value of their shares, then the cost of capital associated with that issue would be

$$C = \frac{1}{25} + 0.12 = 0.16$$

The category of retained earnings is usually treated the same as common shares. That is, the cost of capital associated with retained earnings is set equal to that for common shares. The reason is that the common shareholders will be the prime beneficiaries of the returns to investment of the retained earnings.

Given the interpretation of the cost of capital associated with each of the major sources, it is a simple matter to compute the weighted average. What is needed is the composition of total capitalization attributable to each. This may be obtained from accounting records such as those used for computing the balance sheet. Using the costs of capital of the preceding examples, for a firm having 25 percent of capitalization attributable to debt, 40 percent to common shares, 10 percent to preferred shares, and 25 percent to retained earnings, one would have a weighted cost of capital, CC

$$CC = (0.25)(0.10) + (0.4)(0.16) + (0.1)(0.12) + (0.25)(0.16)$$

$$= 0.141 \text{ or } 14.1 \text{ percent}$$

The weighted cost of capital described above fits a historical cost perspective. Under common industrial practice, this is frequently viewed as satisfactory, since most firms seek a long-term cost of capital as a basis for setting the MARR.

> The fundamental consideration is the average return over a protracted period of time, not the specific rate of return over any particular year or short period of time. This long-time rate of return on investment represents the official viewpoint as to the highest average rate of return which can be expected consistent with a healthy growth of the business, and may be referred to as the economic return attainable.*

---

*A. Bradley, *NACA Bulletin,* Jan. 1, 1927. Current policy is believed to be essentially the same in this regard.

From this viewpoint it would not be appropriate to compute the interest cost of debt based on the current market value of the bonds or notes, since, at any particular moment, market rates might be atypically high or low, and, thus, not representative of the returns needed in the long run.

Capital-asset-pricing Model Approach

For the past ten to fifteen years, the theoretical literature has been dominated by an advocacy of the capital-asset-pricing model for determining the cost of capital. Although it is still not as widely used for that purpose as the weighted-cost-of-capital approach, its application promises to increase, and so is included here. Space does not permit a detailed development of the theory involved. For further reading, consult the references listed.

The capital-asset-pricing model is based on the comparison of the risk involved in a specific security with the most generalized portfolio of investments, called the market portfolio. The model thus recognizes the fact that investments of higher risk justify higher returns. For example, common stocks have historically been recognized as significantly more risky than federal government securities. A fifty-year average of performance indicates that U.S. Treasury bills have yielded about 3 percent annual return, while common stocks have yielded around 12 percent, roughly a 9-percent premium. Thus, it can be said that investments of about the same risk as common stocks over the past fifty years should be discounted at about 9 percent more than risk-free investments. The problem with this last statement is that it does not answer the following question for the particular firms: How is its risk of investment to be evaluated (by comparison to, say, a common-stock average over 50 years)?

The capital-asset-pricing model addresses this issue by examining the historical risks of the firm's securities, as measured by market performance. Here, risk is defined as the variability in the market price of a firm's securities. Such risk can be judged in relation to the variability of a well-diversified portfolio. In other words, is the value of the firm's securities likely to change more or less than the rest of the portfolio in response to economic events?

The portfolio used for such analyses is the so-called market portfolio, as measured by a popular average such as Standard and Poor's Composite Index. Then the risk of the firm's securities can be measured by a parameter called *beta*, the sensitivity of the firm's common stock price in relation to changes in the market portfolio. Beta indicates the degree to which investors believe the indicated security price will vary in relation to a 1-percent change in the market portfolio. A beta greater than 1.0 indicates that a security is expected to be more sensitive to price changes than the rest of the market; it is thus associated with greater risk. A beta less than 1.0 indicates that a security is expected to be less sensitive to price changes than the rest of the market; it is thus said to be less risky.

These days empirical measurement of betas is done commonly, almost routinely, by a number of brokerage firms and investment advisory services for their clients. A typical procedure involves measuring recent market performance and using those measures as an estimate of beta. For example, to estimate betas for common stocks, one brokerage firm uses the monthly change over a period of five years in the market price of the indicated stock in relation to a market portfolio represented by Standard and Poor's Composite Index. This provides 60 observations, which are fitted by linear regression analysis to estimate beta. (The slope of the fitted line is the beta estimate. See Figure 12–1.)

Betas can also be measured for the common stock of industry groups. They show how different industries are viewed with respect to risk. Industry groups with the lowest betas and the lowest perceived risk include electric utilities and telephone companies, which show industry betas around .50. By comparison, manufacturers of electronic components have betas of about 1.50.

Given an estimate of beta for a security, the capital-asset-pricing model indicates the risk premium which is associated with investment in that security. The risk premium is given by

$$r - r_f = \beta(r_m - r_f) \tag{12--1}$$

where r is the expected return indicated for the security, r_f is the risk-free return (given, say, by the Treasury bill rate), r_m is the return rate for the market portfolio, and β is beta for that security. Note that $r_m - r_f$ is the risk premium associated with the market portfolio and $r - r_f$ is the expected risk premium associated with the security in question. The capital-asset-pricing model is thus a simple statement indicating that the risk premium for a particular security is proportional to that for the market portfolio, with the proportionality constant being beta. This is depicted in Figure 12–2, which shows that all investments would plot along the market line.

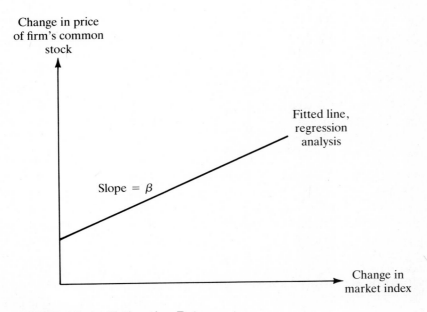

FIGURE 12–1 Estimating Betas

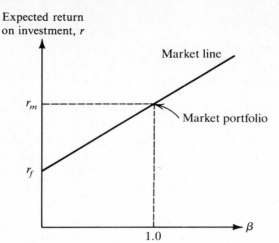

Expected return
on investment, r

FIGURE 12–2 Capital-Asset-Pricing Model

Clearly Equation 12–1 gives the cost of capital for each of the sources of capital for the firm.

$$r = r_f + \beta(r_m - r_f) \tag{12–2}$$

These may be computed for both debt and equity and then combined in a weighted average as was done in the preceding section. Example 12–1 illustrates this.

EXAMPLE 12–1

A corporation listed on the New York Stock Exchange has 60 percent of its capitalization attributable to equity, 40 percent to debt. The estimated beta for its common stock is 0.85, indicating less than average risk. The estimated beta for its debt securities is very low, so they will be assumed to be risk free (i.e., $\beta = 0$). Suppose Treasury bills yield 8 percent (i.e., $r_f = 0.08$), and that the market premium is 9 percent (i.e., $r_m - r_f = 0.09$). Then

$$r_{\text{debt}} = r_f + \beta_{\text{debt}} (r_m - r_f) = 0.08 + 0(0.09) = 0.08$$

and

$$r_{\text{equity}} = r_f + \beta_{\text{equity}} (r_m - r_f) = 0.08 + 0.85(0.09) = 0.1565$$

Since the proportions of capital associated with debt and equity are given, the composite cost of capital indicated by the capital-asset-pricing model would be

$$r = 0.40(0.08) + 0.60(0.1565) = 0.1259, \text{ or } 12.59 \text{ percent}$$

From the Cost of Capital to the MARR

The preceding sections have described two procedures for estimating a company's cost of capital. Although many firms use this "unadjusted" rate for the inflation-free MARR, theoretically this is not quite correct. The estimation of the cost of capital should be viewed as a first step toward obtaining the opportunity cost of capital for that project or an alternative. This opportunity cost of capital is influenced by the degree of risk associated with the project.

Clearly, all firms entertain the possibility of various types of projects with respect to the risk entailed. It would not make sense to use the same MARR for the most and least risky projects. Thus, as part of the project evaluation process, a firm may choose to establish a procedure to adjust the company cost of capital to reflect the risk that the given project entails.

Suppose, for example, that the company cost of capital is estimated to be 18 percent. This may be unadjusted for evaluation of certain types of projects which are viewed as equivalent in risk to operation of the firm as a whole, such as a moderate expansion in production capacity in response to long-term, readily predictable growth in the market. In contrast, a project, such as introduction of a new product, may be somewhat riskier, carrying a higher MARR than the company cost of capital, say 20 percent. A very risky alternative, such as the development, production, and marketing of an entirely new technology for the consumer market, might require an even larger risk premium and result in an MARR of 25 percent. On the other hand, a cost-reduction proposal posing little or no risk, and thus presenting less risk than the average for the operation of the company as a whole, might carry an MARR somewhat less than the (overall) company cost of capital. In this example, that might be about 16 percent.

Of course, there are no mathematical rules indicating how the risk adjustment should be made. It requires sound judgment. A company's effort to systematize this process, while maintaining consistency, would probably utilize a classification system for the risk adjustment described above.

Capital Budgeting

Given that a firm's engineers and managers have developed a large number of proposals for investment, each having been studied in relation to its potential qualitative and quantitative contributions (*PW, AW,* or *IRR*) as against the applicable MARR, how can decisions be made regarding which to fund and implement? From a narrow theoretical perspective, one could argue that all proposals which will give a positive *PW* or *AW* should be accepted. Such a policy will maximize the value of the firm and will result in the greatest long-run return to owners. However, such a policy is almost never practical, for two reasons.

First of all, if overall business conditions are sound and engineers and managers are enthusiastic, aggressive, imaginative, and capable, they are most often likely to generate more investment proposals than the readily available capital can support.

Although it is theoretically true that additional capital can be raised as long as investment returns exceed the marginal cost of capital (MCC)—a capital-rationing viewpoint depicted in Figure 12–3—this is not usually viewed as a pragmatic, workable policy. It is generally felt that there is a limit to the amount of capital that can be raised in the short run without significantly affecting the attitudes of lenders and potential investors toward the risk associated with the firm's operation or without significantly diluting the holdings of the existing owners.

Secondly, for successful implementation, projects require human factors— competent managers and well-trained and motivated employees. The short-run limits on availability of human resources typically constrain the number of projects that can be accepted, even if capital were available. Therefore, a common approach is to determine the amount of capital to be made available in the short run. This is based on many factors, including business conditions, the recent financial performance of the firm, the terms available in the financial markets, strategic requirements, the human resources available, etc., as well as the investment needs as evidenced by the proposals being brought forward. Clearly such decisions require as much "art" as "science" and are rendered at the highest levels of the organization. In this approach, once the amount of capital to be made available in the short run is determined, it becomes an important parameter for the budgeting process.

Ranking

When the total capital is fixed, one way to allocate investment funds is by some ranking mechanism. This is depicted in Figure 12–4. If independent projects are

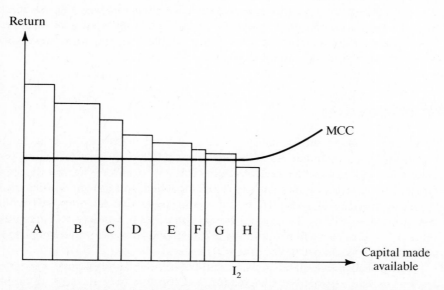

FIGURE 12–3 Capital Rationing—Capital Limited Only by MCC and Investment Return

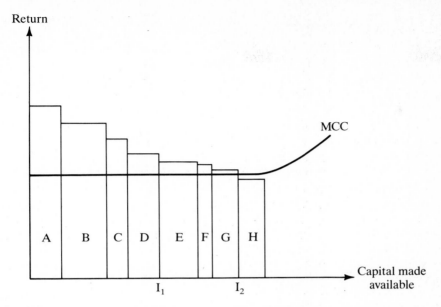

FIGURE 12–4 Capital Rationing—Capital Limited to Some Amount. $I_1, < I_2.$

ranked by some measure of desirability, such as IRR, and I_1 is available, then allocations can be made. Figure 12–4 shows that, conceptually, the last project to be accepted has a higher IRR than the cost of capital. (Obviously, there may be situations in which the last project accepted coincides with the cost of capital. This would imply that the firm had decided to commit and acquire sufficient capital for all its "acceptable" projects, as described in Figure 12–3.) However, in practice, a strict ranking procedure with capital fixed is likely to lead to a mix of projects where the last (or last few) projects included are not as desirable as others excluded from the final choice but must be included to meet the capital availability constraint. Although the collection would certainly be a good one, the procedure can lead to a less than optimal collection of projects, as Example, 12–2a, indicates.

EXAMPLE 12–2a

Suppose a firm has determined that it can make available $10 million for investment in the next fiscal year. The MARR is 16 percent. Seven acceptable projects have been submitted for approval, their total capital requirements being $14 million. The projects' investment requirements and their ranked expected IRR's are given in Table 12–1.

    A ranking procedure with capital fixed at $10 million would select projects I, II, III, and IV, taking $8 million. Since project V, the next ranked, requires $4 million, it would have to be passed over for lower-ranked projects

TABLE 12–1 Investment Requirements and IRR

Project	Investment ($ million)	IRR
I	3.0	30
II	.5	26
III	2.0	23
IV	2.5	21
V	4.0	20
VI	1.0	18
VII	2.0	16

if the $10 million constraint is to be maintained. In this example, one would end up with projects I, II, III, IV, and VII, yielding an average return of

$$\frac{3(0.30) + 0.5(0.26) + 2(0.23) + 2.5(0.21) + 2(0.16)}{10} = 0.2335 \text{ or } 23.35\%$$

That this collection is not optimal can be seen by computing the average return for a different collection, I, II, IV, and V.

$$\frac{3(0.30) + 0.5(0.26) + 2.5(0.21) + 4(0.20)}{10} = 0.2355, \text{ or } 23.55\%$$

The Present Worth Index Although the internal rate of return is frequently used in practice as the yardstick for ranking alternatives so as to allocate investment funds when capital is fixed, it does not always yield the best results. The present worth yardstick may be used alternatively. Then, the net present worth of each independent alternative is computed given the applicable MARR. The collection of projects that maximizes total net present worth, while staying within the limit of available capital would then be optimal. This collection may be identified either by trial and error, if the collection is not too large, or using linear programming procedures as outlined in the next section and in the references.

Optimization Models

The literature stresses the importance of choosing an optimal collection of projects mathematically. Linear programming offers an insight into how this might be approached for our preceding example.

EXAMPLE 12–2b

Suppose that in addition to the information in Table 12–1, one can estimate the additional manpower needed for each project. Furthermore, assume that the personnel department advises that no more than 100 new employees can be absorbed (i.e., recruited and hired, or transferred, trained for the new job, and given proper supervision) in the next fiscal year. Table 12–2 indicates the estimated net new employees needed for each project.

TABLE 12–2   Manpower Requirements

Project	New Employees Needed
I	50
II	10
III	40
IV	15
V	65
VI	12
VII	35

A linear programming formulation of the problem would then be as follows: Let $X_i = 0$ if project i is not assigned. Let $X_i = 1$ if project i is assigned. (Actually, since only 0 or 1 is permitted for the variable, this would be referred to as an integer programming problem, a special case of linear programming.) Then the criterion function would be,

Maximize: $30X_I + 26X_{II} + 23X_{III} + 21X_{IV} + 20X_V + 18X_{VI} + 16X_{VII}$

subject to:
$3X_I + 0.5X_{II} + 2X_{III} + 2.5X_{IV} + 4X_V + X_{VI} + 2X_{VII} \leq 10$
$50X_I + 10X_{II} + 40X_{III} + 15X_{IV} + 65X_V + 12X_{VI} + 35X_{VII} \leq 100$

The criterion (maximization) function seeks a solution with the greatest aggregate return. The first constraint refers to the limits of capital, and the second refers to the limits on new employees.

Our example, being on a small scale, can be solved by trial and error—that is, by choosing project combinations that maximize returns while not violating the two constraints. Clearly, this would not be feasible for problems with more projects and/or more constraints. Although a mathematical procedure which can be implemented with computers exists for finding the optimal solution in a relatively small number of steps, it is beyond the scope of this book to delve into the mathematics of linear programming. For references in this area see the recommended sources for this chapter.

Payback Period

The (nondiscounted) payback period is the number of years needed for the sum of positive cash flows to equal the investment. Taken alone, it is not adequate as a yardstick for measuring the desirability of capital investment. However, it is still commonly used in actual industrial practice as a supplement to the time value measures described in this book. Its use stems from the widely held view that alternatives with shorter payback periods are less risky, other things being equal.

Exercises

12–1 For computing the capital cost for a firm the following information is provided:

Capital Source	Percent of Capital Attributable to That Source	Percent Yield or Rate
Long-term debt	30	10
Short-term debt	10	15
Common stock	60	12

   a. What is the after-tax cost of capital for this firm, if a weighted cost of capital is used?
   b. What would the before-tax cost of capital be if the income-tax rate for the firm is 50 percent?

12–2 The following data are provided for a firm seeking to compute its weighted cost of capital:

Capital Source	Percent of Capital Attributable to That Source	Percent Yield or Rate
Long-term debt	30	12
Short-term debt	10	16
Preferred stock	15	—
Common stock	45	14

The preferred stock pays $6.50 per share; $50 was the net issue price received by the corporation for these shares.
   a. What is the after-tax cost of capital for this firm, if a weighted cost of capital is used?
   b. What would the before-tax cost of capital be if the income-tax rate for the firm is 50 percent?

12–3 Suppose a utility company (enjoying a local monopoly but regulated by a state commission) desires to compute its cost of capital using the capital-asset-pricing model. Suppose 40 percent of capital is attributable to long-term debt; 40 percent is attributable to common shares; and 20 percent is attributable to preferred shares. The estimated betas are: for its common stock, 0.9; for preferred stock, 0.8; for its long-term debt, 0.15. If Treasury bills yield 8 percent and the market premium is 9 percent, what would its aggregate cost of capital be?

12–4 A relatively new manufacturing company has just "gone public"—that is, it has just issued its first common shares tradable on an exchange. It has no bonds on the market, only the new common shares. After a year's trading, the estimated beta of the common stock is 1.7. If a risk-free investment such as Treasury bills pays 9 percent, and the market premium is also 9 percent, what would the cost of capital be, using the capital-asset-pricing model?

12–5 A small company must decide on a number of independent investment alternatives, for which study has revealed the following economic information:

Alternative	Investment	Rate of Return
A	$25,000	20%
B	35,000	16
C	50,000	26
D	10,000	14
E	20,000	24

Assume that the company's capital is investable at 14 percent (if any of the above are not implemented).

a. If the firm has $200,000 of capital available, which projects would be funded?

b. Assuming that only $80,000 can be used, which projects would be chosen from an economic perspective? What rate of return is earned on the least attractive of those chosen?

c. Now suppose that capital can be borrowed at a rate of 16 percent. Which projects would be just fundable, and how much money would be borrowed?

12–6 The following proposals have been studied, with the results as shown:

Alternative	Investment	Rate of Return
A	$125,000	28%
B	50,000	25
C	100,000	23
D	75,000	20
E	50,000	16
F	80,000	12

a. If the MARR is 20 percent and $480,000 is available to invest, which alternatives should be chosen?

b. Suppose that only $300,000 of internal funds is available, which would be chosen now?

c. If, in addition to *b*, it is known that funds can be borrowed over the long term for 14 percent, which collection makes sense now?

12–7 Suppose a collection of independent projects is available and has been studied, but that some of them have more than one variation. These variations are mutually exclusive. For example, the variations may be different models or manufacturers of a given type of equipment. In this example, independent alternatives are designated with Roman numerals; variations or mutually exclusive alternatives with capital letters. A given set of alternatives has been analyzed with results shown below:

Alternative	Investment	Incremental Rate of Return	As Compared With
I A	$25,000	25%	0
I B	10,000	18	IA
II G	50,000	20	0
II H	15,000	22	IIG
II I	10,000	12	IIH
III M	35,000	15	0
III N	15,000	12	IIIM

    a. Assuming that $90,000 is available, which alternatives would be chosen?

    b. If funds can be borrowed at 14 percent, which alternatives would be chosen and how much money would be borrowed?

    Hint: Try ranking the alternatives on the basis of their returns, noting the incremental investment required. This will indicate not only which independent alternatives can be funded but which of the variations (mutually exclusive alternatives) for an independent alternative is best as well.

12–8 A cost-cutting proposal promises to save $20,000 per year. The equipment needed is expected to have an economic life of 10 years with no appreciable salvage value. Suppose the firm's cost of equity capital is estimated to be 15 percent, and that of debt is 12 percent. The proportion of the company's capital attributable to debt is about 30 percent. If the cost-cutting investment can be justified if its return is twice the cost of capital, what is the most the firm will invest in the new process? (Do this on a pretax basis. You may then compare the result obtained to an after-tax analysis, treating the investment as being depreciable over 5 years with no salvage value. Choose the depreciation method you prefer, and use an income-tax rate of 50 percent.)

12–9 A company raises additional needed capital by issuing $1 million in bonds due in 10 years. The investment banking firm which handles the transaction charges a fee of $30,000, so that the firm realizes only $970,000. The coupon rate is 12 percent and, since the interest is paid semiannually, the firm must pay $60,000 of interest every 6 months. Suppose that the firm is required, as part of its covenant in issuing the bonds, to establish a sinking fund to repay them in 10 years. The company will make semiannual deposits into the fund, earning 9 percent per year.

    a. Compute the amount payable into the sinking fund; add this to the interest payment on the bonds to arrive at the total amount payable every 6 months.

b. Contrast the above arrangement with the following: Suppose that the firm retires one-tenth of the face value of the bonds outstanding after each of the 10 years. Thus, it pays interest on the entire initial amount the first year, on $900,000 the second year, and so on. What is the schedule of interest and principal payment over the 10 years?

c. Other things being equal, which repayment schedule seems preferable? How did you arrive at your conclusion?

12–10 The common stock of a publicly traded company now sells for $10 per share. Dividends amount to $0.40 per share per year.

a. Assuming investors expect a 12-percent per year increase in the value of their shares, what would the company's cost of equity capital be?

b. On what basis do you suppose such an estimate of stockholders' expectations can be made?

c. Using the capital-asset-pricing model, determine the cost of equity capital. Use a reported beta of 1.4, a market premium of 9 percent, and a risk-free interest rate of 7 percent. How does this estimating procedure differ from that in *a*?

Selected References

Barges, A. *The Effect of Capital Structure on the Cost of Capital,* Englewood Cliffs, NJ: Prentice-Hall, 1963.

Bierman, H. and C. Alderfer, "Estimating the Cost of Capital: A Different Approach," *Decision Sciences,* Vol. 1, No. 1, January, 1970, pp. 40–53.

Bierman, H. and S. Smidt, *The Capital Budgeting Decision,* 5th ed. New York: Macmillan, 1980.

Brealey, R. and S. Myers, *Principles of Corporate Finance,* New York: McGraw-Hill, 1981.

Brigham, E. and J. Pappas, *Managerial Economics,* 2nd ed. Hinsdale, IL: Dryden Press, 1976.

Freeland, J. and M. Rosenblatt, "An Analysis of Linear Programming Formulation for the Capital Rationing Problem," *The Engineering Economist,* Vol. 24, No. 4, Fall, 1978, pp. 49–61.

Hayes, J. "Discount Rates in Linear Programming Formulations of the Capital Budgeting Problem," *The Engineering Economist,* Vol. 29, No. 2, Winter, 1984, pp. 113–126.

Hillier, F. and G. Lieberman, *Operations Research,* 3rd ed. San Francisco: Holden-Day, 1980.

Langer, H. "Valuation, Gains from Leverage, and the Weighted Cost of Capital as a Cutoff Rate," *The Engineering Economist,* Vol. 29, No. 1, Fall, 1983, pp. 1–11.

Lintner, J. "The Cost of Capital and Optimal Financing of Corporate Growth," *Journal of Finance,* Vol. 18, No. 2, May, 1963, pp. 292–310.

Miller, M. and F. Modigliani, "The Cost of Capital, Corporation Finance and the Theory of Investment," *American Economic Review,* Vol. 48, No. 3, June, 1958, pp. 261–297.

Oso, J. "The Proper Role of the Tax-Adjusted Cost of Capital In Present Value Studies," *The Engineering Economist,* Vol. 24, No. 1, Fall, 1978, pp. 1–12.

Rubinstein, M. "A Mean-Variance Synthesis of Corporate Financial Theory," *Journal of Finance,* Vol. 28, March, 1973, pp. 167–182.

Schall, L., G. Sundem, and W. Geijsbeek, "Survey and Analysis of Capital Budgeting Methods," *Journal of Finance*, Vol. 33, March, 1978, pp. 281–287.

Stevens, G. *Economic and Financial Analysis of Capital Investment,* New York: John Wiley & Sons, 1979.

VanHorne, J. *Financial Management and Policy,* 5th ed. Englewood Cliffs, NJ: Prentice-Hall, 1980.

Weingartner, H. *Mathematical Programming and the Analysis of Capital Budgeting Problems,* Englewood Cliffs, NJ: Prentice-Hall, 1963.

Weston, J. and E. Brigham, *Essentials of Managerial Finance,* Hinsdale, IL: Dryden Press, 1974.

13

Risk in Engineering Economic Analysis

Introduction

To a certain extent, all the material of the text up to this point has dealt with situations involving some degree of risk. Usually this was due to an inability to predict the economic parameters of the problem with absolute precision. This was handled by attempting to make the best estimates of those parameters and by conducting sensitivity analyses to indicate the degree of risk if the actual future parameters were to vary substantially from the estimates. Although every decision situation involving future outcomes embodies some risk, some situations are far more riskful than others.

In this chapter, risk is defined as a condition in which a variety of outcomes is possible, for which an estimate may be made of the relative likelihood of each of those outcomes. How choices may be made under those circumstances—i.e., where there is risk that the most economically desirable outcome may not come to pass—will be the theme of the chapter. The relative likelihood will be measured in terms of probabilities. A brief introduction to probabilistic concepts will be presented at a practical and nonformal level, partly because of the limitations of space and partly because the fundamental aim is to provide the reader with a pragmatic framework for analyzing risk situations. (For a more detailed and formal presentation of probability theory and its application to decision theory and analysis, see the references at the end of the chapter.)

This definition of risk leaves two other possible situations, certainty and uncertainty. Certainty is defined as a condition for which the outcomes of decisions are known for sure. In practice, there is no such thing as complete certainty. As mentioned earlier, it is frequently sufficient to treat the future as almost certain, resulting in an outcome for which numerical estimates of the result can safely be made. Uncertainty is defined as a situation in which the relative likelihood of the outcomes cannot be estimated, or worse still, the outcomes cannot even be unambiguously identified. Conditions of uncertainty typically present themselves when new ground is broken, so that there is little experience on which decisions may be based.

Probability Concepts

Probability is a decimal measure of relative likelihood, with values ranging between zero and one. That it is a decimal measure stems from the fact that the probability is a ratio formed from

$$p = \frac{x}{N} \tag{13-1}$$

where p is the decimal probability measure, x is the number of ways a particular outcome may occur, and N is the total number of ways that all the outcomes may occur. Thus, probability measures close to 1 indicate high relative likelihood; measures close to 0 indicate low relative likelihood; a probability of exactly 1 indicates a certain outcome; a probability of 0 indicates impossibility.

In practice, probability measurements may be categorized as abstract, historical, or subjective. Abstract probability measures are those which may be evaluated from Equation 13–1 by directly analyzing the situation. For instance, the probability of tossing a coin and having a head come up is one-half. This is because there is a total of two outcomes, head and tail, head accounting for one of them. Another example is that of a die. The probability of a 2 landing face up when a die is rolled is one-sixth. Here there is a total of six outcomes, one of which is the 2. Finally, one may consider a deck of cards. If one does an experiment in which the deck is shuffled and then cut, the probability of, say, a heart turning up is one-fourth. This is because there are 52 cards in all, 13 of which are hearts. The ratio of 13/52 is 1/4.

Historical probabilities are based on some examination of past experience because the probabilities cannot be reasoned out from the context of the situation. An example would be the probability that a certain piece of manufacturing equipment will not be usable because of breakdown. This may be estimated from the ratio of recorded downtime to total available time for a representative recent period. Implicit in the use of historical probabilities is the assumption that the forces at work in the period for which the measurement was made will continue to apply in the future in more or less the same way. When this assumption is valid, the historical probabilities are reliable estimates on which to make judgments about the future.

Subjective probabilities are measurements based on human judgment. They may be relied upon when it is not possible or practical to carry out abstract or historical probability computations. Ordinarily, they would be estimated by individuals having a wealth of personal experience to lean on, and they should, additionally, take into consideration any relevant facts bearing on the probability measurements. An example would be an engineer's judgment that there is a .8 probability that a long-used vendor will be able to deliver a newly designed part on time.

A Basic Axiom

A fundamental property of probabilities is that the sum of the probabilities for all the basic outcomes must equal one.

$$\sum_i P_i = 1$$

To see what this means, refer to the previous examples for abstract probabilities. For the coin-tossing illustration, the basic outcomes are head and tail; the probability for each outcome is $1/2$; and the sum is 1. For the die-rolling case, the basic outcomes are the six faces numbered 1 through 6; the probability for each outcome is $1/6$; and the sum is 1. For the card example, the basic outcomes were the 52 cards; the probability for each outcome is $1/52$; and the sum is 1.

The simplicity of this axiom belies its usefulness. Having computed the probabilities in a particular case, it is a good idea to make some checks. One way to do so is to apply the axiom above. Even though the fact that the sum equals 1 does not guarantee the absence of error, the failure of the sum to equal 1 certainly does indicate an error. Another practical use of the axiom is for situations where the probabilities of all the basic events except one have been evaluated. It may be cumbersome or tedious to compute the probability of the last event. In that case, one may prefer to compute

$$1 - \sum_{i-1} P_i \tag{13–1'}$$

An example of the attractiveness of this approach will be provided by Examples 13–1 and 13–5.

The Rule of Addition

Often it is necessary to be able to evaluate the probability of compound events. Two compound events of special interest are the union of two basic events and the intersection. The union corresponds to either one, or the other, or both. Given that the basic events are labeled A and B, the union is denoted as either $(A \text{ or } B)$ or $(A \cup B)$. Some examples of the unions of basic events follow:

Suppose a deck of cards is shuffled and a single card is drawn. Denote the basic outcome "ace" as A and the basic outcome "club" as C. Then the compound event "ace or club or both" can be denoted as $(A \text{ or } C)$ or $(A \cup C)$. [In this book the notation $(A \text{ or } C)$ will be used.]

As a second illustration, suppose that a die is rolled. The basic outcome "one" can be denoted as O, and the basic outcome "two" can be denoted as T. Then the compound event "one or two or both" can be denoted as $(O$ or $T)$. (The fact that in such an experiment it is impossible for a one and a two to occur simultaneously means that here the union refers to a "one or a two." More will be said of this later.)

The probabilities of these compound events are denoted as $P(A$ or $C)$ or $P(O$ or $T)$. If it is desired to evaluate these probabilities, they may be reasoned out as follows.

For $(A$ or $C)$. There are 52 cards in all. Of the total, there are 4 aces, including the ace of clubs. Thus, there are 12 additional clubs to be included in the compound event. The number of cards corresponding to "ace or club or both" is, therefore, $4 + 12$ or 16. The probability of the compound event is then $16/52$ or .308.

For $(O$ or $T)$. There are six outcomes to the roll of a die. One outcome corresponds to a "one," another to a "two." The number of outcomes corresponding to "one or two" is 2. The probability of the compound event is then $2/6$ or .333.

A mathematical rule giving the same results is easily formulated. Note that in computing $P(A$ or $C)$ it was necessary to avoid "double counting" the ace of clubs. That card corresponds to the intersection compound event, both ace and club, denoted by $(A$ and $C)$ or $(A \cap C)$. [In this text, the notation for intersection will be $(A$ and $C)$]. The probability of the intersection event is denoted by $P(A$ and $C)$. In this case, since there is only one ace of clubs, $P(A$ and $C) = 1/52$. If the number of aces, 4, were added to the number of clubs, 13, the total, 17, would be one more than is needed, because of the double counting of the ace of clubs. Therefore, a mathematical rule based on addition would have to subtract the double-counted outcome to be correct. In this example, such a rule would give

$$P(A \text{ or } C) = 4/52 + 13/52 - 1/52 = 16/52$$

In general, such a rule can be written as

$$P(A \text{ or } B) = P(A) + P(B) - P(A \text{ and } B) \qquad (13\text{–}2)$$

Another way to visualize this rule is by analogy to a Venn diagram. Figure 13–1 shows such a diagram. The area of shape A can represent $P(A)$, and the area

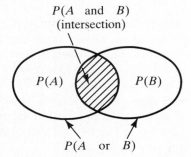

$P(A \text{ and } B)$
(intersection)

$P(A)$ $P(B)$

$P(A \text{ or } B)$

FIGURE 13–1   Venn Diagram $P(A \text{ or } B)$

of shape B can represent $P(B)$. The area of the joint shape of A and B would then correspond to $P(A$ or $B)$. That joint area can be obtained by adding area A to B and then subtracting the double-counted area or intersection, which is shaded.

A special case of the rule of addition exists when $P(A$ and $B) = 0$. That means that it is impossible to have events A and B occur simultaneously. When that happens, A and B are said to be mutually exclusive events. Of course, this means that the rule of addition becomes purely additive.

$$P(A \text{ or } B) = P(A) + P(B)$$

An example of this case is the compound event $(O$ or $T)$ referred to previously. Since it is impossible to roll a die and have both a one and a two as a single outcome, $P(O$ and $T) = 0$. Thus, the events "one" and "two" are mutually exclusive, and

$$P(O \text{ or } T) = 1/6 + 1/6 = 1/3 = .333$$

as before.

The Rule of Multiplication

The rule of multiplication is used to evaluate the probability of the compound event $(A$ and $B)$ under certain conditions. One of those is the existence of conditional probabilities. A conditional probability refers to an event the probability of which is affected by the occurrence of a related event. As an illustration, consider a small urn containing three marbles, two black and one white as in Figure 13–2. Suppose an experiment is conducted in which the contents of the urn are mixed and two marbles are withdrawn at random, without replacement of the first marble after it has been removed. Further suppose that immediately after the first draw, the result is announced, and it is a black marble. It should be clear that the evaluation of the probability of drawing either a white or a black marble on the next draw is dependent on what came out first. Since black was withdrawn first, two marbles remain, one black and one white. Thus, the probability of drawing either a white or a black on the second draw would be one-half. If, on the other hand, the first draw was a white marble, the second draw would have to be black.

A conditional probability is denoted by $P(A/B)$, where A is the event for which the probability is stated, and B is the related event that influences it. Returning to the urn example above, if B_i denotes withdrawal of a black marble on the ith draw, and W_i denotes withdrawal of a white marble on the ith draw, then one can write for the example,

$$P(B_2/B_1) = 1/2, \quad P(W_2/B_1) = 1/2$$

$$P(B_2/W_1) = 1, \quad P(W_2/W_1) = 0$$

Given the meaning and notation for conditional probabilities, it is possible to write the rule of multiplication.

$$P(A \text{ and } B) = P(A)P(B/A) \tag{13–3}$$

To illustrate it with the urn example,

$$P(B_1 \text{ and } W_2) = P(B_1)P(W_2/B_1) = (2/3)(1/2) = 1/3$$

The reader should use the rule to verify that $P(B_1 \text{ and } B_2)$ and $P(W_1 \text{ and } B_2)$ are both 1/3. Note, also, that these sum to 1, as one would expect from Equation 13–1.

Although the following is not a proof, it does serve to give credence to the rule of multiplication. Suppose the three marbles are labeled as in Figure 13–2: x, y, and z. Then there are six ways in total that two marbles may be withdrawn in succession:

$$x, y \qquad x, z \qquad y, x \qquad y, z \qquad z, x \qquad z, y$$

$P(B_1 \text{ and } B_2)$ correspond to two out of those six ways, x,z and z,x. Thus, $P(B_1 \text{ and } B_2) = 2/6$, or 1/3, which agrees with that obtained under the rule of multiplication. The reader may verify $P(B_1 \text{ and } W_2)$ and $P(W_1 \text{ and } B_2)$ in a similar fashion.

The rule of multiplication has a special case also. If $P(B/A) = P(B)$ (i.e., event B is not influenced by A), then

$$P(A \text{ and } B) = P(A)P(B)$$

In such cases, A and B are said to be independent events. The urn example can be converted to one illustrating independent events if the experiment is modified so that the first marble withdrawn is replaced before proceeding. With replacement, then,

$$P(B_1 \text{ and } W_2) = P(B)P(W) = (2/3)(1/3) = 2/9$$

Once again, the reader may make the rest of the probability computations for sampling two marbles, this time with replacement. From Equation 13–1, these should, of course, add to one. Don't forget that there is now one more compound event than before. (This should be verified before proceeding.)

The use of the multiplication rule can be illustrated with some simple examples.

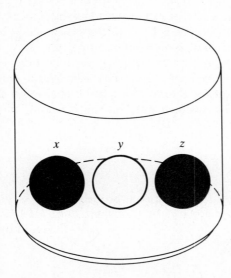

FIGURE 13–2   Urn, Two Black Marbles, One White Marble

EXAMPLE 13–1

In a production lot of 100 integrated circuits, suppose 4 percent, or four chips, are defective. If three chips are drawn randomly from the entire lot without replacement, what is the probability that one or more of the sampled items are defective? Denote the event, drawing a defective, as D, and the event, drawing a good item, as G. Represent the conditional probability of drawing a good chip on the ith draw as $P(G_i/B)$, where B is a conditional event, and, similarly, the probability of drawing a defective, as $P(D_i/B)$. Also denote the probability of receiving a given number of defectives or good items in three draws as $P(D = a)$ and $P(G = b)$, where a, $b = 0, 1, 2, 3$. Then $P(D \geq 1) = 1 - P(D = 0)$, from Equation 13–1′. Since $P(D = 0)$ equals the probability of drawing three good chips in succession,

$$P(D \geq 1) = 1 - P(G_1)P(G_2/G_1)P(G_3/G_1 \text{ and } G_2)$$

$$= 1 - (96/100)(95/99)(94/98)$$

$$= 1 - .884 = .116$$

EXAMPLE 13–2

Suppose the preceding problem is modified so that each chip is returned to the lot before sampling again (i.e., there is sampling with replacement). Now what would be the probability of more than one defective?

Here the successive draws are independent of the preceding ones. Thus,

$$P(D \geq 1) = 1 - P(G)P(G)P(G)$$

$$= 1 - (96/100)^3$$

$$= 1 - .885 = .115$$

Note that the probabilities obtained under both sets of conditions are close to one another. This is generally true when small samples are taken from large populations. Thus, in such cases, practitioners often approximate the conditional probabilities desired with results obtained assuming independence.

Discrete Probability Distributions

A discrete probability distribution is made up of a finite set of events, denoted by X_i, $i = 1, \cdots, n$. The probabilities associated with each event are denoted by $P(X_i)$

and the events are defined so that

$$\sum_i P(X_i) = 1$$

A few examples follow.

EXAMPLE 13—3

Suppose a die is painted over so that three of the six faces on the cube are labeled "one"; two faces are labeled "two"; and one face is labeled "three." Then the probability distribution would appear as

X_i	1	2	3
$P(X_i)$	1/2	1/3	1/6

EXAMPLE 13—4

A certain part, manufactured for many years, is monitored for cosmetic defects or flaws. As many as four flaws per unit have been observed. The historical data indicate the proportion of production exhibiting a given number of flaws, and these are being used as probabilities.

X_i	0	1	2	3	4
$P(X_i)$	.90	.04	.03	.02	.01

EXAMPLE 13—5

One of the first state lotteries in modern times worked as follows: The player purchased a prenumbered ticket for $0.50. The number contained six digits. After the weekly close of ticket sales, a six-digit number was randomly drawn. A different prize was awarded for each of the following outcomes: (1) matching all six digits, (2) matching the last five digits only, (3) matching the last four digits only, (4) matching the last three digits only, and (5) matching the last two digits only. What were the probabilities associated with each of the winning outcomes? What was the probability of not winning a prize? Write this as a discrete probability distribution.

The probability of matching the first digit is .1. Since the generation of the subsequent digits of the winning number is independent of the previous ones, the probability of matching any two digits is $(.1)^2$. Similarly, the probability of matching all six digits is $(.1)^6$ or one in a million.

The probability of matching the last five digits only might at first seem to be $(.1)^5$. However, this overlooks the fact that one in ten of such matches would also have matched all six digits, and, thus, would be in a different prize category. Therefore the probability of winning the prize for matching the last five digits only would be $(10-1)/10^6$ or nine in a million. The rest of the "winning" probabilities may be reasoned out in a similar pattern.

Finally, the probability of not winning may be most easily evaluated by Equation 13–1′, or one minus the sum of the "winning" probabilities. Thus, the probability distribution may be written as

X_i	$P(X_i)$
Match all six.	$1 \times 10^{-6} = .000001$
Match last five only.	$9 \times 10^{-6} = .000009$
Match last four only.	$90 \times 10^{-6} = .000090$
Match last three only.	$900 \times 10^{-6} = .000900$
Match last two only.	$9000 \times 10^{-6} = .009000$
Don't win.	$1 - \Sigma$ above $= .99$

Since it paid prizes so infrequently (about one time in a hundred, overall) it should not be surprising that this lottery was not popular and was subsequently dropped in favor of a different format.

Expected Values

A measure of central tendency of a probability distribution is provided by its expected value, defined as

$$E(X) = \sum_i X_i P(X_i) \tag{13–4}$$

The expected value, or expectation, as it is sometimes called, gives the long-term tendency for the process if the experiment described by the probability distribution is repeated many times. In other words, the more the experiment is repeated, the more the tendency would be for the long-run average of the process to approach the expected value.

The computation and interpretation of expected value is illustrated by returning to the probability distributions of Examples 13–3 through 13–5.

EXAMPLE 13–6

For Example 13–3, the distribution was

X_i	1	2	3
$P(X_i)$	1/2	1/3	1/6

Its expected value would be

$$E(X) = 1(1/2) + 2(1/3) + 3(1/6) = 5/3 = 1.67$$

The interpretation of that expected value is that if such a die were rolled repeatedly, about half the rolls would come up "one," about one-third would come up "two," and about one-sixth "three." Those proportions would tend to give a long-term average approaching 5/3.

EXAMPLE 13—7

The distribution of Example 13—4 involved the cosmetic blemishes of a particular manufactured part. The distribution was

X_i	0	1	2	3	4
$P(X_i)$	.90	.04	.03	.02	.01

This would give an expected value of

$$E(X) = 0(.9) + 1(.04) + 2(.03) + 3(.02) + 4(.01) = .2$$

The interpretation here is that the average number of blemishes per part would approach .2 over the long term.

EXAMPLE 13—8

The lottery Example 13—5 did not indicate the "payoffs" for winning. These must now be given if the expected value is to be computed. The prizes for that lottery were as follows:

X_i Description	X_i Payoff
Match all six.	$50,000
Match last five only.	$5,000
Match last four only.	$500
Match last three only.	$50
Match last two only.	$2.5
Don't win.	0

Recall that each ticket cost $0.50. Thus, to compute the expected value, the ticket price would need to be "netted out." That would give an expected value computation of

$$E(X) = (49,999.5)(10^{-6}) + (4,999.5)(9 \times 10^{-6}) + \cdots$$
$$+ (2.0)(9,000 \times 10^{-6}) + (-.5)(.99)$$

Since this is awkward, some may prefer to use the following theorem to simplify the computation: If Y is a random variable expressed in terms of the random variable X as $Y = X + k$, where k is a constant, then $E(Y) = E(X) + k$. Thus, since each ticket must be purchased for $0.50,

$$E(X) = -.5 + (50,000)(10^{-6}) + (5,000)(9 \times 10^{-6}) + \cdots$$
$$+ (2.5)(9,000 \times 10^{-6}) + (0)(.99)$$

In either case, the expected value turns out to be $-$0.2925. The interpretation of that result is that the long-run average of outcomes for a player would tend to approach $-$0.2925 times the number of plays ever more closely, the more the person plays. (That the result is negative may at first be surprising to the reader. However, negative expected values for the players are common in such games of chance. That people play is due to a number of factors, including the low economic value of the $0.50 ticket price as compared to the relatively high psychic value of the prospects of winning a big prize, no matter how low the chances of such an outcome.)

Furthermore, upon reflection it should be clear that the expected value for the lottery commission is $+$0.2925 per ticket! The expected value measurement is particularly useful for the lottery commission, since it sells many tickets per week. Thus, the expected value gives a very close indication of the net ticket revenues (i.e., revenues less prize payouts), based on the number of tickets sold. For example, sales of a million tickets per week would yield an approximate net weekly revenue of $(1,000,000)(0.2925)$ or $292,500.

Applications of Probability Concepts

In the sections that follow, risk situations are analyzed using the probability concepts outlined to this point.

Decision Matrices

Often where decision-making is undertaken with risk, one can identify four components to the situation: the decisions or actions, the states of nature, the probabilities of the states of nature, and the payoffs. These are described as follows:

Decisions or Actions These refer to the array of alternatives that the decision-maker identifies. Identifying and describing the alternatives are crucial, since it is possible to overlook promising actions. Therefore especial care is often taken to identify the alternatives to be considered—for instance, by conducting preliminary studies or arranging "brainstorming" sessions, including people in the organization who are directly and indirectly involved in the decision. Examples of actions include the amount to bid on a particular contract, and the various design configurations to be considered in a flood-control project.

Sometimes the process of determining the actions resembles aspects of the modeling process in that some simplifications are needed. For example, where a continuum of values for an action is possible, one may distill those to a finite set of actions, as in the bidding problem alluded to above. Although any dollar figure within a given range is possible, it is conceivable that one might want to consider a small number of bids.

States of Nature These are the external factors bearing on the eventual outcomes that are not under the control of the decision-maker. In fact, this is the major source of risk in the decision. For example, there may be three firms competing for a design, fabrication, and delivery contract for a certain product. In that situation, one might identify three states of nature: SN_1 referring to the decision-maker winning the contract; SN_2 and SN_3 referring to each of the other competitors winning.

Here, too, it is often necessary to simplify a complex assortment of states of nature in order for an analysis to be practical.

Probabilities of the States of Nature Although the meaning is self-explanatory, estimating the probabilities poses great difficulties. Most often, historical probabilities are used, although sometimes there is no other choice but to rely on subjective probabilities, which have gained more acceptance in recent years. As an example, suppose one categorizes five states of nature in relation to potential flooding of a particular waterway:

SN_1 = no flood,
SN_2 = up to 2 feet above flood stage,
SN_3 = greater than 2 feet, less than or equal to 5 feet above flood stage,
SN_4 = greater than 5 feet, less than or equal to 10 feet above flood stage, and
SN_5 = greater than 10 feet above flood stage.

The probabilities of these states of nature are often expressed in terms of the chance of occurrence in a particular year. Historical data over 100 years may indicate that in 80 years (corresponding to 4 years out of 5 or .8 probability) there was no flooding at all. In 9 instances in 100 years, flooding occurred at a height of up to 2 feet above flood stage. That would correspond to a probability of .09. The rest of the probabilities can be estimated similarly and are presented below.

SN_i	1	2	3	4	5
$P(SN_i)$	.80	.09	.05	.04	.02

Payoffs The payoffs are estimates of the quantitative results associated with each action, conditional upon the outcome of a particular state of nature. Thus, for example, the construction of a retention channel of a given configuration may be considered for the flood-control illustration above. It should be possible to make estimates of the damage (damage being a negative payoff) associated with each of the previously enumerated states of nature, given the construction of the retention channel. A table of such payoffs is frequently made. An example of one is shown below.

Payoffs (Damage, $000s)

	SN_i				
	1	2	3	4	5
Action:					
Retention					
Channel	0	500	700	1000	1500

Once all the components are described, it is possible to evaluate the various decisions or actions. Of course, there must be a criterion to which each may be compared. One such criterion is to choose the action that maximizes the expected value of the payoffs. Example 13–9 illustrates how this may be done.

EXAMPLE 13—9a

A magazine publisher mails most of its printed copies to subscribers, with a relatively small proportion of production being sold on local newsstands. It is this incremental aspect of production which is "discretionary" and which is to be analyzed. Data are available on newsstand demand (simplified so that only a small number of events need to be considered). Newsstand demand in thousands over the past 24 months is as follows:

Newsstand Sales

Sales (in thousands)	5	6	7	8	9
Frequency (in months)	9	6	4	3	2

Assuming that future sales patterns are expected to be about the same as were recently experienced, these can be converted to probabilities for the state of nature, "newsstand demand."

Newsstand Demand

Demand (in thousands)	5	6	7	8	9
Probability	.375	.25	.167	.125	.083

After careful study, the industrial engineering department indicates that the incremental cost of production and distribution is $0.40 per copy ($400 per thousand). The unit revenue from newsstand sales is $1.10 per copy ($1,100 per thousand). Unsold copies (to the newsstands) may be disposed of by selling them in bulk to a paper recycler for the equivalent of $.02 per copy ($20 per thousand).

Assuming that the publisher wishes to have a standing production order for newsstand sales for the near future, how much should it be?

A decision matrix indicating the possible actions (production levels for newsstand sales), states of nature (newsstand sales), probabilities of states of nature, and payoffs appears in Table 13–1. Since sales vary from 5,000 to 9,000, production of no fewer than 5,000 and no more than 9,000 is being considered. The reader should verify all of the computations of the payoffs in the cells of the matrix. Some sample computations are:

For producing 5,000 copies, the payoff is 1,100(5) − 400(5), or $3,500. Since larger demand represents only potential and not actual sales, and because back-order sales are not permitted in this example, the payoff is the same under all states of nature for the decision: "produce 5,000."

For producing 6,000, with a corresponding state of nature: demand is 5,000. Here the payoff is 1,100(5) − 400(6) + 20(1), or $3,120. Note that this payoff is $380 less than that for producing 5,000 when demand is 5,000. This is because 1,000 must be disposed of at a loss of 400 − 20, or $380.

For producing 6,000 with a corresponding demand of 6,000, the payoff is 1,100(6) − 400(6), or $4,200.

TABLE 13–1   Decision Matrix, Payoffs

Probability		.375	.25	.167	.125	.083
SN$_i$, Demand (in thousands)		5	6	7	8	9
A						
C	5	$3,500	$3,500	$3,500	$3,500	$3,500
T	6	3,120	4,200	4,200	4,200	4,200
I	7	2,740	3,820	4,900	4,900	4,900
O	8	2,360	3,440	4,520	5,600	5,600
N	9	1,980	3,060	4,140	5,220	6,300
S						

As the reader verifies the entries of the payoff table, he should notice that it forms a pattern. The diagonal elements are (1,100 − 400) times sales; vertical entries decline by $380 for each increment of production, as indicated in one of the sample computations; horizontal entries to the left of the diagonal increase by $1,080 for each increase in sales, owing to increase in revenue of $1,100 with a corresponding decline in "salvage revenue" of $20; horizontal entries to the right of the diagonal remain the same for each increase in potential demand, as shown in one of the sample computations.

Given the payoffs in the table and the probabilities associated with each state of nature, it is possible to compute the expected values corresponding to each production action or decision. For example, the expected value for the decision to produce 7,000 copies would be

$$E(X = 7) = 2,740(.375) + 3,820(.25)$$

$$+ 4,900(.167 + .125 + .083) = \$3,820$$

The computation of the expected values for the other possible actions is similar and appears in Table 13–2.

TABLE 13–2   Decision Matrix, Payoffs, Expected Values

Probability		.375	.25	.167	.125	.083	
SN$_i$, Demand (in thousands)		5	6	7	8	9	E(X)
A							
C	5	$3,500	$3,500	$3,500	$3,500	$3,500	$3,500
T	6	3,120	4,200	4,200	4,200	4,200	3,795
I	7	2,740	3,820	4,900	4,900	4,900	3,820
O	8	2,360	3,440	4,520	5,600	5,600	3,666.5
N	9	1,980	3,060	4,140	5,220	6,300	3,374.3
S							

In this example, the highest expected value corresponds to a decision to produce 7,000 copies per month for newsstand distribution. The value of the approach and the result may be better appreciated after considering the following facts:

A naive approach might have considered demand only. The expected value of demand is

$$E(D) = 5(.375) + 6(.25) + 7(.167) + 8(.125) + 9(.083) = 6.29 \text{ thousand}$$

Thus, rounding down to 6,000 would have been a likely conclusion if only demand were taken into account. It is clear from the expected values of the payoffs that such a decision would yield a less profitable result in the long run. In other words, analysis of the payoffs includes the important consideration of the revenue- and cost-generating characteristics of each possible action. Also, it is worth noting that maximizing expected value is quite a good criterion in this case, since the situation will be repeated often. Recall that expected value indicates the long-term tendency of a repetitive process, and it is thus an apt device here.

Other Criteria   The analyst may choose to base his decision on a criterion other than maximizing expected value. The following section outlines several possibilities that have been described in the decision analysis literature. Since these criteria do not consider the probabilities involved, they may be applied more generally, as under the condition of uncertainty, where the probabilities of the states of nature cannot be estimated.

Maximin criterion. This one considers the worst outcome for each decision or action, selecting that action with the most advantageous "worst outcome." For Table 13–2, the maximin criterion would lead to a decision to produce 5,000 units, since the payoffs of the 5,000 demand–state-of-nature column give the "worst outcomes," and producing 5,000 yields the best of these results, a payoff of $3,500.

This criterion leads to a very conservative approach, one that focuses on the worst that can happen. When it is applied to a situation where the payoffs are costs, it may be referred to as the minimax criterion, where the maximum cost or loss is to be minimized.

Maximax criterion. This one considers the best outcome for each decision or action, selecting that action with the most advantageous "best outcome." For Table 13–2, the maximax criterion would lead to a decision to produce 9,000 units, since the payoffs of the 9,000 demand–state-of-nature column give the "best outcomes," and producing 9,000 yields the best of these results, a payoff of $6,300.

The use of this criterion is indicative of a very optimistic viewpoint, one that focuses on the best that can happen.

Hurwicz criterion. This criterion falls between the extremes of the two above. A linear combination of the best and worst payoff of each decision or action is computed, according to the following rule:

$$\alpha BP_i + (1 - \alpha)WP_i$$

where BP_i is the best payoff for the ith decision or action, and WP_i is the worst payoff for the ith decision or action. Then the decision or action with the largest value for $\alpha BP_i + (1 - \alpha) WP_i$ is chosen.

The coefficient α is referred to as the coefficient of optimism. The closer α is to 1, the more the decision approaches that yielded by the maximax criterion; the closer it is to 0, the more it approaches that yielded by the maximin criterion. The reader should verify that if $\alpha = 1$, the Hurwicz criterion becomes the maximax criterion; if $\alpha = 0$, the Hurwicz criterion becomes the maximin criterion.

An Alternative Decision-matrix Approach   Some analysts prefer to construct the decision-matrix payoffs in terms of the applicable opportunity costs, referred to here as opportunity losses. In this context, the opportunity cost is the amount forgone if the action taken is less than the most advantageous.

EXAMPLE 13—9b

One can convert Table 13–2 to one containing opportunity losses, which follows. In the first column, which shows the payoffs for the actions associated with the outcome "demand is 5," the action "produce 5" yields the best economic result. Therefore, the opportunity-loss table assigns zero cost to that action–state-of-nature combination. The other actions, if taken when the demand is 5, yield less desirable economic returns, and thus show positive opportunity losses. Note that the opportunity losses for a given action–state-of-nature combination are the difference between that payoff and best payoff for that state of nature. Table 13–3 shows both the opportunity losses associated wih each action and their expected values.

To illustrate the computation of the expected values, consider production of 7,000 copies.

$$E(X=7) = 760(.375) + 380(.25) + 0(.167)$$
$$+ 700(.125) + 1,400(.083) = \$583.7$$

TABLE 13—3   Decision Matrix, Payoffs, Expected Values   (Opportunity-cost or -loss Approach)

Probability		.375	.25	.167	.125	.083	
SN, Demand (in thousands)		5	6	7	8	9	E(OL)
A							
C	5	$ 0	$ 700	$1,400	$2,100	$2,800	$ 903.7
T	6	380	0	700	1,400	2,100	608.7
I	7	760	380	0	700	1,400	583.7
O	8	1,140	760	380	0	700	739.06
N	9	1,520	1,140	760	380	0	1,029.42
S							

The optimal action, produce 7, corresponds to the lowest expected value of opportunity losses. Note that the opportunity-loss approach is consistent with the result obtained when profit contributions were analyzed.

Expected Value of Perfect Information

The decision-matrix model of a risk problem can be used to explain an important concept in risk analysis: the expected value of perfect information. Example 13–9 will be used for that purpose.

To suppose that the publisher has perfect information is tantamount to saying that he can predict weekly newsstand demand with certainty. Of course, knowing demand for the coming week, he would not be content to simply maintain a "standing order" for a constant number of copies; instead, if profits are to be maximized, he would produce exactly the number of copies to be demanded, as the diagonals of Tables 13–2 and 13–3 indicate. However, the probabilities of a given demand for a week remain the same—that is, the proportion of weeks in which demand would be 5,000 would continue to be .375, and so on. Thus, the expected value of the payoffs with perfect information (i.e., the long-term tendency for the average payoff with certain knowledge of the future), would be

$$E(PI) = 3,500(.375) + 4,200(.25) + 4,900(.167)$$
$$+ 5,600(.125) + 6,300(.083) = \$4,403.7$$

where $E(PI)$ is the expected value with perfect information. This results from the fact that 37.5 percent of the time demand would be 5,000; he would put in a production order for 5,000; and the payoff would be $3,500. In 25 percent of the time demand would be 6,000; he would put in a production order for 6,000; and the payoff would be $4,200, and so on.

Considering that without such perfect information the best policy would be to put in a standing order for 7,000 with a corresponding expected value (i.e., long-term payoff) of $3,820, it can be argued that in the long run, perfect information is worth the difference between $E(PI)$ and $E(X)$, or $583.7.

This example can be generalized. The expected value of perfect information, *EVPI,* can be written as

$$EVPI = E(PI) - E(X) \tag{13-5}$$

where $E(PI)$ is as defined above, and $E(X)$ is the expected value corresponding to the best policy in the absence of perfect information.

This concept can be used to evaluate the usefulness and value of information in real situations. Thus, for example, it can be safely stated that the absolute limit on what the publisher would be rationally willing to pay for information is provided by the *EVPI.* In fact, he would pay less for imperfect information about future demand.

Decision Trees

Another approach in analyzing risk situations with expected values utilizes decision trees. These can be helpful when there is a sequence of decisions to be made. Such would be the case when a choice of one action in combination with a particular state of nature, yielding a given outcome, presents a number of other decision options. In such situations, the decision-matrix approach would be inadequate.

Decision trees schematically represent the sequence of decisions, states of nature, and other subsequent decisions and states of nature. This is done in a network or graph-node format, which will be illustrated with some examples. Probabilities of the states of nature and payoffs are depicted also. Decisions are represented by decision nodes or points as small boxes from which emanate branches representing the available choices or decisions. For Example 13–9, the five production choices would be represented by a decision node with five branches, one for each production decision.

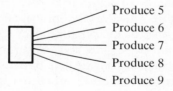

The possibility that different states of nature can occur is represented in a chance node. It contains a circle with branches emanating from it, each branch corresponding to a different state of nature. The probabilities can be indicated along the branches or with the state-of-nature label.

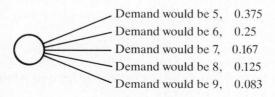

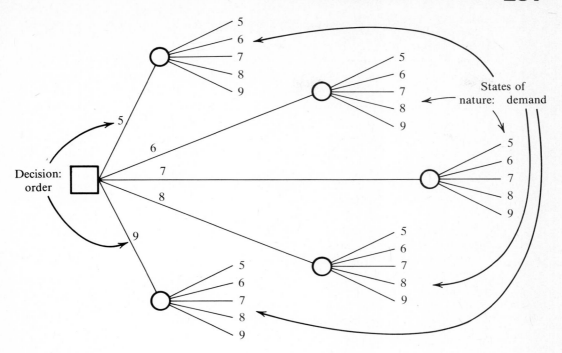

FIGURE 13–3

The most common approach is to begin with the first decision node at the left, laying out the states of nature for the first decision to the right, proceeding with subsequent decisions and states of nature further to the right. This gives rise to a treelike structure: hence the name. Payoffs are generally indicated at the extremities of the branches. When the payoffs are listed at the extremities, they represent net amounts, based on all the investments, costs, and revenues specified by the entire sequence of decisions and states of nature leading to that payoff. Furthermore, since decision trees typically deal with time horizons of several years or more, the payoffs should also be computed using the concepts developed in this book. In other words, they should be net present values (or equivalent annual worths) computed on an after-tax basis and adjusted for inflation, where applicable.

A decision tree for Example 13–9 would look like Figure 13–3.

The computation of expected values in a decision tree is often referred to as "folding back." Although Example 13–9 can be folded back, it is a single-stage situation, and so does not provide a representative framework for the generalized decision-tree problem. Therefore, another example will be used.

EXAMPLE 13–10

A small automobile manufacturer will have sufficient resources to do product design, testing, and production design and planning for only one new car at a time over the next six years. The process will take about three years for each new car introduced. Thus, it can introduce two new cars over the next six years. It is felt that a new compact, (C), and subcompact, (S), should be introduced to satisfy the changing market. The decision being faced is whether to introduce the compact first, followed by the subcompact, or vice versa. A study is done in which a decision tree is used to depict the choices faced, the states of nature thought to be applicable, the probabilities of those states of nature, and the payoffs associated with the decision—state-of-nature combinations. The analysis of the decision tree (Figure 13–4) involves "folding it back" or determining which decision yields the highest expected value of the payoffs.

In Figure 13–4 both the decision nodes and chance nodes are numbered for identification purposes. Node 1 represents the initial decision, whether to introduce a compact or a subcompact first. Node 2 represents the states of nature possible after three years if a compact is initially introduced: sales may be above target (T^+), on target (T), or below target (T^-).

The estimated probabilities are indicated along each branch. Although nodes 4, 5, 7, and 8 are treated as decision points, the fact that only one branch emanates from each indicates no real choice at that point. For example, node 4 shows that, having introduced a compact first, and having indications that sales will exceed targets, the firm would go ahead with development of the subcompact. On the other hand, nodes 6 and 9 show that, with indications that sales will be below target for the configuration chosen initially, there are two choices that would be entertained. For instance, node 6 shows that, having first introduced a compact with initial sales below target, the firm could choose between planning a subcompact or a newly designed compact car.

The states of nature for nodes 10 to 17 are structured like those for nodes 2 and 3. However, in this example, the probabilities and payoffs depend on the previous conditions and their ordering. In other words, $P(T^+)$ at node 10 is not the same as $P(T^+)$ at node 11 or node 2; it is a conditional probability thought to be applicable to this situation. (In other situations the probabilities of the states of nature at the branch extremities may be the same. See Figure 13–5 in Example 13–11.) Also, the payoff of 100 corresponding to branch 1–2–4–10 is not the same as the payoff of branch 1–3–7–14. In this case, it was felt that a compact, being more profitable than a subcompact at a given target level of sales, would provide a greater payoff if it preceded the introduction of the subcompact. Recall that each payoff represents a discounted cash flow, either a present worth or periodic worth, computed on an after-tax basis, and possibly adjusted for the anticipated effects of inflation.

Given the decision tree, the next task is to compute the expected values of the decisions at node 1 by folding back. Thus, the expected value of node 10 would be given by,

$$E(10) = 100(.3) + 90(.5) + 50(.2) = 85$$

where $E(10)$ represents the indicated expected value. The reader should verify the expected values for nodes 11 to 17. The expected values appear in the boxes adjacent to the nodes. Folding back amounts to assigning the expected

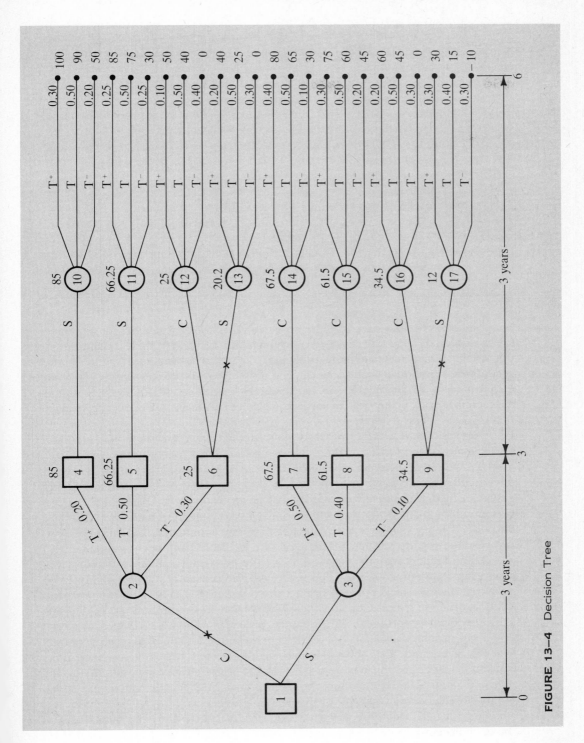

FIGURE 13—4 Decision Tree

value of 85 to decision node 4. Similarly, the values for nodes 5, 7, and 8 may be filled in. As for node 6, it is a decision point with two choices or branches. One has an expected value of 25, the other, 20.5. A decision-maker wishing to maximize the expected value of the payoffs would choose the decision branch with the higher expected value. He then assigns its expected value to the decision node. Given the data here, this is equivalent to saying that having first introduced a compact, and having experienced below-target initial sales, one would design another compact rather than a subcompact. This is emphasized in the diagram by "crossing out" the decision branch with the lower expected value(s). The reader should verify this procedure by examining node 9.

Once this has been done, the folding back may continue to the preceding stage. The expected value of node 2 is found from

$$E(2) = 85(.2) + 66.25(.5) + 25(.3) = 57.625$$

Similarly, $E(3)$ is 61.8. Thus, since the branch corresponding to producing the subcompact first has the higher expected value, it would be chosen by an expected-value maximizer.

Since the decision tree uses the expected-value criterion, results can sometimes be unexpected. Here, for example, although the compact car is thought to be potentially more profitable, the probability of its achieving satisfactory market acceptance offsets that feature. In other words, although the subcompact may not contribute as much profit, given the most desirable market conditions, those conditions are such that the subcompact, having greater chances of satisfactory market acceptance, is the sensible first choice.

Note that the decision tree does not purport to map out the entire sequence of decisions at the outset. If the decision-maker is an expected-value maximizer, the tree would be used for the initial decision only. Then, after time has passed, the situation would need to be re-evaluated, on the basis of the information then available. In this example, this means that no decision for the second three-year period will be made now. That will be re-examined after the subcompact has been introduced, after a sense of the market at that time is obtained, and after the costs and revenues of the various alternatives then available have been ascertained.

Some readers may be skeptical about the practicality of this tool. However, where large investments are to be made and the pitfalls are consequently large, industrial decision-makers try to bring analytical methods to bear on the problems faced in order to gain the greatest possible insights. The industrial use of decision trees has been reported in the literature. (See the references for this chapter.) Computer programs have been written to perform the folding back. (See Appendix E.) (However, note that the computation of expected values is the least time-consuming aspect of the whole process. Studying the alternatives, identifying the states of nature, and estimating the probabilities and payoffs under the stated assumptions take the vast bulk of the analyst's time.)

Even where decision trees are not practical, owing to the difficulties of estimating probabilities or payoffs, it is possible that the schematic representation of decisions, states of nature, and ensuing decisions and states of nature, may be helpful in clarifying the possibilities and identifying potential pitfalls.

Bayesian Analysis

Bayes's Rule permits computation of conditional probabilities based on other probabilities. This can sometimes be useful in the context of a decision-tree analysis and will be illustrated later in this section. Before this can be done, however, it will be necessary to introduce Bayes's Rule. As with the other material in this chapter, the approach will be informal, and will draw on illustrative examples to develop the concepts.

Bayes's Rule   Bayes's Rule may be thought of as an extension of the rule of multiplication. From Equation 13–3,

$$P(A \text{ and } B) = P(A)P(B/A)$$

One may also write it as,

$$P(A \text{ and } B) = P(B)P(A/B)$$

(It was not also stated this way earlier in the chapter because the illustrative example of drawing marbles from an urn used events defined sequentially. The reader should verify to himself that the "flipped" condition, (A/B), makes no sense in the context of the urn example.)

Permitting $P(A \text{ and } B)$ to be written both ways allows the statement

$$P(A)P(B/A) = P(B)P(A/B)$$

Suppose that one wishes to compute $P(A/B)$. One can easily obtain that from the statement above by dividing both sides by $P(B)$.

$$P(A/B) = \frac{P(A)P(B/A)}{P(B)} \tag{13–6}$$

Finally, the denominator, $P(B)$, can be "segmented," as the following illustration will suggest:

Suppose a manufacturer produces three models of a given product, X, Y, and Z. Quality control inspections classify the output as defective, D, or as nondefective, N. All three models are produced simultaneously in the plant. Six hundred items are sampled for a quality control study. Although the items have not yet been inspected, suppose that the distribution of defective and good items for the different models are as specified in Table 13–4.

Designate the probability of randomly drawing a defective item as D, and a good item as N. Designate the random selection of models X, Y, or Z with those letters. Conditional events and probabilities may also be designated. For example, $P(D/X)$ represents the conditional probability of drawing a defective, given that it is of model type X. From the table it is easy to visualize that $P(D/X)$ would be $3/200$. In order to see what is meant by "segmenting" the probabilities, first look

TABLE 13–4 Quality Distribution by Model

	Model			
	X	Y	Z	Total
Defective	3	2	5	10
Nondefective	197	298	95	590
	200	300	100	600

at the table to see the unconditional probability of drawing a defective. $P(D) = 10/600$. Then one may reason as follows: the total number of defectives is contributed by each model, and each model is present in differing proportions. The probability of drawing a defective, then, may be thought of as a weighted average, depending on the relative weight of each model and the percent defective in each. Thus, the proportion of defectives may be alternatively written as

$$P(X)P(D/X) + P(Y)P(D/Y) + P(Z)P(D/Z)$$

Evaluating this for our example,

$$\frac{200}{600} \times \frac{3}{200} + \frac{300}{600} \times \frac{2}{300} + \frac{100}{600} \times \frac{5}{100} = \frac{10}{600}$$

Of course, this agrees with the result obtained before, dividing the total number of defectives by the total number of items.

This example may be generalized so that one can write

$$P(B) = P(N_1)P(B/N_1) + P(N_2)P(B/N_2) + \cdots + P(N_n)P(B/N_n)$$

which is the "segmentation" of $P(B)$ referred to previously. (For our example, B corresponds to the event, drawing a defective; N_i, $i = 1, 2, 3$, corresponds to models X, Y, and Z.)

Using the above "segmentation" of $P(B)$ in Equation 13–6, one obtains

$$P(A_i/B) = \frac{P(A_i)P(B/A_i)}{P(A_i)P(B/A_i) + \cdots + P(A_n)P(B/A_n)} \qquad (13\text{–}7)$$

which is Bayes's Rule. As can be seen from Equation 13–7, Bayes's Rule can be used to compute "flipped" conditional probabilities, when conditional probabilities are estimatable in the "unflipped" direction. In other words, given the conditional probabilities, $P(B/A_i)$, and the unconditional probabilities, $P(A_i)$, one can find $P(A_i/B)$.

In order to see how Bayes's Rule can be put to practical use, consider the following petroleum exploration problem. Although it is an oversimplified portrayal of the general problem, it provides a suitable illustration of the underlying probability concepts.

Example 13–11

Suppose that an oil-discovery division of a large company acquires exploration and exploitation rights to a large tract of land. Although the tract is deemed promising, this does not mean that the chances are high that a productive well can be found at any site. In fact, it is felt that the chances that a productive well will be drilled at any particular spot, labeled O, is only .1. The probability of a nonproductive or dry hole, labeled D, is .9. Simplifying the payoffs, it is felt that the average productive well would have a payoff of 400, while a dry hole would involve a payoff of -50. If no drilling takes place the payoffs are represented as -1, whether or not there is oil in the ground. (This negative payoff is due to the expense in acquiring the rights.) Table 13–5 provides a summary as a decision matrix.

TABLE 13–5   Decision Matrix, Payoffs

| | States of Nature | | Expected Value |
	O	D	
Drill	400	-50	-5
Don't Drill	-1	-1	-1
	.1	.9	
	Probabilities of States of Nature		

The reader should verify that the expected value of the payoff for drilling is less than for not drilling. Thus, an expected-value maximizer would not venture to drill. Of course, one way to look at this situation is that the very nature of the business involves risk-taking, and high risk at that. So one might say that oil drillers are not expected-value maximizers. But is this really so? To answer this question we shall explore the possibility that additional information will help make a better decision. In doing so, Bayes's Rule will be used. To begin that process, the expected value of perfect information can be computed, which will give an upper limit for what one would be willing to pay for real, imperfect information:

$$EVPI = [400(.1) + (-1)(.9)] - (-1) = 40.1$$

Now suppose that some topographical/geological testing may be done at individual sites which will provide additional helpful but imperfect information about the likelihood of finding oil there. The cost of the testing is 5, in the units of the problem. Since this is less than the *EVPI*, it surely is worth considering. Table 13–6 gives data on the historical reliability of the testing procedure.

The interpretation of the conditional probabilities may be seen from column O as follows: where such testing has been done before, in all instances where oil has been discovered, the test correctly predicted oil 70 percent of the time; it incorrectly predicted no oil 30 percent of the time.

TABLE 13—6   Previous Experience with the Test

		Actual States of Nature	
		O	D
Survey Predicts	O	.7 $P(O_p/O) = .7$	.1 $P(O_p/D) = .1$
	D	.3 $P(D_p/O) = .3$	.9 $P(D_p/D) = .9$

O_p = outcome, "oil is predicted"
D_p = outcome, "oil is not predicted"

These conditional probabilities indicate that the test has been a fairly good indicator of whether oil is present in exploitable quantities. Using the test might involve a simple decision rule, such as "if the test predicts *O,* drill; if not, don't." Assuming that the test continues to predict as well, it would be desirable to evaluate, in advance, the risks involved in conducting the test and in acting on its recommendations, and the expected values of the new testing-decision process. In other words, conditional probabilities like $P(O/O_p)$ are wanted. Since these conditions are "flipped" in relation to the given historical conditional probabilities of the test, Bayes's Rule can be used to compute them.

Figure 13—5 shows a decision tree that incorporates an evaluation of the test and computation of the flipped conditional probabilities.

Decision node 1 indicates a choice between expending the funds (5) for additional information and deciding without the information. The latter is equivalent to the decision matrix of Table 13—5 and has an expected value of -1, shown at node 2. The lower branch of decision node 1 depicts the option of testing. Node 5 is a chance node, the branches representing the outcomes that oil may or may not be predicted, and labeled O_p and D_p. Node 6 represents the choice that would exist, given a test predicting oil: one could drill or not. If one drills, one faces the same states of nature as before, oil or a dry hole. Although the payoffs for both of those states of nature are unchanged, one now needs to compute the probability associated with each of the branches. For branch O, it is $P(O/O_p)$. For branch D, it is $P(D/O_p)$. These may be computed from Bayes's Rule (Equation 13—7) as follows:

$$P(O/O_p) = \frac{P(O)P(O_p/O)}{P(O)P(O_p/O) + P(D)P(O_p/D)}$$

$$= \frac{.1(.7)}{.1(.7) + .9(.1)} = .4375$$

$$P(D/O_p) = \frac{P(D)P(O_p/D)}{P(O)P(O_p/O) + P(D)P(O_p/D)}$$

$$= \frac{.9(.1)}{.1(.7) + .9(.1)} = .5625$$

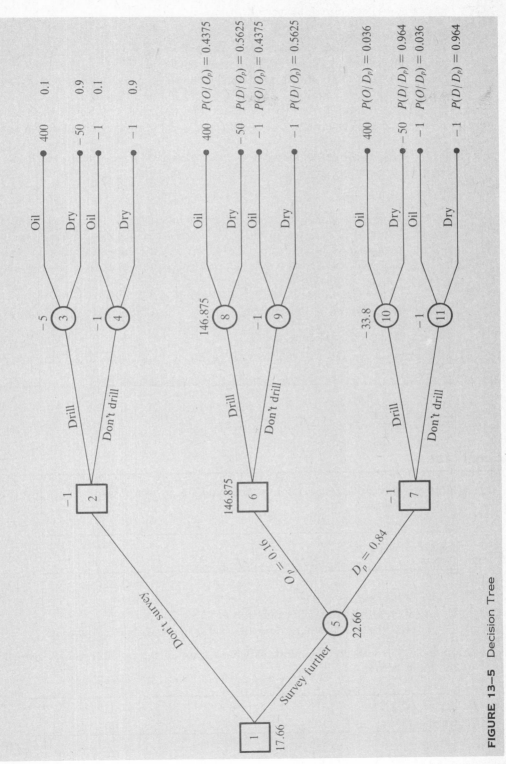

FIGURE 13–5 Decision Tree

In evaluating Equation 13–7, $P(O)$ and $P(D)$ are referred to as the a priori probabilities, in that they relate to the initial estimates provided. $P(O_p/O)$ and $P(O_p/D)$ come from the historical probabilities recorded for the test. $P(O/O_p)$ and $P(D/O_p)$ are referred to as revised probabilities. The reader should verify the values for the revised probabilities, $P(O/D_p)$ and $P(D/D_p)$. Note that the denominator, $P(O)P(O_p/O) + P(D)P(O_p/D)$, is $P(O_p)$, which will be put to use later.

The decision tree can now be folded back. For example, node 8 has an expected value of

$$E(8) = 400(.4375) + (-50)(.5625) = 146.875$$

The comparison of this value with that for not drilling, -1, shows that the decision rule "drill if the test predicts oil" is consistent with maximizing expected value. In analogous fashion, it can be shown that the expected value of node 7 is -1, and that the decision rule "don't drill if the test fails to predict oil" is similarly consistent with maximizing expected value. Continuing to fold back, the expected value of node 5 is 22.66 [based on $146.875(.16) + (-1)(.84)$]. Since the test costs 5, the expected value of node 1 would be the maximum of $(22.66 - 5, -1)$, or 17.66. Thus, the decision tree indicates that it is wise to get additional information and test at a cost of 5; that if the test indicates the presence of oil, to drill; and that if the test indicates no oil, then not to drill.

Exercises

13–1 Suppose X is a discrete random variable taking on the possible values given in the table below:

Random variable	X	0	1	2	3	4
Probability	$P(X)$	.1	.2	?	.2	.1

    a. What is $P(X = 2)$?
    b. Sketch the probability function $P(X)$ vs. X.
    c. Find the $E(X)$. Where does that fall in your sketch of *b* above?

13–2 Suppose Y is a discrete random variable taking on the possible values given in the table below:

Random variable	Y	0	1	2	3	4
Probability	$P(Y)$	.1	.15	?	.25	.30

    a. What is $P(Y = 2)$?
    b. Sketch the probability function $P(Y)$ vs. Y.
    c. Find the $E(Y)$. Where does that fall in your sketch of *b* above?
    d. How does this illustration differ from Exercise 13–1?

13–3 Given the data below concerning a production run, complete the table and use it to answer the questions that follow:

		Model				
		A	**B**	**C**	**D**	**Total**
C O L O R	Red		300	200	100	800
	Blue	100		200	250	700
	White	500	400	—	200	
					550	2,900

Suppose an experiment is done in which items are randomly drawn from this completed production run.
    a. What is the probability of drawing a red item?
    b. What is the probability of drawing one of model type *D*?
    c. What is the probability of drawing a blue one of type A?
    d. What is the probability of drawing one that is either red or model *D* or both?
    e. What is the probability of drawing one that is either white or model *B* or both?
    f. Given that one having just been drawn is of model *B*, what is the probability that it is red?
    g. Given that one having just been drawn is blue, what is the probability that it is model *D*?
    h. Are model and color independent characteristics? Explain.
Now suppose that two items are drawn in succession, without replacement.
    i. What is the probability of drawing two model *A*s? Two blues?
    j. What is the probability of drawing one of model *A* followed by one of model *B*?
    k. What is the probability of drawing one of model *A* followed by a white item?

13–4 Suppose five items have been removed from a production lot for testing. The items are classified as either good or defective. Given that one in the group is defective, what is the probability that the first item tested is good? The first two? Three? Four?

13–5 Long experience with stamping a particular sheet metal part indicates that the long-term average result for the length and width for every hundred items produced is the following:

		Length			
		Short	OK	Long	Total
W	Narrow	1	2	1	4
I	OK	6	86	2	94
D	Wide	0	2	0	2
T		7	90	3	100
H					

a. What is the probability of producing parts that are too long? Too short?

b. What is the probability of producing parts that are too narrow? Too wide?

c. What is the probability of producing acceptable parts?

d. What is the probability of producing parts that are either too long or too wide or both?

e. What is the probability of producing parts that are either too short or too narrow or both?

f. What is the probability of producing parts that are either too short or too wide?

g. Given that a produced part is OK in length, what is the probability that it is too narrow?

h. Given that a produced part is too short, what is the probability that it is too narrow? Too wide? OK in its width?

13–6 Suppose one has two dice. One of them is conventional (i.e., its faces are numbered 1, 2, 3, 4, 5, 6). The other is different, having its faces numbered 1, 1, 1, 2, 2, 3. Both dice are rolled simultaneously.

a. What is the probability that both show 1s? That both show 3s?

b. What is the probability that the conventional die, C, shows a 2, while the different one, D, shows a 3?

The random variable Z = the sum of the outcomes of the two dice.

c. What is the probability that $Z = 2$?

d. Find the probabilities for the remaining outcomes (i.e., $Z = 3, \ldots, 9$). Sketch the probability function $P(Z)$ vs. Z.

e. Use d to compute $E(Z)$.

f. Compute $E(C) + E(D)$. (This should agree with part e.)

13–7 A company uses a large number of similar machines. It examines its historical records to see how long they lasted. The information follows:

Years of use	5	6	7	8	9	10
Number giving that use	12	15	22	18	13	6

a. What is the probability that such a machine will last at least 9 years?

b. What is the expected life of such a machine?

13–8 Do this problem on a before-tax basis. *a.* Suppose your firm faced a one-percent chance of suffering a single $100,000 insurable loss in five years. If funds are worth 12 percent and insurance decisions are based purely on expected values, at most how much would the firm be willing to pay now to insure against such a loss?

   b. Suppose that the potential $100,000 loss is viewed as possible at any time over the five years. However, for simplicity, assume that a potential $100,000 loss exists each year with a probability of .002. If the insurance premium is prepaid on an annual basis, with funds worth the same as before, at most how much would it be worth now to protect against such a loss? Again base the calculations on expected losses.

13–9 (This is a variant of exercise 4–13.) Data for a deferred-investment alternative vs. a full-investment one is depicted in Cash Flow Diagrams D13–1 and 13–2. The time at which new capacity would be needed is represented by *t.*

D13–1

Full ($000):

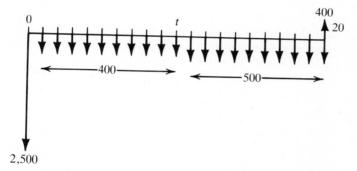

   This may be interpreted to mean that investment is $2.5 million, operating and maintenance expenses are $400,000 per year for the first *t* years, $500,000 per year for the last (20 − *t*) years, with a salvage value of $400,000 at year 20.

D13–2

Deferred ($000):

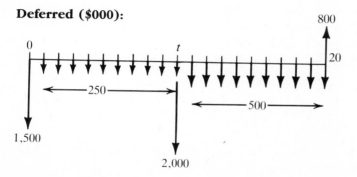

In D13–2 there is a subsequent investment of $2 million in year *t*. The other parameters of initial investment, operating expense, and salvage value also differ from the preceding diagram. If the MARR is 12 percent, which has a more desirable expected value of present worth? The probability distribution for *t* is:

t	8	9	10	11	12
P(*t*)	.40	.25	.15	.10	.10

13–10 (This is a variant of exercise 4–14.) A bridge project is being considered under the following two configurations: Under the first, a four-lane bridge can be constructed for $8 million. It is expected to serve the anticipated needs for the next 20 years. Maintenance would be $80,000 per year for the first 5 years, going up by $25,000 for the next 5 years and similarly for the remaining 10 years. The salvage value at year 20, to be used for studying this alternative, is $2 million.

The other alternative is to construct a two-lane bridge now at a cost of $5 million. Maintenance would be $55,000 per year for the first 5 years, going up to $70,000 per year until two more lanes are added, as needed (at time *t*). At the time two more lanes would be added to accommodate the anticipated growth in traffic, an additional $6 million would have to be invested. Maintenance of the combined four-lane facility would then be $130,000 after time *t* and up to year 16, and $155,000 in years 16 to 20. Salvage value for the second alternative would be $2.5 million at year 20. If funds are worth 10 percent, which alternative is superior, based on expected present worth? The probability distribution for *t* is

t	8	9	10	11	12
P(*t*)	.10	.10	.15	.25	.40

13–11 In designing a dam for flood control, a number of configurations are being considered. They are listed below in order of increasing protection against flooding. Such protection is expressed for the given structure as the probability per year of being breached, with ensuing floods. The required investment for each structure is also given, as is the expected financial loss if flooding occurs. (That the losses are not the same is due to the differential effect of a given flood under each configuration.) Assume that funds are worth 7 percent; the structures have a life for study purposes of 40 years; and a design decision is based on minimizing the total of annualized capital cost and expected yearly financial flood losses.

Configuration	P(Flood)	Expected losses	Required investment
A	.05	$ 8,500,000	$2,000,000
B	.03	11,666,667	3,000,000
C	.02	15,000,000	3,500,000
D	.01	27,500,000	4,000,000

13–12 The plant manager wants you to study the economic feasibility of purchasing a standby machine which would be part of an important process, regularly in use. A standby machine would prevent production downtime losses attributable to a breakdown. Such breakdowns have averaged $2,500 of lost production per incident. In analyzing the worth of the standby proposal, the capital cost of the standby machine will be considered but not its operating cost (since only one machine would be operating at any one time). Also to be considered is the likelihood of breakdown of the existing machine. A table of the probability distribution of breakdowns is provided below:

Number of breakdowns per year	2	3	4	5	6	7	8	9	10	11	12	13
Probability		.03	.05	.09	.13	.15	.15	.13	.10	.07	.05	.03	.02

If the pretax MARR is 18 percent, how much, at most, would the firm be willing to pay for the replacement machine? Assume that it would last for ten years with no salvage value. Assume also that the probability that the replacement machine breaks down in its intended use is negligible.

13–13 A clothing manufacturer must decide on initial production for the fall season. Although many styles are produced, a new line of suits will constitute a major part of the line. The distribution of sizes has been determined on the basis of historical patterns; however, the total number of suits to produce initially is the problem. Only a single advance production run will be made, and it must be planned 4 months ahead of time; subsequent production will be based on reorders, if the initial run is all sold. Assume that the suits will be sold to retailers in lots of 200. On the basis of available evidence for this year and past data, the manufacturer expects demand to be between 1,000 and 2,000 suits. The manufacturing cost per suit is $75; the wholesale price has been set at $125 per suit. Units not sold during the fall season will be disposed of at a price of $35 per suit. The probability that orders will be the amounts given are:

Demand	1,000	1,200	1,400	1,600	1,800	2,000
Probability	.10	.15	.25	.25	.20	.05

    a. How many suits should be made to maximize expected profit contribution?
    b. Repeat *a* with the probabilities all the same (i.e., 1/6).
    c. What would be the value of getting additional information in *a*?

13–14 An airline runs a regular flight between two cities using an aircraft with 100 seats, all with one class and fare. Although advance purchasers of tickets generally do use them for the intended flight, in some instances they do not. It has been customary for the line to refund passengers' fares for unused tickets, and this remains company policy. That being the case, the airline faces a dilemma concerning its booking policy. If it books a maximum of 100 seats, then in the long run some revenue will be lost due to no-shows. If it overbooks, there will be less of a tendency to suffer financial loss from no-shows, but there will be instances where more than 100 passengers show up for a specific flight. Suppose that the one-class fare for this flight is such that the net revenue per additional passenger is $200. If a passenger showing up with a ticket cannot be seated, the airline's analyst estimates the penalty of direct payments to the passenger and loss of goodwill at $400. (The net loss is thus $200.) In this problem, consider that no more than 102 seats are sought by passengers. Evaluate the policy of selling 100, 101, or 102 seats if that many are sought. Assume that no fewer than 98 will show up in any case. The probabilities that a given number of passengers will show up for given ticket sales were estimated from previous experience when overbooking was commonly done. These are shown below.

		STATE OF NATURE—NUMBER SHOWING UP				
		98	99	100	101	102
S	100	.08	.80	.12		
E	101	.04	.10	.82	.04	
L	102	.02	.06	.84	.06	.02

a. Use the criterion of maximizing expected value to determine what the ticket-selling policy should be if 100 seats are sought; if 101 are sought; if 102 are sought.

b. Indicate the assumptions you used in *a*.

c. How would you evaluate the expected value of perfect information?

13–15 For the decision tree given in Figure 13–6, discount the indicated values, compute the expected values at the relevant nodes, and complete the folding-back process. What is the expected value of perfect information? Capital is worth 12 percent.

13–16 For the decision tree given in Figure 13–7, discount the indicated values, compute the expected values at the relevant nodes, and complete the folding-back process. What is the expected value of perfect information? Capital is worth 12 percent.

13–17 A civil engineer is studying a residential development proposal to determine the optimal pattern of development. Two possibilities are being considered. One is to move rapidly ahead with development of the entire property over the next 2 years. This would require completion of all the roads and streets, water, sewer, and electrical lines. The other alternative is to proceed more slowly, developing only a part of the property in the first 2 years. Thus, the initial road, water, sewer, and electrical construction requirements would be less. Under this arrangement, the balance of the property could be developed later if conditions warrant it. However, partial development in the first 2 years, followed by completion in the next 2 years, would result in greater construction costs for the above-mentioned services, because of lost opportunities for economies of scale.

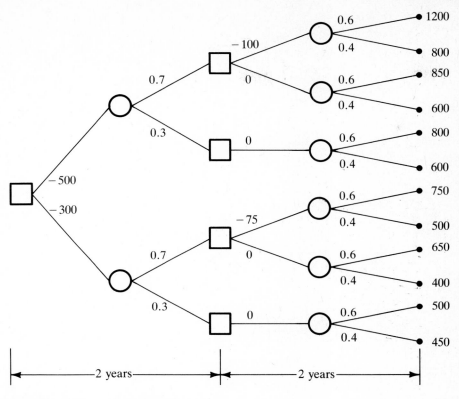

FIGURE 13—6   (Exercise 13—15)

Of course, the possibility also exists that partial initial development will be followed by sale of the remaining undeveloped property. In fact, since the developer wants to stay busy building and selling homes, assume that the study period is 4 years. In this case, that amounts to two 2-year periods.

Clearly, the relative desirability of the two alternatives depends on the demand for homes and on the economics of completing the construction in 2 years as opposed to 4. Estimating the demand for houses in the first 2-year period was simplified by categorizing it as brisk, medium, or poor. The estimates of profits in the following table were based on these categories. Also, it was felt that the demand in the second 2-year period might vary from that in the first. The assumptions were made that if demand in the first period was brisk, the second period would not be poor; if demand in the first period was medium, the second period could be brisk, medium, or poor; if demand in the first period was poor, demand in the second period would not be brisk.

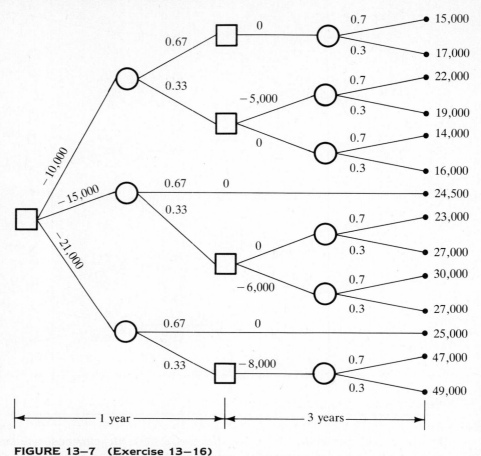

FIGURE 13–7 (Exercise 13–16)

Outcomes

Partial development, First Period:

Home Demand	Net Profit, Home Sales	Market Value, Residual Property
Brisk	$ 500,000	$300,000
Medium	375,000	300,000
Poor	250,000	200,000

Full Development, First Period:

Home Demand	Net Profit, Home Sales
Brisk	$1,200,000
Medium	750,000
Poor	300,000

Partial Development, Second Period:

Home Demand	Net Profit, Home Sales
Brisk	$ 400,000
Medium	300,000
Poor	200,000

Probabilities

First Period:

Brisk	.2
Medium	.5
Poor	.3

Second Period:

	Given First Period Demand		
	Brisk	**Medium**	**Poor**
Brisk	.4	.2	0
Medium	.6	.5	.6
Poor	0	.3	.4

a. Given the above data, construct a decision tree describing the situation. Use the data to fold it back and determine the alternative with the greater expected value.

b. What would perfect information about demand be worth?

13–18 The chief engineer of a design department of an aircraft manufacturer faces the following problem: Performance specifications have been laid out for a component of a new aircraft. It must be designed and fully tested in 2 years. In the past, certain materials have been used for similar components. If these familiar materials are used now, it is estimated that the design and testing can be completed in 1 year at a cost of $750,000. An alternative has been suggested, based upon newly developed materials and alloys. Although these offer potential benefits, they are not assured in advance. The department does not have the manpower or equipment to pursue both alternatives simultaneously. If it attempts the new approach, the resulting prototype will be either superior, inferior, or equivalent to the component arrived at by conventional means.

A determination is made that, if the new approach is pursued, progress will be evaluated in exactly 1 year. At that time its prospects will be rated "superior," "inferior," or "equivalent" to the results expected for the conventional approach. Development work for the new approach is also estimated to require $750,000 of expense over the next year. If the new approach is followed, then after the first year, a choice will be faced: either to continue development of the new approach or revert to the conventional. It is felt that continuing to develop the new approach would then cost an additional $1,000,000 over the second year. Reverting to the conventional approach would require only $500,000 of development costs for the second year.

The suitability of the new approach will be determined from the estimated savings it would provide over its life cycle in manufacturing-cost savings and reductions in maintenance in the field. The conventional approach will be used as a standard. Thus, if found to be "superior," the new approach would provide a benefit of $2,000,000 over its life cycle; if found equivalent, there would be no difference in cost; if inferior, it would add $700,000 to manufacturing and field-maintenance costs over its life cycle. Assume that there is a .3 probability that the new approach will prove superior; .5 that it will prove equivalent; and .2 that it will prove inferior.

a. Construct a decision tree for this situation, and fold back the branches to see which alternative has the highest expected value.

b. Suppose the advantage of a superior component under the new approach was worth a $3,000,000 savings. How would the analysis change?

13–19 Refer to Exercise 13–17. Suppose a home-marketing consulting firm offers to analyze the prospects for selling homes in the first two-year period described in that exercise. The property owner requests information concerning the marketing firm's track record and transforms the data into the following table, applicable to the given situation.

Consultant's Track Record

		Actual State of Nature		
		Brisk	Medium	Poor
Study	Brisk	.6	.3	.1
predicted	Medium	.3	.5	.2
	Poor	.1	.2	.7

If the study is undertaken so as to provide a prediction of either brisk, medium, or poor sales and if it would cost $35,000, should it be undertaken? If so, how should the results be used? Prepare a decision tree incorporating this new possibility and use Bayes's Rule to generate the required probabilities.

Selected References

Chernoff, H. and L. Moses, *Elementary Decision Theory,* New York: John Wiley & Sons, 1959.

Holloway, C. *Decision Making under Uncertainty: Models and Choices,* Englewood Cliffs, NJ: Prentice-Hall, 1979.

Kenney, R. and H. Raiffa, *Decisions with Multiple Objectives: Preferences and Value Tradeoffs,* New York: John Wiley & Sons, 1976.

Larson, H. *Introduction to Probability Theory and Statistical Inference,* 2nd ed. New York: John Wiley & Sons, 1974.

Moore, W. and S. Chen, "The Value of Perfect Information in Capital Budgeting Decisions with Unknown Cash Flow Parameters," *The Engineering Economist,* Vol. 29, No. 1, Fall, 1983, pp. 41–51.

Oakford, R., A. Salazar and H. DiGuilio, "The Long Term Effectiveness of Expected Net Present Value Maximization in an Environment of Incomplete and Uncertain Information," *AIIE Transactions,* Vol. 13, No. 3, September, 1981, pp. 265–276.

Rose, L. *Engineering Investment Decisions: Planning Under Uncertainty,* Amsterdam: Elsevier, 1976.

Schlaifer, R. *Analysis of Decisions under Uncertainty,* Vol. 1, New York: McGraw-Hill, 1967.

14

Analysis of Public Projects

Introduction

Thus far this text has addressed itself primarily to problems of economic decision-making in the private sector. These may be characterized by the following features:

1. The private sector consists of firms owned by individuals or partners and by corporations chartered for the purpose of conducting manufacturing, service, or trade with the aim of earning profit for their owners. Since under private ownership decisions are made with the view of ultimately earning a profit (or maximizing profit), economic yardsticks and decision-making tools were sought that would enable the most profitable choices to be made. In such an environment, overall economic performance can be measured readily by such yardsticks as the rate of return on invested capital.

2. Alternatives to be analyzed in the private sector generally have a singular, dominant, or unambiguous purpose. For example, a new manufacturing process may be developed to improve labor productivity and lower unit production costs. In principle, the economic evaluation of such a process is relatively straightforward, since the net benefits owing to reduced labor costs can be weighed against the capital investment required.

3. When private-sector economic decisions are made, the element of controversy or conflict of purpose, although sometimes present, is not generally paramount. For instance, investments to improve la-

bor productivity are usually noncontroversial. (However, in certain situations in which immediate, large-scale layoffs directly attributable to the investment will occur, such decisions have sometimes been quite controversial, especially in situations where organized labor is powerful. In such cases, labor has construed its interests as opposite to those of owners and managers.)

In this chapter, attention will be turned to the problems of economic decision-making in the public sector. This may be compared, item for item above, with the characteristics of the private milieu:

1. The public sector refers to government: federal, state, and local. It refers also to agencies of government and to pseudogovernmental corporations established for satisfying some public need or purpose. Examples of pseudogovernmental corporations include the Port of New York and New Jersey Authority, established to provide public facilities and services to encourage the use of that port in national and international commerce; the Tennessee Valley Authority, organized to supply hydroelectric power, flood control, water, and recreational facilities; and the Triborough Bridge and Tunnel Authority of New York City, intended to provide bridge and tunnel transportation connections between the various boroughs of the city separated by waterways.

   In the public sector, project decisions are often needed to protect life and property and to provide an economic infrastructure and services that will facilitate the performance of the private sector. In some instances the user pays a fee for the service provided by the public sector; sometimes the fees are intended to defray all the costs of providing the service, including capital costs. Such projects are referred to as "self-liquidating." The tolls collected by the Triborough Bridge and Tunnel Authority for vehicular crossings, for example, are intended to make those projects self-liquidating. However, in no event is profit, per se, sought in the public sector. In other cases the fees only partially offset the cost of service or there are no fees at all. In such cases, government pays part or all the costs through borrowing or taxation. (Borrowed funds must be repaid through future taxes or user fees.)

   Since decisions in the public sector are often made with the aim of protecting life and property, as well as facilitating the performance of the private sector, it may not be possible to employ a profit-maximization criterion. Hence, readily quantifiable economic yardsticks are not so easily developed for general use in the comparison of alternatives, and economic performance cannot be measured as easily in the public sector as in the private sector.

2. Alternatives to be analyzed in the public sector often have multiple objectives. For example, construction of the World Trade Center by the Port of New York and New Jersey Authority was originally intended to satisfy a variety of purposes: to provide a headquarters for the Port Authority; to provide centralized space for international trading companies as an inducement to conduct such busi-

ness in and through the Port of New York and New Jersey; to provide space for the U.S. Customs Service; and to provide office space for the government of New York State.

Multiple-purpose projects, more typical in the public sector, complicate the quantitative evaluation of the economic desirability of the projects. The more purposes, the more likely that the benefits of some of them are not so readily quantifiable.

3. The element of controversy or conflict of purpose is more likely to be present in public-sector projects than in the private sector. For example, a proposal to construct a roadway, dam, or public housing project may elicit community opposition.

Furthermore, public projects may be differentiated from those in the private sector by their method of financing, which includes the possibilities of taxation, low-interest loans (as with tax-exempt bonds), self-liquidating bonds, and guarantees of private loans. This distinction in financing the projects tends to confuse the issue of the true cost of capital, or opportunity cost, to be associated with public-sector projects. More will be said of this later. Until that point, the discount rate or MARR will be treated as given. (In the public sector, MARR is not as apt a label as discount rate. Thus, from here on in this chapter, the phrase "discount rate" will be used.) Public projects also differ from private ones in that politics is more apt to be involved in the decision-making process.

Even though there are significant differences between the public and private sector, the need for attempting to provide economic efficiency in public projects is unquestionable. Resources are scarce in society as a whole, and this obviously includes government and its agencies. Thus, resources must be used to their best advantage. Even though there may be substantive weaknesses in an analytical procedure to evaluate a public project, it is generally agreed that some rational attempt should be made, since an economic study, though not without flaws, is better than sole reliance on snap judgment, experience, politics, or expediency.

Engineers and managers working in the public sector often find themselves in situations where such analytical studies are called for, and in fact, may be required by legislation or executive order. Civil engineers are involved with such studies for almost all types of public works: roads, bridges, dams, irrigation systems, flood-control systems, waste water systems, government office buildings, schools, and so on. Increasingly, as government operations come under close public scrutiny with the aim of reducing inefficiency, waste, and/or corruption, ways are sought to evaluate more objectively the desirability of contemplated projects.

In many instances, the conduct of public-sector economic studies is similar to that in the private sector. For example, an agency may be faced with a decision that can be described as one of cost minimization (as outlined in Chapter 4). To illustrate, a sanitation department may be evaluating two types of truck for potential large-scale purchase: a small one requiring two operators, which would have to travel for unloading more frequently than a larger model that has greater capacity but requires larger investment, more fuel per mile, and three operators. These two alternatives can be compared by computing the annual worth (equivalent annual cost) of each truck over its economic life. The only differences between such a study and one done for a private carting firm are:

1. The public study would be done on a pretax basis, since the governmental agency is not subject to income taxes. The study for the private carting firm would be done on an after-tax basis, as indicated in Chapter 8.
2. The question of what discount rate to use would arise. This is an issue that will be addressed separately, at the end of the chapter. As indicated, until that point is reached, the discount rate will be treated as a given parameter.

Benefit-cost Ratios

Since public projects often may not be evaluatable or comparable on the basis of costs alone, some consideration must be given to the benefits as well. In the private sector, this is done by accounting for the revenue associated with a project (or sometimes by the implied revenue provided by cost savings). In the public sector it is sometimes possible to identify explicit revenues of a project (or implied revenue provided by cost savings) without including other benefits of a less quantitative nature. Such is often the case with self-liquidating projects where the fees for services are the revenues. Where this occurs, it is possible to evaluate the alternatives in the same way as in the private sector. Thus, the computation of equivalent annual worth or present worth may be sufficient to compare the available alternatives. As with cost-minimizing problems, there still exists the problem of choosing the discount rate, which will be addressed later.

However, in public projects it is frequently impossible to estimate accurately a financial revenue associated with a particular benefit or there may be some benefits of a less easily quantifiable nature. This stems from the very nature of public projects, where the benefits may be multifaceted, as well as communal or societal in nature, and hence not easily evaluated by a price in the marketplace.

In such cases, analysts resort to a cost-benefit (or benefit-cost) analysis, which attempts to estimate the benefits and compare them to the costs of a particular project. The difficulty is that substantial judgment must be used and specific and somewhat arbitrary assumptions made in order to identify and quantify the various benefits. Given the evaluation of benefits, a simple ratio is formed between the benefits and costs, as is given by Equation 14–1:

$$B/C = \frac{B}{C + O + M} \tag{14-1}$$

where B/C is the benefit-cost ratio; B designates benefits, C is for capital cost, O for operating cost, and M for maintenance costs; and B, C, O, and M are computed similarly, either in terms of the equivalent annual worth or present worth of the period flows. Under this formulation, a public project is considered acceptable if the benefits exceed the costs (i.e., if the B/C ratio exceeds 1.0).

Example 14–1, which is adapted from an article by David Willer,* illustrates the use of the B/C ratio.

EXAMPLE 14–1

A municipality with a small river running through its borders having an existing dam at a strategic location over which it has responsibility, is considering the installation of a small-scale hydroelectric generator. The energy generated would be sold to the local utility and would serve to defray, to a small degree, the town's considerable electricity expenses. Because the town's financial resources are limited, such an investment can be justified only if the benefits of such a project outweigh the costs, including investment. Factors affecting the generator's economic feasibility include physical factors, such as the head and flow of water at the dam site, and economic factors, such as the amount of the investment required, the financing costs, operating and maintenance costs, and the value of the energy generated.

For now, suppose that a 5,000-kw power plant is being considered. The hydrology conditions indicate that such a power plant is expected to experience a 50-percent capacity factor. It would require a \$5 million investment. Operating and maintenance costs are estimated at \$75,000 per year. The economic life of the project is felt to be 35 years. The discount rate to be used for the study is 8 percent. Electricity is worth 50 mills per kilowatt hour (\$0.05). With this data a B/C ratio can be computed as follows: The annual benefit, treated here as an annuity, is

$$B = (24 \text{ hrs./day})(365 \text{ days/yr.})(5,000\text{kw})(\$0.05/\text{kwh})(0.50)$$

$$= \$1,095,000/\text{yr}.$$

The annual costs are

$$C + O + M = \$5 \text{ million}(A/P \ 8\%, 35) + \$75,000 = \$504,016$$

Thus, such a project would give a favorable B/C ratio and would be viewed as desirable.

$$B/C = \frac{B}{C + O + M} = \frac{1,095,000}{504,016} = 2.17$$

This simple example shows how the B/C ratio may be computed and interpreted. As with many analytical tools, however, the simplicity of the example fails to indicate the full extent to which such tools may be used in real design situations. As has been indicated, the process of design is frequently iterative, in that a design

*David Willer, "Determining Feasibility of Small Scale Hydropower," adapted from *ASCE Journal of Energy Engineering,* vol. 107, December, 1981; pp. 210–212.

can be tested or evaluated with respect to a certain characteristic and, if it is found lacking in any respect, can be modified accordingly. Then it can be tested or evaluated again, with the process continuing until a satisfactory design is arrived at. From this viewpoint, the situation of Example 14–1 may be considered more generally. With regard to small-scale hydroelectric plants, at least one design issue is economic: choosing an acceptable power-plant capacity. The greater the generator capacity, the greater the capital cost, and to a certain extent the operating and maintenance costs. As Willer pointed out,

> . . . if the capital cost, value of power, and the interest rate are known, then for the selected size of power plant, the energy the power plant must generate (plant load factor) for a B/C ratio of 1.0 can be determined. If the hydrology will not permit the generation of the energy required, the power plant must be reduced in size until the ratio is more favorable. Certain sites, obviously, may never produce a *B/C* ratio greater than 1.0.*

Modified Benefit-cost Ratio Some analysts and public agencies compute the *B/C* ratio somewhat differently than in Equation 14–1. In such cases the numerator is used for net benefits and the denominator for capital costs only. Such an expression of the B/C ratio would be given by

$$B/C = \frac{B - (O + M)}{C} \qquad (14\text{–}2)$$

and is then referred to as the modified benefit-cost ratio. The expression given by Equation 14–1 is called the conventional *B/C* ratio. The modified *B/C* ratio is consistent with the conventional one in that both will give a ratio greater than or less than 1.0 for the same set of flows. However, if mutually exclusive alternatives are considered, the two *B/C* ratios may not give the same ranking of projects. In such situations, an incremental analysis should be conducted. This is illustrated in the next section.

Incremental Benefit-cost Analysis When mutually exclusive alternatives exist, the decision as to where to apply scarce capital in the public sector should be justified economically under the same principles as in the private sector. Thus, as indicated in Chapter 4, each additional increment of capital investment should be warranted, based on the benefits provided, the additional costs incurred, and the discount rate. In line with that principle, a *B/C* ratio may be computed for each incremental amount of capital called for under each successively larger investment.

The decision rule to be employed, then, works the same way when applied to public projects as indicated more generally in Chapter 4. Each incremental amount of capital, up to the maximum amount available for investment, should have an incremental *B/C* ratio greater than 1.0 in order to be justified. If a given capital increment is not warranted, the next alternative should be compared incrementally to the previously acceptable one, using the aforementioned criterion.

Ibid.

Example 14–2 will help to clarify this concept and to show how incremental B/C ratios are computed. The conventional B/C ratio is used.

EXAMPLE 14–2

Suppose that for the situation of Example 14–1, involving the economic generation of electric power from a small stream, a variety of alternatives are considered. These are mutually exclusive alternatives, in that only one will be chosen. The estimated characteristics of the alternatives are depicted in Table 14–1.

TABLE 14–1   Alternatives for Generating Energy at Existing Dam

| | Generator Alternative | | | |
	A 1,000 kw	B 3,000 kw	C 4,000 kw	D 5,000 kw
Investment, I ($000,000)	3	4	4.5	5
$O + M$ ($000)	45	60	67.5	75
Load Factor (percent)	70	65	60	50

The table indicates that, for the anticipated hydrology conditions, increasing investment gives greater generating capacity but lower load factors. With the same value of electric energy as was used in Example 14–1 (50 mills, or $0.05 per kwh) the same economic life for the generator (35 years), and the same discount rate (8 percent), Table 14–1 can be converted to a table of benefits and costs.

TABLE 14–2   Alternatives for Generating Energy at Existing Dam

| | Generator Alternative | | | |
	A 1,000 kw	B 3,000 kw	C 4,000 kw	D 5,000 kw
Annual Equivalent				
Inv. Cost, C	$257,410	$342,213	$ 386,115	$ 429,016
$O + M$	45,000	60,000	67,500	75,000
$C + O + M$	302,410	403,213	453,615	504,016
Annual Benefit, B	306,600	854,100	1,051,200	1,095,000
B/C ratio	1.01	2.12	2.32	2.17

The reader should verify that the equivalent annual capital costs are simply $I(A/P\ 8\%,\ 35)$ and the benefits are

$$24(365)(\text{generator kw})(\text{electricity value})(\text{load factor})$$

Note that all four alternatives, when considered independently, show a B/C ratio greater than 1.0. Thus, one might conclude, incorrectly, that if $5,000,000 is available, alternative *D* is best because it "uses all investable funds satisfactorily." That this is incorrect may be seen from an incremental analysis, as prescribed previously. Each successive increment of capital must be justified on the basis of the economic benefits it provides. With this in mind, Table 14–2 can be converted to one showing the incremental benefits and costs. The first comparison is between *B* and *A,* since *A* would be at least minimally acceptable.

TABLE 14–3   Alternatives for Generating Energy at Existing Dam, Incremental Analysis

	Generator Comparison		
	B − A	*C − B*	*D − C*
Incremental *C + O + M*	$100,803	$ 50,402	$50,401
Incremental *B*	547,500	197,100	43,800
Incremental *B/C*	5.43	3.91	0.87

From this table it is clear that *C* is best, since an additional investment of $500,000 to obtain generator *C* as compared to *B* provides more additional benefits than costs, while an additional investment of $500,000 to obtain *D,* as opposed to *C,* would provide more additional annual costs than benefits. (Here the optimal choice under an incremental analysis is the same as that with the maximum *B/C* ratio. However, because this is a coincidence not always duplicated, the analyst must perform the incremental analysis in order to choose the best alternative.) Recall, too, that the correct comparison is between an alternative and its next smaller acceptable one. Thus, had *B* turned out to be undesirable, *C* would have been compared to *A.*

Risk-benefit Analysis

The concept of risk has already been discussed in Chapter 13. Frequently, it is necessary to consider the economic impact of a variety of public works designs, each affected differently by risk; or the impact of additional public expenditure to reduce public risk regarding some loss of life, health, or property may need to be considered. The extension of benefit-cost analysis to include public-sector risk situations has become known as risk-benefit analysis.

To begin with, project costs are treated the same as in benefit-cost analysis. Investment, operating, maintenance, and administrative costs can be evaluated as before. Now, however, the additional expected (annual) cost associated with the risk factor can be added to the project costs to give a total expected cost for a given design. Then the decision-maker can choose the design that minimizes total cost (including that for the risk factor). Alternatively, an equivalent conclusion can be reached under an incremental approach. Here, reductions in expected risk losses attributable to an investment are treated as incremental benefits, which can be weighed against the incremental total cost needed to provide that benefit.

To illustrate these issues, consider the problem of the design of highway stream crossings in areas designated as flood-prone.* In such situations there are a number of design objectives. Among them is the need to design structures so that backwater will not exceed some arbitrary limit for large floods having low probability of occurrence during the life of the project. Another goal is to provide operational use of the structure for most hydrological conditions by elevating the highways so that there is a low probability of traffic interruption due to flood waters.

Designing structures to satisfy such performance criteria, however, ignores the overall economic performance attributable to a particular design. As a result, adherence to performance criteria alone can often lead to costly overdesign as a direct result of considering one design factor as opposed to another.

> The ability to pass a given design flood requires the overdesign of the structure in relation to all floods of a lesser magnitude. This also requires a considerable additional investment in the initial construction costs for an event with low probability of occurrence.

By contrast, although the design of a structure able to perform satisfactorily only for lesser floods requires smaller investment, such a design would be more susceptible to repair and maintenance costs. Risk-benefit analysis applied to this situation involves an evaluation of the total costs for each alternative design with the aim of choosing the design that minimizes total cost.

Total costs are composed of construction costs (those required for building the indicated bridge, approach roads, and embankments) and the expected costs of flood occurrence (including bridge damage from inundation, embankment erosion and bridge pier scour, traffic-related losses, and upstream-flooding losses). For each configuration of bridge length and embankment height under consideration, then, the analysts must estimate not only construction costs but also those costs associated with floods of known return interval and weight them by the flood's probability of occurrence to obtain the expected values. Then the designs may be compared with respect to total costs.

To demonstrate their contention that without risk-benefit analysis costly overdesign can occur, Knepp et al. analyzed a highway stream-crossing near Tallahalla, Mississippi. Representative data on hydrologic, hydraulic, traffic, and construction characteristics were obtained. Nine designs were analyzed: the existing bridge structure of 500-foot length, 315-foot embankment elevation; and structures with 1,000-foot, 500-foot, and 300-foot lengths; and embankment elevations of 310 feet, 315 feet, and 320 feet. The result of the study, duplicated in Figure 14–1, shows that the most cost-effective design was not the existing structure but one with a 300-foot bridge length and a 315-foot embankment elevation. This design had expected annual costs that were $20,000 less than the existing configuration.

---

*This illustration was adapted from A. Knepp et al., "Benefit-Risk Analysis in the Design of Highway Stream Crossings," *Public Roads,* September, 1976, pp. 66–69. The quotations in this section and Figure 14–1 appeared in that article.

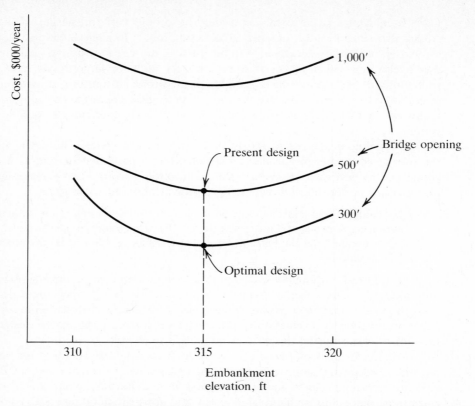

FIGURE 14–1 Risk-Benefit Analysis for a Stream Crossing (from Knepp et al., "Benefit-Risk Analysis in the Design of Highway Stream Crossings")

The Discount Rate To Be Used in Public-sector Analysis

Thus far, the discount rate used when computing B/C ratios was treated as given. For the engineer or decision analyst conducting a study, this is not an unrealistic situation, for rarely will the analyst have discretion over a wide range of rates to be used. For example, a federal directive issued by the Office of Management and Budget indicated an unadjusted rate (for inflation) of 10 percent for a large variety of federal projects.

That analysts should not have unilateral discretion to set the discount rate should be clear from the following two points, taken together:

1. Variation in the discount rate would affect the computation of the B/C ratios and would therefore affect the apparent desirability of a project. The dependence of the B/C ratio on the discount rate is analogous to the dependence of other yardsticks, such as PW, AW, or IRR, on the discount rate, as has already been illustrated in Chapter 3. Figure 14–2 makes this explicit. It shows the (modified) B/C ratio as a function of the discount rate for the following cash flows: a 25-

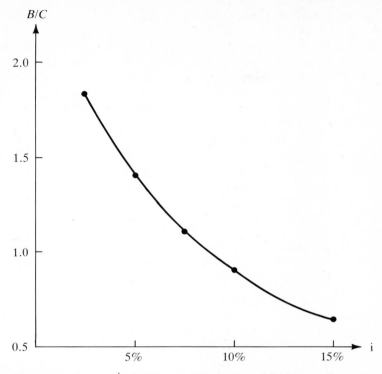

FIGURE 14–2 B/C Ratio as a Function of Discount Rate

year project, requiring net investment of $1 million, and net benefits, $B - (O + M)$, of $10,000 per year.

2. Since approval of public projects is influenced by political factors to a far greater degree than in the private sector, it is best to rely on studies that are not affected by charges of special-interest influence or political meddling to make projects seem more or less attractive.

Therefore, in the public sector, the rate is typically set at the highest bureacratic or administrative levels. As mentioned previously, federal project studies use rates set by OMB. State projects, and those of special agencies or pseudogovernmental corporations, use rates set by fiat from the highest administrative authority. The administrative bureaucracy of local government, which is of smaller scale than that for state or federal gvernment, allows the technical staff conducting the study to have greater influence in setting the discount rate. Even there, however, ultimate authority for setting rates rests outside the realm of those conducting the study. In local government, rates may be set by authority of the executive branch through the mayor's office or one of the specialized departments, such as the finance department.

Even though the discount rate is not usually decided by the analyst conducting the study, professionals should understand the basis on which the rate is set. A discussion of those issues follows.

To begin with, there has always been a considerable amount of debate surrounding the subject of the appropriate discount rate for public projects. The dominant side on this debate has drawn on a strict interpretation of orthodox economic theory, while the other has been more willing to be concerned with social needs or political will in setting rates. In the debate those who subscribe to the conservative economic view have concentrated on the real opportunity cost of capital to the public sector. Such fiscal conservatives generally favor relatively high discount rates.

The group advocating greater consideration of social needs has been concerned with the usefulness or importance of public investment and the fact that, in the absence of such investment, social needs may not be fulfilled adequately. Such advocates prefer to see relatively low discount rates, since evaluation of public-sector projects is more positive at lower rates. Clearly, the degree to which discount rates should be set below the opportunity cost of capital by such advocates is arbitrary; similarly, the determination of adequate funding of social needs is also arbitrary; thus, the debate is never-ending.

The theoretical economic basis for setting discount rates revolves about three possibilities:

1. When borrowed funds are earmarked for the particular project being studied, the rate should be based on the interest rate paid by the public body for the borrowed funds. This is a popular and widely held viewpoint, particularly in local government and for self-liquidating projects. Its popularity stems from the fact that the cost of borrowed capital can be directly related to the project being considered and that it is easily understood by citizens. Also, it does not heavily penalize the consideration of public works in a B/C analysis, since bonds of states, municipalities, and their agencies are exempt from federal income taxes and consequently bear low rates of interest. Thus, discount rates based on such borrowing are also low.

2. It should be based on the opportunity cost of capital. Theoretically, the discount rate should be based not on the cost of borrowing for a particular project but on the relevant opportunity cost of capital. In this context, the opportunity costs may be viewed from two perspectives. One is that of the government body in which the study is conducted. If one considers that the agency ranks all its projects with respect to their economic desirability and then rations them so that, for the available capital, only the n most desirable are approved, then the rate of return associated with the best alternative forgone is the relevant opportunity cost to the agency and is the proper discount rate to be applied for study purposes. Although this perspective is satisfactory from a theoretical point of view, it lacks practicality. The reason for this is that rates of return are usually not computed for public projects (part of the rationale for B/C ratios) and, thus, no such ranking of projects is feasible.

3. Another perspective from which to measure the opportunity cost is that of the taxpayer. Here the rationale is that government competes with the private sector for funds and that, for optimal societal economic efficiency, government should not preempt more desirable private investment, as determined in the capital markets. This view sets the discount rate by examining the opportunity costs

of taxpayers, considering the productivity of capital in the private sector were it not taxed by government. E. B. Staats illustrated that perspective using a set of time-related interest and income-tax assumptions.* His calculations were based on assumptions and interest rates that were prevalent in the late 1960s. His estimate then was that such an opportunity cost rate would be close to 10 percent. (Note that the OMB directive of 1972, mentioned previously, was for 10 percent.) At the time this book was written, a similar computation would yield a result substantially higher, about 14 percent, owing to a large increase in interest rates for all types of borrowing, including governmental, and to higher private-sector rates of return.

Other Uses of Benefit-cost Analysis

Another application for benefit-cost analysis in the public sector has emerged in recent years. As government—federal, state, and local—has become more concerned with regulations governing health and safety, there has been a movement to measure the desirability of proposed or existing regulation. Benefit-cost analysis has been advocated and used for this purpose. That movement is motivated by the growing realization that public law should be practical, as well as meeting some avowed social purpose, and that society's resources, which are limited, must be put to use with maximal effectiveness. From such a view, regulation should be promulgated only after a thorough analysis indicates that its benefits outweigh the costs.

As an example of benefit-cost analysis applied to government regulation, consider the following: A 1981 study by the Organization for Economic Cooperation and Development indicated that a 50-percent reduction in sulfur emissions could be achieved in Europe by 1985, resulting in cleaner air and less acid rain. The cost was an estimated $4.6 billion in annual investment, which was expected to be within the estimated range of savings attributed to improved lake quality, health, crops, and less building damage.†

A study published in 1982 evaluated by benefit-cost analysis a cotton-dust regulation applied to the textile manufacturing industry by the Occupational Safety and Health Administration in 1978.‡ The study concluded that the costs far outweighed the benefits for the forty-year study period.

Benefit-cost analysis applied to regulatory policy is even more difficult than for specific public works projects. Large amounts of data are required to conduct a B/C analysis to evaluate a specific regulation. Numerous assumptions need to be made, many of which may be characterized by critics as arbitrary and biased. Consideration of costs should include the issues of who incurs them, their magnitude, and how burdensome the regulation will be for the affected industry, locale, or

---

*E. B. Staats, "Survey of Use by Federal Agencies of the Discounting Technique Evaluating Future Programs," in H. H. Hinricks and G. M. Taylor, eds., *Program Budgeting and Benefit-Cost Analysis*, Pacific Palisades, CA: Goodyear, 1969.

†"OECD Study Develops Cost Benefit Methodology," *Journal of the Air Pollution Control Association*, June, 1981, pp. 680–696.

††R. Dardis et al., "Benefit-cost Analysis of the Cotton Dust Regulation," *Textile Research Journal*, June, 1982, pp. 380–388.

sector of society. Similarly, for the benefits: who will gain and how much? Finally, what are the alternatives? Are there more cost-effective ways of achieving satisfactory results?

Exercises

14–1 The project engineer for a county-sponsored water carrier has examined three proposals which are intended to foster irrigation:

	Project		
	A	**B**	**C**
Initial Investment	$350,000	$500,000	$750,000
Net Annual Benefits	60,000	90,000	115,000

Assume a 15-year study period for the projects and that funds are worth 10 percent. Examine the present worth of benefits in doing an incremental benefit-cost analysis to determine which alternative is best.

14–2 A government project takes 2 years to complete. The development and construction expenses for the first year are $500,000; they are $300,000 for the second year. If benefits start accruing in the third year, what would the benefit-cost ratio be if:

    a. The benefits were $140,000 per year for 10 years?

    b. The benefits were $120,000 in the third year, increasing by $5,000 per year for 9 years (10 years of benefits in total).

    c. The benefits were $120,000 in the third year, increasing by 4 percent per year for 9 years (10 years of benefits in total). Assume funds are worth 10 percent.

14–3 The chief engineer in a city sanitation department is heading a project to determine the best site for a resource recovery plant, using refuse. Two acceptable sites have been identified, but they differ in their suitability for construction and, hence, in their investment requirements. Because they differ also with respect to distance from the area to be served, the net benefits for the two sites are not the same. The data are summarized as follows:

	Project	
	A	**B**
Initial Investment	$10,000,000	$12,000,000
Net Annual Benefits	1,100,000	1,200,000

If the projects have a 25-year study period and funds are worth 8 percent, which site is the more economic?

14–4 Three mutually exclusive alternatives are being examined by a state transportation department engineer:

	Project		
	A	**B**	**C**
Initial Investment	$75,000	$100,000	$125,000
Annual Benefits	34,000	45,000	60,000
Annual Costs	10,000	15,000	20,000

If funds are worth 10 percent and the projects are examined over a 10-year study period, analyze the incremental benefit-cost ratios to determine which is superior.

14–5 The public works department of a small town is considering the purchase of a computer system to improve its scheduling, handling of materials inventory, equipment maintenance records, and the like. Two systems are being evaluated on a benefit-cost basis. The data are as follows:

	System	
	A	**B**
Initial Investment	$12,000	$15,000
Annual Benefits	3,800	5,000
Salvage Value	1,000	1,200

Assuming a 6-year study period with salvage values for the systems as indicated, which should be recommended? Funds are worth 8 percent.

14–6 A public works department must choose between two types of equipment to be used for plowing snow, spreading salt and sand, and general hauling. After study it is felt that one piece of equipment offers somewhat more benefit because its greater range requires less frequent return for refueling. Also, the maintenance and operating costs differ. These are summarized in the table following:

	Equipment	
	A	**B**
Initial Investment	$60,000	$75,000
Annual Benefits	14,000	15,000
Annual Costs	5,000	4,000
Salvage Value	3,000	2,000
Useful Life	10 yrs.	12 yrs.

Examine the incremental benefit-cost ratio to determine which should be purchased if funds are worth 6 percent.

14–7 An industrial park located in a small town currently has few tenants, because of transportation problems. At present the only road to the park passes through the town center and already has heavy traffic. Furthermore, there is no direct route from the town center to major highways. The town's traffic engineer, in conjunction with the economic development commission, has studied the feasibility of providing a second link to the park, which would avoid the town center. In order to effect this link to existing roads and thence to the entrance of a major interstate highway, a bridge and short stretch of roadway would need to be built.

Their study has led to the following data: Cost of land for right of way, $100,000; construction costs, $2,500,000; annual maintenance costs, $50,000; additional town services, $25,000 per year. The economic development commission studies indicate that the improved transportation link would result in full occupancy of the park. This would provide a net increase of $10,000,000 per year in business activity for the town; $200,000 per year in added property and miscellaneous taxes; and a reduction in vehicular accidents in the congested town center, saving $75,000 per year in property damage.

If the study period is 25 years and $i = 8$ percent, compute the benefit-cost ratio. Would you expect any opposition from the town council, which has bonding authority for the construction costs?

14–8 A long-abandoned small airport, 25 acres in area, is located in a town outside a port city. The land which the airport occupies has not been developed by its private owners, because of a complex history of zoning and financial problems. The town wishes to see the land developed for productive use and, after many years of fruitless waiting, proposes to acquire the land under an economic development program partly sponsored by the state. This has become a controversial issue, since many citizens feel that the town "should not be in the development business." The following proposal is submitted to the legislative council for approval:

The land would be acquired by the town. The town's cost would be $250,000, the remaining $500,000 to be paid by a state grant. The land would be divided into small lots to be sold to business firms for light industrial or commercial use at $25,000 per acre (gross). The town would make the necessary general improvements, such as water, sewers, electrical supply, and a small amount of additional road construction. Specific development needs of the individual sites would be the responsibility of the lot owners. The general improvements are estimated to cost $6,000 per acre and would be made during a period of one year after the land is acquired.

Once the individual lot owners take title (after the improvements are completed) the town's property and miscellaneous taxes are expected to increase by $75,000 for the first year of operation and $150,000 for the second year of operation, rising by $5,000 per year thereafter. Since the lots would be individually owned, there would be no maintenance expense for the town. However, additional town services are estimated to cost $20,000 per year, beginning with the first year of operation.

a. What is the benefit-cost ratio for this proposal? Assume a 20-year study period and that capital is valued at 8 percent.

b. What are the risks in the proposal? Would these be substantial if the going rate for industrial land is greater than $25,000 per acre?

14–9 Gloucester County contains a small city as well as an extensive rural area. In the last 15 years it has experienced rapid suburban growth and development along its eastern end. Auto and truck traffic has become very congested along the existing road and bridge connecting the eastern part of the county to the city. Since the existing bridge is very old, a new one is proposed that will better serve the current and projected traffic flows. It will cost $25,000,000 to construct and will require $350,000 in annual maintenance. It is required that the users of the facility pay a toll covering all costs of construction and maintenance. Suppose that there is an average of 12,000 passenger-car crossings per day. Of these, 2/3 pay the regular passenger fare while 1/3 pay a commuter fare, which is half the regular amount. In addition, there are 2,000 trucks and buses per day which pay an average of 2.5 times the regular passenger-car fare. Given this information, what would the fare structure be? Assume that construction costs are financed with 25-year bonds carrying an 8-percent interest rate.

14–10 A northeastern town's new school superintendent is given the responsibility of studying the desirability of replacing a school constructed only 10 years ago. Although in excellent condition, it has an electric heating system which has become very expensive as rates have risen dramatically. The school was financed by 12-percent, 20-year bonds. Electric-heating costs are now about $350,000 per year. The school can now be sold for $1 million and a new school constructed for $2.5 million. The new school would then require only about $150,000 for heating, using fossil fuel. It would be financed with 20-year bonds at 8 percent per year. In all other respects the maintenance costs of the two schools would be similar.

    a. Given only these two alternatives, what should be done?

    b. Assuming that the existing school can be renovated to install a fossil-fuel heating system with operating costs as given above, at most how much could the renovation expense be and still be economic?

    c. What would happen to your recommendation in *b* if electric rates are expected to rise faster than fossil-fuel prices in the years to come?

14–11 Two sites are being considered by a regional authority for a new airport, intended primarily to serve a certain city. Although the closer site is more convenient, requiring less travel time and expense, the land for that site is more costly. In addition, the terrain for the closer site is not as suitable for construction, leading to higher costs. The relevant data are presented as follows:

	Site	
	Closer	**Farther**
Land Cost	$ 3 million	$ 1 million
Construction Cost	30 million	25 million
Maintenance Cost	600,000	500,000

In addition to airport construction and operating expense, the authority wishes to consider the relative travel costs of the two sites. They differ by 10 miles in their travel distance from the city to be served. Operating costs are $0.20 per mile for automobiles, $0.75 per mile for buses. Travel time for passengers is thought to be worth $5 per hour per car and $100 per hour per bus.

    a. Assume 5 times as many auto trips to the airport as bus trips and that all vehicles average 25 m.p.h. How many daily trips of each type make the total cost of both sites the same? Use a 25-year life for the airports, with funds being worth 8 percent.

    b. If 1,000 auto trips and 200 bus trips are projected daily, which site is superior?

14-12 A state legislative study panel is investigating the need for government regulation to foster worker safety on construction projects. Regulations have been proposed which, if implemented, will cost an estimated $3 million per year to contractors and subcontractors. It is estimated that the regulations will result in a 15-percent reduction in accidents of all sorts, uniformly distributed. Currently settlements for accidental deaths are about $1,000,000 per year. Settlements for permanent injuries are about $2,000,000 per year. Indirect economic losses due to disruption and lost production time because of accidents is about $600,000 per year. Also to be considered is the fact that the regulations are expected to reduce insurance premiums by $500,000 per year. What does a benefit-cost ratio applied to this risk situation indicate about the desirability of the proposed regulations?

14-13 Investigate a public project in your local area to see how the costs and benefits were identified and estimated.

Selected References

American Association of State Highway Officials, *Road User Benefit Analysis for Highway Improvements,* Washington, D.C., 1960.

Dasgupta, A. and D. Pearce, *Cost-Benefit Analysis,* New York: Barnes and Noble, 1972.

Frost, M. *How To Use Benefit-Cost Analysis in Project Appraisal,* 2nd ed. New York: Macmillan, 1979.

James, L. and R. Lee, *Economics of Water Resource Planning,* New York: McGraw-Hill, 1971.

McKean, R. *Efficiency in Government Through Systems Analysis,* New York: John Wiley & Sons, 1958.

Mishan, E. *Cost-Benefit Analysis,* New York: Praeger, 1976.

Sassone, P. and W. Schaffer, *Cost-Benefit Analysis: A Handbook,* New York: Academic Press, 1978.

U.S. Congress, Joint Economic Committee, Subcommittee on Priorities in Government, *Benefit Cost Analysis of Federal Programs,* 92nd Congress, 2nd Session, 1973.

U.S. Government, Subcommittee of Evaluation Standards of the Inter-Agency Committee on Water Resouurces, *Proposed Practices for Economic Analysis of River Basin Projects,* Washington, D.C.: Government Printing Office, 1958.

Winfrey, R. *Economic Analysis for Highways*, Scranton, PA: International Textbook Co., 1969.

15

Project Scheduling and Management: Critical Path Methods

Introduction

The practice of engineering has always involved projects of large scale, that require many workers, both skilled and unskilled, that are composed of many interdependent functional components or activities, and that take a long time to complete. This was even true in ancient times, as evidenced by the construction of the pyramids of Egypt, the temples of Greece, and the aqueducts and public buildings of Rome. Such large-scale projects required competent human organization and management to coordinate the various activities for timely completion of the projects with minimum waste of resources.

Although large-scale projects have been known throughout history, they occurred rarely in the everyday life of society. Thus, exceptional men could be chosen to lead and manage them. In modern times, projects have become increasingly large and complex, with multiple interdependent activities, some of which are crucial as prerequisites to others. Examples of these appear in Table 15–1.

What makes any of these undertakings complex is that the mix of resources they require—equipment, energy, and especially manpower—can be varied in order to affect their completion date. But although it is true that any increase in resources will be likely to speed up the completion of a particular activity (with an increase in its cost), it does not follow that such an action will necessarily affect the completion time of the whole project. Many activities are often done in parallel, and many of these have slack insofar as they affect the project's completion date.

323

TABLE 15–1 Projects with Interdependent Components

Construction:
    Roads, bridges, tunnels, buildings, factories, processing plants
Aerospace:
    Development and production of new aircraft
Military:
    Development and production of new weapons systems
Industrial:
    Development and production of new products or processes
Miscellaneous:
    Organization of a large-scale professional meeting
    Development and production of new large-scale computer software programs

There are also likely to be bottlenecks presented by certain sequences of activities. It is these that are of particular concern, since they determine the earliest completion time for the project; delays there will affect the final outcome, and resources brought to bear there will help to avoid or shorten any overall delays. These bottlenecks, referred to as critical paths, are sufficiently important to be the focus of this chapter.

As a result of the increasing complexity of modern engineering projects, even the smallest of them now requires special attention to the issues of proper scheduling, management, and control. Therefore, the well-trained engineer needs to be familiar with the available tools that facilitate good project planning, scheduling, and control. This chapter concerns itself with some network tools that have become very widely used for economic planning, scheduling, and management of projects with interdependent components. These are referred to as critical path scheduling models. Their common feature is that they are based on network graph models (described in a following section) and that they identify the critical path (or paths) which is the bottleneck sequence of activities for the project as a whole. Once bottlenecks are identifiable, the expected final completion date can be estimated. In addition, attention can be given to the bottleneck activities as they are under way, so that delays in those crucial activities can be anticipated and remedial action taken to avoid overall delay in the project's completion date.

The Gantt Chart

Before network models are described, a procedure developed earlier will be delineated. Although it is not a network model, it is commonly used as a scheduling and control tool in its own right, as well as in conjunction with network models. In fact, one can readily purchase board kits from office supply firms to depict this sort of chart for operational use and display.

    The Gantt chart is attributed to Henry Gantt, who for many years was a working colleague of Frederick Taylor, considered by many the father of scientific

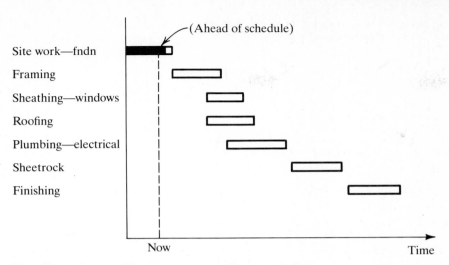

FIGURE 15–1 Gantt Chart, Simple Construction

management in the United States. The chart was intended to improve the ability of managers to schedule and control projects with multiple activities (see Figure 15–1). It displays the various activities along the vertical axis and time on the horizontal axis. The scheduled time for each activity is represented by a horizontal bar whose length corresponds to the required time in units of the horizontal axis. The scheduled beginning and end times for the activity are depicted by the end points of the bar. The plan, as depicted by the chart, can be monitored once the project begins by using a pointer to indicate the current date and by filling in or coloring the bars to indicate the proportion of each activity that has been completed by that time. A thoughtful examination of the chart indicates that it is most helpful in identifying activities that are falling behind schedule. However, the prerequisites of future activities are not specified explicitly in the chart (although presumably they were taken into account implicitly when the initial schedule was constructed). Thus, the chart does not indicate which activities can be permitted to fall behind schedule, and by how much, without affecting the overall completion date. Similarly, it does not distinguish which activities cannot be allowed to fall behind schedule without extending the completion date. Finally, the Gantt chart would not be practical for a project with a very large number of activities, since the vertical axis would have to be very long.

Network Models

Network models represent events as nodes and activities as their connecting arrows. Activities such as those described on the vertical axis of the Gantt chart are the functional components of projects. They are the elemental, differentiable work elements of the project, such as "build footings" or "design landing gear." Events (nodes) are the beginning or end points of the activities; they are critical points in time, or milestones. Corresponding events for the activities mentioned above might be "begin design work for landing gear," "complete landing-gear design," or "complete footings." Frequently the nodes representing events are labeled with numbers. Activities can then be denoted by the nodes that contain them (see Figure 15–2). The activity "design landing gear" can be represented by 109–110.

Begin
design

Complete
design

FIGURE 15–2

A small-scale network is shown in Figure 15–3. Convention calls for the first event, 1, to be on the extreme left, with activities (arrows) and succeeding events (nodes) proceeding to the right. Nodes with more than one activity (arrow) emanating from them (1 and 2) indicate events or milestones that are prerequisites for the start of more than one activity; nodes with more than one activity (arrow) pointing to them (5 and 6) indicate events that cannot occur until each of the indicated activities is finished.

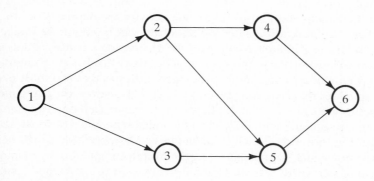

FIGURE 15–3

Critical Paths

In order to gain more familiarity with network models, to see the significance of their critical paths, and to understand how to compute them, consider the abstract Example 15–1a, illustrated by Figure 15–4. After it has been analyzed and a critical path identification procedure has been introduced, a realistic example will be presented.

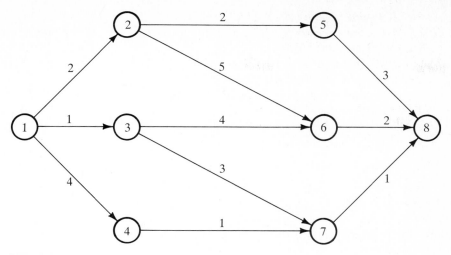

FIGURE 15–4

EXAMPLE 15–1a

Note that there are only eight events, labeled 1 to 8, and 11 activities: 1–2, 1–3, 1–4, 2–5, 2–6, 3–6, 3–7, 4–7, 5–8, 6–8, 7–8. Event 1 is the beginning time and event 8 is the end time of the project. Events 2, 3, 4, and 5 each have only one prerequisite activity; events 6, 7, and 8 have more than one. Events 4, 5, 6, and 7 are succeeded by only one activity; events 1, 2, and 3 are succeeded by more than one. Assume that the time required (in months) for each event is known. In Figure 15–4 these times are shown alongside the activity arrows. Since this is a small network, it is easy to enumerate the different paths through the network and to add the cumulative times associated with each path. These are as follows:

Path	Cumulative Time (months)
1–2–5–8	7
1–2–6–8	9
1–3–6–8	7
1–3–7–8	5
1–4–7–8	6

The network represents a situation in which each path is "traversed" in carrying out the project. Thus, the longest path is critical in that it determines the minimum time that the entire project can take. It is also true that if delays occur in activities along that path (e.g., 1–2, or 2–6, or 6–8) then the project would inevitably be delayed. For these reasons it is referred to as the critical path. In the example, the critical path is 9 months.

Although the critical path was easily found in this example by enumeration, that would take too long for larger networks. A much more efficient procedure involves finding the path(s) for which there is no slack. This is accomplished by computing the earliest start time (EST), and latest finish time (LFT) for each event. The EST for each event, j, is found by traversing the network from left to right. The EST for event 1 is defined as 0. From there on, it is given by

$$EST_j = Max_{(i,j)} (EST_i + t_{i,j}) \qquad i < j$$

where EST_j is the earliest start time for event j, (i,j) is the set of activities from event i to event j, and $t_{i,j}$ is the time required for activity i to j. Thus, the earliest start time for an event is determined by the EST for preceding events plus the maximum time required to complete the activities for that event.

Example 15–1b shows how EST's are computed.

EXAMPLE 15–1b

$EST_1 = 0$

$EST_2 = EST_1 + t_{1,2} = 0 + 2 = 2$

$EST_3 = EST_1 + t_{1,3} = 0 + 1 = 1$

$EST_4 = EST_1 + t_{1,4} = 0 + 4 = 4$

$EST_5 = EST_2 + t_{2,5} = 2 + 2 = 4$

$EST_6 = Max [(EST_2 + t_{2,6}); (EST_3 + t_{3,6})] = Max [(2 + 5); (1 + 4)] = 7$

$EST_7 = Max [(EST_3 + t_{3,7}); (EST_4 + t_{4,7})] = Max [(1 + 3); (4 + 1)] = 5$

$EST_8 = Max [(EST_5 + t_{5,8}); (EST_6 + t_{6,8}); (EST_7 + t_{7,8})]$

$\qquad = Max [(4 + 3); (7 + 2); (5 + 1)] = 9$

Note that the EST for the final event (project completion) agrees with the total time for the critical path obtained by enumeration. In other words, the project can take no less time than the EST for the final event.

The next step is to compute the LFT for each event. This is done by traversing the network backward, from the last to the initial event. The LFT for the last event, project termination, is equal to its EST. Since the LFT is the latest finish time for all activities leading to a particular event, it is given by

$$LFT_i = Min_{(i,j)} (LFT_j - t_{i,j}) \qquad i < j$$

EXAMPLE 15–1c

Example 15–1c illustrates the computation of the LFT's:

$LFT_8 = EST_8 = 9$

$LFT_7 = LFT_8 - t_{7,8} = 9 - 1 = 8$

$LFT_6 = LFT_8 - t_{6,8} = 9 - 2 = 7$

$LFT_5 = LFT_8 - t_{5,8} = 9 - 3 = 6$

$LFT_4 = LFT_7 - t_{4,7} = 8 - 1 = 7$

$LFT_3 = Min[(LFT_6 - t_{3,6}); (LFT_7 - t_{3,7})] = Min[(7 - 4); (8 - 3)] = 3$

$LFT_2 = Min[(LFT_5 - t_{2,5}); (LFT_6 - t_{2,6})] = Min[(6 - 2); (7 - 5)] = 2$

$LFT_1 = Min[(LFT_4 - t_{1,4}); (LFT_3 - t_{1,3}); (LFT_2 - t_{1,2})]$

$\qquad = Min[(7 - 4); (3 - 1); (2 - 2)] = 0$

After the earliest start times and latest finish times are computed, the critical path can be identified. Since it is the path with no slack, it is composed of a complete, continuous chain in which the EST and LFT for each event are the same. If more than one path connects two points along this chain, the critical path is the one for which the total activity time equals the difference between EST (or LFT) for those events.

EXAMPLE 15–1d

Example 15–1d displays a table in which the slack (or lack thereof) is depicted for each event. With an example as small as 15–1 it is easy to trace the network by inspection, using the results in Table 15–2 to trace the critical path, 1–2–6–8. This result is the same as that obtained previously by enumeration.

TABLE 15–2   Event Slack Times*

Event	EST	LFT	Slack
1	0	0	0
2	2	2	0
3	1	3	2
4	4	7	3
5	4	6	2
6	7	7	0
7	5	8	3
8	9	9	0

*Sometimes referred to as float time.

Dummy Activities Sometimes situations are such that events should not be connected by a standard activity. In such cases a dummy activity may be appropriate. For example, one may need to indicate precedence of one event over another, even though no real activity takes place between them. Or, when two or more events are prerequisites to one activity, it may not make sense to connect each of them with a separate activity arrow, because doing so would imply differentiable activities when, in fact, only one activity may be taking place. Consider the case of a product made up of 3 subassemblies. Each of the subassemblies must be completed before final assembly can take place. The advantage to using dummy activities in this case is illustrated by Figure 15–5, where nodes 1, 2, and 3 represent completion of each subassembly, and node 4 designates the completion of final assembly. Clearly Figure 15–5b is more apt. Dummy activities are assigned zero times. In all other respects, they are treated the same way as standard events.

The Critical Path Method

Many scheduling methods have been developed that utilize the network concepts described in the preceding section but differ with respect to some aspect of application. The critical path method of scheduling, usually referred to as CPM, is one of the most widely used. It was developed by the DuPont Company in 1957–1958 to aid in the scheduling of a large project. Example 15–2 gives the flavor of a realistic scheduling problem and its CPM analysis.

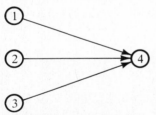

Without a dummy, one
would need to show the
same activity 3 times.

(a)

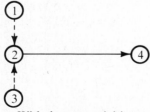

With dummy activities
showing zero times.

(b)

FIGURE 15–5 Dummy Activity. Multiple Event
Prerequisites for One Activity

EXAMPLE 15—2

Consider a construction project, a simple one-unit electrical pumping station without standby power, enclosed by a block-wall building.* As a first step in conducting a CPM analysis, all the elemental activities required to carry out the project must be identified. In addition, the expected time required to perform the activities must be estimated. Clearly this requires certain assumptions about the intensity with which resources will be put to use on that activity. A common approach in first estimating the activity times is to choose the method and manpower staffing levels that are reasonable and cost-minimizing for that activity. These are shown in Table 15—3.

TABLE 15—3 Activities and Expected Times

Activity	Expected Time (days)
Clear and grade site	5
Form and place rebars	5
Place concrete for footings and floor	8
Construct walls	6
Erect crane	4
Build roof	4
Install power supply	5
Paint exterior	3
Install pump and motor	5
Install controls	8
Test run	2
Landscape	4
Pave	3
Install suction and discharge lines	6
Deliver controls	1
Deliver crane	1
Deliver pump and motor	1
Install interior lighting	2

Next, in order to prepare the network diagram, the preceding, succeeding, and concurrent activities must be determined for each of the elemental activities. This can be done as a list, which becomes quite long for a large project. Even for Example 15—2, the list would be longer than the needs required by this text. Consequently, Table 15—4 shows the preceding, succeeding, and concurrent activities for only one activity.

Once the precedence relationships have been established, the network diagram can be drawn. This appears in Figure 15—6. Note that event numbers have been indicated in the node circles and expected activity times along the

*This example was adapted from R. Olson, "Critical Path Method of Work Scheduling," *Journal of the American Water Works Assoc.,* Vol. 61, No. 9 (September, 1969), by permission. Copyright © 1969, The American Water Works Association.

TABLE 15–4   Activity Precedence Relationships

Elemental Activity	Preceding Activity(ies)	Succeeding Activity	Concurrent Activity(ies)
.	.	.	.
.	.	.	.
.	.	.	.
Build roof	Erect crane	Paint	Install power supply
	Deliver pump & motor		Deliver controls
			Install pump & motor
			Install suct. & disch. lines
.	.	.	.
.	.	.	.
.	.	.	.

arrows, and that some dummy activities have been used. The reader should verify that the critical path is along 1–2–3–4–5–6–9–10–11–13, with the minimum time for completion being 43 days.

Applying CPM   Although the CPM network is ideal for planning in general and identifying crucial activities in particular, a diagram such as Figure 15–6 is not very helpful for monitoring and managing the project's progress. Therefore, some writers recommend that a modified Gantt chart be set up for that purpose. A sample of such a chart as applied to Example 15–2 appears in Figure 15–7. The added feature of this chart is the inclusion of slack times on the horizontal bars. Thus, the actual progress of the activities can be checked against the total of their expected and slack times. The Gantt chart should be used in conjunction with the network diagram, however, since the precedence relationships will not be apparent without it.

Once the project gets under way, it is unlikely that all the activities will progress as originally scheduled. Therefore, a periodic, routine reporting system is necessary. It should include project, activity, and time-period identification; it should indicate whether the activity in question is on the critical path; and it should show which subactivities (activity components) were begun, continued, and/or completed during that period.

Such a reporting system, when used in conjunction with CPM, permits remedial action to be taken before a project gets delayed. Critical path methods are helpful in that consequential delays (i.e., delays in activities on the critical path and in activities approaching their limit of slack) can be readily identified. Corrective measures that can be adopted include the use of additional manpower and equipment, overtime, additional shifts or crews, subcontracting some of the work, and so on.

It should be clear that critical path methods are useful for developing economic schedules as well as for dealing with existing ones. Trial schedules can be modified to see what the expected results would be if the total project time were shortened. When this is done it should be borne in mind that the time of some activities cannot be shortened as easily as that of others; for example, concrete cures at a rate independent of staffing levels.

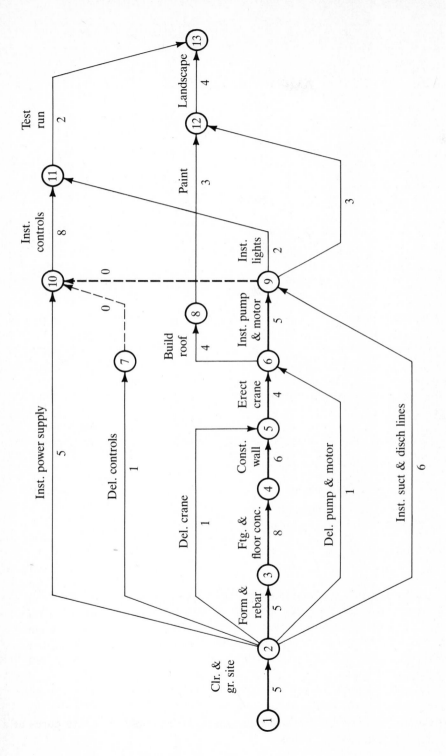

Network diagram to which event numbers and activity durations have been added.

FIGURE 15–6

333

Activity			Working Days				
Name	Code	CP*	10	20	30	40	50
Clr. gr. site	1-2	X					
Form & rebar	2-3	X					
Ftg. & fir. conc.	3-4	X					
Const. walls	4-5	X					
Erect crane	5-6	X					
Inst. power sply.	2-10						
Del. controls	2-7						
Del. crane	2-5						
Del. pump & motor	2-6						
Inst. suct. lines	2-9						
Inst. pump & motor	6-9	X					
Inst. controls	10-11	X					
Test run	11-13	X					
Build roof	6-8						
Paint	8-12						
Landscape	12-13						
Interior lights	9-11						
Pave	9-12						

* Critical Path   Duration   Free Float   Interfering Float

FIGURE 15—7

Another aim of a trial schedule might be to evaluate the cost of a particular staffing level for given activities. Alternative staffing levels, resulting in different expected activity times and activity costs, can then be intelligently compared. Other trial schedules can indicate the total cost or manpower ramifications of uniformly changing the way that the component activities are executed—doubling crew sizes, for instance.

When the number of events is large, it is tedious and time-consuming to compute critical paths and slacks by hand. Computer programs are available for this purpose and are particularly helpful in the "design stage," when schedules are being modified (and networks recomputed) to meet cost, manning, or total project time requirements.*

PERT

PERT, which stands for Program Evaluation and Review Technique, was developed by the U.S. Navy at about the same time the DuPont Company was developing CPM. It was necessitated by the scheduling and management problems posed by a complex, large-scale project: the first submarine-launched ballistic missile system.

PERT differs from CPM mainly in that it treats the time for an activity as a random variable. PERT allows estimates of the pessimistic, optimistic, and most likely times associated with each. Thus, PERT is better suited than CPM to situations in which the activities are novel or where previous experience is not so sure a guide to the anticipated activity durations. This was precisely the case for the Polaris program, for which PERT was developed—that is, it was a research, development, and production project for which substantial technological innovation and application were necessary. PERT-type methods are now routinely used for scheduling and managing engineering R&D. They are also used for complex computer software development projects.

One way to apply the PERT procedure is to assume that the beta probability distribution describes the likelihood of completing an activity in a given time. This is depicted in Figure 15–8.

The beta distribution was chosen because its shape could be skewed to the right, which was believed to be descriptive of most activities. Therefore, if t_o represents the optimistic time (almost no chance that the activity can be finished earlier under the assumed *modus operandi*), t_{ml} the most likely time (the time that would occur most often for the activity if it was a repeatable experiment),

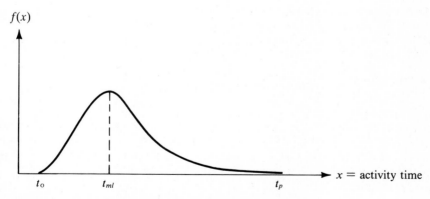

FIGURE 15–8 Beta Distribution Density Function

*Such a program was published in *Industrial Engineering*, January, 1981, pp. 20–22 and is reprinted in Appendix F.

and t_p the pessimistic time (almost no chance that the activity will take longer under the assumed *modus operandi*), then the expected value for activity duration for the beta distribution can be approximated. If $E(t)$ is the expected activity time, that approximation is given by,

$$E(t) = \frac{t_o + 4t_{ml} + t_p}{6} \qquad t_o \leq t_{ml} \leq t_p \tag{15-1}$$

Once the expected times have been computed for each of the activities, the expected project duration and critical path can be computed as before. This is because a random variable $X = A + B + \cdots + N$, has an expected value $E(X) = E(A) + E(B) + \cdots + E(N)$. In addition, the PERT approach allows one to estimate the probability that an activity can be completed by a given time. To do so, it is necessary to have the variance of the time associated with each activity. For the assumed beta distribution, the variance for activity i,j would be approximated by σ_{ij}^2, where

$$\sigma_{ij}^2 = \left(\frac{t_p - t_o}{6}\right)^2 \tag{15-2}$$

Then, if the number of activities and events in the network and between events p and r is large, one can assume that the random variable, duration time of events p through r, is normally distributed. The expected value of that duration time is given by the sum of the expected activity times of events p through r, and its variance is given by the sum of the activity variances. Although Example 15-3, illustrated by Figure 15-9, involves only three activities, the normality assumption is employed to illustrate the idea.

EXAMPLE 15-3

The reader should first verify that for each activity the optimistic, pessimistic, and most likely times (in weeks) shown below the arrows in Figure 15-9 give the expected times and variances (in parentheses) above the arrows. Then the

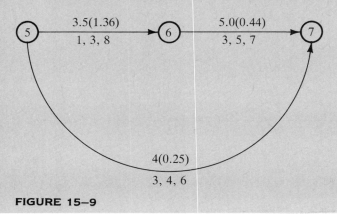

FIGURE 15-9

duration of 5,7 is treated as though it is determined by the critical path of its component activities; expected times, here being 5–6–7. (Since this is a probabilistic situation, it is clearly possible for 5–7 to actually take longer than 5–6–7, although that is not as likely.) The expected time for 5,7 is given by

$$E(5 - 7) = E(5 - 6) + E(6 - 7) = 3.5 + 5 = 8.5$$

Its variance is

$$\sigma_{5,7}^2 = \sigma_{5,6}^2 + \sigma_{6,7}^2 = 1.36 + 0.44 = 1.8$$

Thus, if the project were now at event 5, one could estimate the probability of completing event 7 in a given time interval. For example, if there were a penalty for not completing event 7 in 10 more weeks, one would be interested in the probability of suffering that penalty. The probability of being penalized is 1 − the probability of completing (5,7) within 10 weeks. That, in turn, is given by the normal distribution with mean 8.5 weeks and variance of 1.8 weeks^2 (standard deviation of 1.34 weeks) and is given by

$$P(X \leqslant 10) = P\left(Z \leqslant \frac{10 - 8.5}{1.34} \right) = P(Z \leqslant 1.12) = .8686$$

Therefore, there is only about a 13-percent chance that the activities will not be completed in time to avoid the penalty.

Exercises

15–1 Given the following information, construct a network diagram incorporating the precedence relationships indicated. Label the nodes any way you wish.

Activity Precedence Relationships

Elemental Activity	Preceding Activity(ies)	Succeeding Activity(ies)	Concurrent Activity(ies)
A		D	B,C
B		E	A,C
C		F,G,H	A,B
D	A	I	E
E	B	I	D
F	C	J	G,H
G	C	K	F,H
H	C	L	F,G
I	D,E	M	J
J	F	M	I
K	G	N	L
L	H	N	K
M	I,J		N
N	K,L		M

15–2 Given the following information, construct a network diagram incorporating the precedence relationships indicated. Label the nodes any way you wish. You may need dummy activities in certain instances.

Activity Precedence Relationships

Elemental Activity	Preceding Activity(ies)	Succeeding Activity(ies)	Concurrent Activity(ies)
A		C,D	B
B		E,F	A
C	A	G,H	D
D	A	I	C
E	A,B	J,K	F
F	A,B	L	E
G	C	M	H
H	C	N	G
I	D	N	H,J
J	E,F	N	I,K
K	E,F	O	J,L
L	F	O	K
M	G		N,O
N	H,I,J		M,O
O	K,L		M,N

15–3 The following table shows a set of activities and their event nodes, as well as the activity times for a project. Construct its network diagram; calculate the earliest start and latest finish times as well as slack times for the events, using that information to determine the critical path.

Activity	Duration
1–2	3
1–3	5
1–4	2
2–5	6
2–6	1
3–6	2
4–6	5
4–7	6
5–8	3
6–8	1
7–8	3

15–4 The following table shows a set of activities and their event nodes, as well as the activity times for a project. Construct its network diagram; calculate the earliest start and latest finish times as well as slack times for the events, using that information to determine the critical path.

Activity	Duration
1–2	3
1–3	2
1–4	3
2–5	2
2–6	4
3–6	4
4–7	4
4–8	2
5–9	5
6–9	2
7–9	6
7–11	4
8–10	4
10–11	3
9–11	2

15–5 Given the following information, construct a network diagram incorporating the precedence relationships indicated. Label the nodes any way you wish. You may need dummy activities in certain instances. Then use the indicated durations to calculate the earliest start and latest finish times as well as slack times for the events, using that information to determine the critical path.

Activity Precedence Relationships

Elemental Activity	Preceding Activity(ies)	Succeeding Activity(ies)	Concurrent Activity(ies)	Duration
A		C,D,E	B	3
B		F,G	A	5
C	A	H	D,E	4
D	A	I	C,E	2
E	A	J,K	C,D	5
F	B	J,K	E,G	2
G	B	K,L	F	4
H	C	M		5
I	D	N	J	6
J	E,F	N	I,K	2
K	E,F,G,	N,O	J,L	3
L	G	N,O	K	3
M	H	N	I	3
N	M,I,J,K,L	O		4
O	K,L		N	3

15–6 The following diagram is a PERT-type network showing the optimistic, pessimistic, and most likely duration times for the activities.

  a. Using the assumptions described in the text, find the expected duration for the entire project.

  b. What is the probability that the project will require less than 14 time units?

  c. What is the probability that event 5 can begin by time 10?

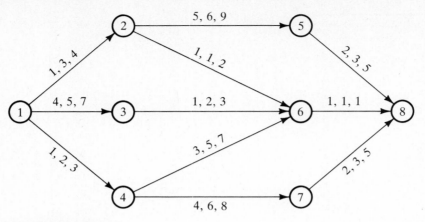

FIGURE 15–EX-1

15–7 A PERT-type network showing the optimistic, pessimistic, and most likely duration times for the activities is diagrammed below.

   a. Using the assumptions described in the text, find the expected duration for the entire project.

   b. What is the probability that the project will require less than 15 time units?

   c. What is the probability that event 7 can begin by time 8?

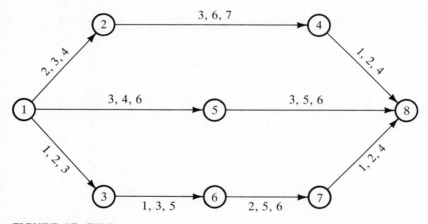

FIGURE 15–EX-2

Selected References

Hillier, F. and G. Lieberman, *Operations Research,* 3rd ed. San Francisco: Holden-Day, 1980.

Moder, J. and C. Phillips, *Project Management with CPM and PERT,* 2nd ed., New York: Van Nostrand, 1970.

Whitehouse, G. *Systems Analysis and Design using Network Techniques,* Englewood Cliffs, NJ: Prentice-Hall, 1973.

Whitehouse, G. and B. Wechsler, *Applied Operations Research,* New York: John Wiley & Sons, 1976.

Whitehouse, G. and D. Washburn, "Critical Path Analysis," *Industrial Engineering,* December, 1980.

16

Conclusion

Introduction

This chapter will tie together many of the concepts explored throughout the text. Since all the preceding examples and exercises were
designed to demonstrate particular points, there has been no opportunity to see how these tools are commonly used in industry. Therefore, the first part of this chapter will provide a case study for that
purpose. Then another illustration, based on a student project, will
show how engineering economics concepts may be applied in practice. A short discussion of pitfalls in the application of these tools will
conclude the chapter.

Case Study

This case reports on a preliminary feasibility study (including economic analysis)
conducted at a multidivision international company. The location is a manufacturing
plant where health and beauty products are made. Prior to the study, all output
of the three main production departments (accounting for about 85 percent of
plant production) was manually palletized for shipment. Of that amount, only a
small portion of the loaded pallets received a stretch wrapping. (Stretch wrapping
is the application of a film covering the entire loaded pallet. The advantages to
this process are outlined later.) The purpose of the study was to determine whether
automatic palletizing of the entire output of the departments mentioned could be

accomplished, and if so, how this could be done; to determine whether all the output could be stretch-wrapped; and to perform an economic analysis, given the design requirements for automatic palletizing and stretch-wrapping, to see whether investment for these purposes could be justified.

Existing Conditions

Prior to the study, all output of the three major departments, packed in shipping cartons, was loaded onto pallets by hand. Materials-handling employees performed this operation by removing cases from the line, stacking the cases according to a set pattern, then ticketing the completed pallet. This operation had a history of causing numerous accidents, from minor lacerations to major back injuries. In the few instances when stretch-wrapping was called for, an additional materials-handling employee was needed. The completed pallets would then be transported by a forklift truck to the holding area in the warehouse, where further ticketing would be done. Finally the completed pallet would be transported to the distribution center.

After several weeks of careful study of the output flow requirements, research into available equipment, and trial layouts of the equipment given the existing building configuration, two similar proposals emerged for conducting the automatic palletizing and stretch-wrapping. They both required the purchase of two centrally located palletizers, two portable palletizers, and one centrally located stretch-wrapper. In addition, conveyor and material flow control equipment would be needed.

Proposal 1

The first system proposed would be located in the holding area of the warehouse, occupying about 4,000 square feet of space. It would require the service of only one materials-handling employee.

The processing sequence was planned as follows:

- Finished goods would leave each production line and enter an overhead conveying system by means of incline conveyors. Electronic photocells would be positioned so as to prevent any jam-ups which might occur as a carton enters the system. The cartons would then be conveyed to the holding area.
- Upon entering the holding area, cartons would be diverted to their respective holding lines through the use of electronic photocells. Decline conveyors would then be used to bring the cartons to floor level.
- A counter would allow the appropriate number of cartons to enter the conveyor feeding into the palletizing machines. A computer would be used to orchestrate the flow of cartons from the seven holding lines to the two palletizers.
- The cartons would enter the palletizer, be stacked following a programmed pattern, and finally be deposited onto a conveyor feeding into the automatic stretch-wrapper. A shuttle car permits three pallets to be formed at a time.

- Pallets would then be transported by conveyor into an in-line automatic stretch-wrapper. Pallets would be stretch-wrapped, with film securing the top as well as the sides of the load. The wrapped pallet would then exit the stretch-wrapper and await transportation to a shuttle by means of a forklift truck.
- The preceding description applies to most, but not all, production of the three departments in question. Owing to limitations in the facilities structure, a few products cannot be processed as described above but would require portable palletizers located on the production floor. These pallets would then be taken by forklift truck to the warehouse for stretch-wrapping.

The investment requirements for this system appear in subsequent tables.

Advantages/Disadvantages, Proposal 1 The following advantages were identified for the first proposal:

- *Direct labor savings*
  A reduction of 8 or 9 materials-handling personnel was projected as a result of the automatic palletizing. However, one line-service employee and one materials-handling employee would be needed for the system.
- *Indirect labor savings*
  Since the forklift trucks would pick up most of the finished pallets in the warehouse as opposed to the production floor, there would be a reduction in forklift run time. This provides savings in forklift operators' hours and in fuel usage, and a reduction in traffic.
- *Reduced injuries*
  By eliminating manual palletizing, a reduction in injuries was projected. This would lead to reduced downtime and reduced insurance payments attributable to a lower injury incidence rate.
- *Reduced product damage*
  The stretch-wrapping procedure was expected to reduce the incidence of goods damaged in transit by breakage and dirt contamination.

The following disadvantages of the system were identified and considered:

- *Ticketing requirements*
  New procedures would have to be developed for ticketing the pallets, since the current procedures would not be suitable.
- *Space usage*
  Additional warehouse floor space would be needed for the system, preempting other uses. Alteration in the warehouse layout would be a response to this need.

The net savings (or costs) attributable to each of the above was estimated and is reported in the tables that follow.

Proposal 2

The methods and procedures for Proposal 2 are essentially the same as for Proposal 1. However, it differs with respect to location. In Proposal 2, the system would be located in the warehouse near the truck dock rather than in the holding area. As a result, the system would require slightly more space than Proposal 1.

Advantages/Disadvantages, Proposal 2 The same advantages and disadvantages would accrue to Proposal 2, with the following differences:

- *Materials flow reduction*
  With the proposed system located adjacent to the truck dock, finished goods would be deposited closer to the shipping point. Also, raw materials could be held closer to the production floor (since the holding area would not be needed for the palletizing and wrapping system). The net result would be a further reduction in forklift truck run time with a concomitant reduction in indirect labor requirements.
- *Investment cost increase*
  A greater investment cost is needed for Proposal 2, since a longer conveyor is required.

The net savings (or costs) attributable to each of the above was estimated and appears in the tables that follow.

Economic Analysis

Table 16–1 shows how the capital costs for the two proposals were developed, and Table 16–2 does the same for net annual savings. These data, developed as a

TABLE 16–1 Capital Costs

	Proposal 1	Proposal 2
Conveyor	$110,500	$138,000
Palletizers	267,000	267,000
Stretch-wrapper	67,500	67,500
Photoelectric system	50,000	50,000
Installation	21,500	21,500
Contingency	41,320	41,320
Total	$557,820	$587,520

TABLE 16–2 Annual Savings

	Proposal 1	Proposal 2
Reductions:		
Direct labor	$175,000	$175,000
Indirect labor	11,862	19,642
Accident costs	15,000	15,000
Product damage costs	84,300	84,300
Total	$286,162	$293,942
Cost increases	80,180	81,807
(See Table 16–3)		
Net annual savings	$205,982	$212,135

result of careful, detailed study and design work, are sufficient to permit evaluation of the proposals as outlined throughout this text.

Before turning attention to the economic analysis, however, the reader may find it interesting and useful to get a glimpse of how the study process has been formalized in a major American corporation such as this one. (Such a formalization is actually quite common.) As in most big firms, the process has been formalized in order to provide a reliable planning process, while insuring uniformity among divisions and departments in the analytical procedures.

It should be pointed out that, although an attempt is being made here to depict the nature of the details of the study, it is not practical to duplicate it entirely. Such a study might have many scores of pages of documents and working papers, which it is not feasible to present here. Thus, not all of the documents used for this study appear in the text, but included are facsimiles of the most pertinent ones that were used for the economic analysis.

Figure 16–1, which shows the basis of support for the appropriation requests for each of the two proposals, presents data for the first full year of operation. Noteworthy in the calculation of benefits is the itemization of categories for potential savings and increases in costs attributable to the proposals. Lines 1 to 5 address those categories most likely to be affected in the typical study: direct and indirect labor, materials, maintenance and repair, and local insurance and taxes. Lines 6 and 7 are for categories associated with particular proposals—in this case energy use and costs related to accidents. It is the accumulated figures of lines 1 to 7 that appeared in Table 16–2, Annual Savings. Clearly a good deal of work went into these estimates.

The highlights of the assumptions and the concluding computations in arriving at the estimates for lines 1 to 7 are also reported on the form shown in Figure 16–2, Supplemental Material and Overhead Information.

In this company, once data have been developed in the form shown above, the after-tax return on investment for the proposal(s) must be computed. This is done by hand and is facilitated by a form shown in Figure 16–3, Schedule of Discounted Cash Flow. For proposals 1 and 2, depreciation was under ACRS (1984 percentages: 15, 22, 21, 21, 21). A 15-year study period was employed, assuming no market value of the investment after 15 years. Although the format of Figure 16–3 is somewhat different from that used in Chapter 8 of this text, the reader will no doubt see that it yields results that are algebraically equivalent. Columns 7 to 9 of the schedule provide trials for the IRR computations. For Proposal 1 the author obtained a 25.3 percent return using the computer programs described in the text, and for Proposal 2 the result was 24.8 percent. These are very close to the results obtained by the study analyst using linear interpolation for the data in Figure 16–3.

Recommendations

The firm's analyst suggested some further study before a final recommendation could be made. However, the preliminary results embodied in the existing study were deemed sufficient to indicate the apparent desirability of automating the

APPROPRIATION REQUEST - SUMMARY OF RECOVERY C103-B

	Location/Country
	Local Project Number

All figures in U.S. $ Exchange Rate: FORM 6086245 B

Date:

OJECT TITLE: CENTRALIZED PALLETIZING AND STRETCHWRAP - PROPOSAL I

PREPARED BY

Calculation of Recovery	Year of Installation 19 __ Increase	Decrease	First Full Year 19__ Increase	Decrease	Average First Full 3 Years Increase	Decrease
1. Direct Labor Including Benefits			32,000	175,000		
2. Indirect Labor Including Benefits				11,862		
3. Raw and Packing Materials			12,510	84,300		
4. Maintenance and Repair			18,800			
5. Local Insurance and Taxes			9,020			
6. Others (Explain) Energy Usage			7,850			
7. Disability Savings				15,000		
8. Total Changes in Cost						
9. Gross Saving (To Form C104 Column 1)			205,982			
10. Subtract: Depreciation	(	)	(83,673)		(	)
Pre-Tax Savings			122,309			
12. Subtract: Income Taxes a __46__ %	Above (Line 11) To Form C102 Section 1V **Profit (Loss) Effect**		(56,262)		(	)
13. Net Savings or Increased Profit (To Form C102 Section 111H)			66,047			
14. Add: Depreciation			83,673			
15. Annual Cash Flow (To Form C102 Section 111J)			149,720			

COMMENTS:

FIGURE 16—1a

palletizing operation and conducting stretch-wrapping for all output. Proposals 1 and 2 were both viewed as effective, but Proposal 2 was thought to be qualitatively more desirable because it recommended a preferable location. Both systems were viewed as capable of adaptation for further automation or for expansion for meeting future needs of increased production.

Comments

Although this study did incorporate many of the tools discussed in this text, a number of important issues were not touched on at all. Included in these are the following:

In comparing Proposals 1 and 2 it is possible not only to evaluate the qualitative difference between them (i.e., location) but also to compute the incremental benefit provided by Proposal 2 in relation to its increased investment re-

APPROPRIATION REQUEST - SUMMARY OF RECOVERY C103-B	Location/Country

FORM 6066245 B

All figures in U.S. $ Exchange Rate:

Local Project Number

Date:

PROJECT TITLE: CENTRALIZED PALLETIZING AND
STRETCHWRAP - PROPOSAL II

PREPARED BY

Calculation of Recovery	Year of Installation 19 __		First Full Year 19 __		Average First Full 3 Years	
	Increase	Decrease	Increase	Decrease	Increase	Decrease
1. Direct Labor Including Benefits			32,000	175,000		
2. Indirect Labor Including Benefits				19,642		
3. Raw and Packing Materials			12,510	84,300		
4. Maintenance and Repair			19,900			
5. Local Insurance and Taxes			9,547			
6. Others (Explain) Energy Usage			7,850			
7. Disability Savings				15,000		
8. Total Changes in Cost						
9. Gross Saving (To Form C104 Column 1)			212,135			
10. Subtract: Depreciation	(	)	(88,125)		(	)
Pre-Tax Savings			124,010			
12. Subtract: Income Taxes @ __46__ %	Above (Line 11) To Form C102 Section IV "Profit (Loss) Effect"		(57,045)		(	)
13. Net Savings or Increased Profit (To Form C102 Section 111H)			66,965			
14. Add: Depreciation			88,125			
15. Annual Cash Flow (To Form C102 Section 111J)			155,090			

COMMENTS:

FIGURE 16—1b

quirement. The author conducted such an exercise, which appears in Tables 16–3 and 16–4. The transformation of these incremental benefits to after-tax flows appears in Table 16–4 (see p. 354). This yields an internal rate of return on the incremental investment of 13.96 percent. There was no indication in the study documents as to whether this would be satisfactory. However, such an analysis should be done to see whether the incremental investment can be justified economically.

No sensitivity analysis was done for any of the variables of the study.

No attention was given to the potential effects of inflation over the fifteen-year study period. The greatest annual flow, by far, was the annual savings in labor expense, which was offset to a small degree by other costs. Thus, it is likely

TABLE 16—3   Incremental Analysis, Proposal 2 vs. 1

Incremental Investment	$29,700
Incremental Savings	$6,153/yr.

SUPPLEMENTAL MATERIAL AND OVERHEAD INFORMATION

FORM C103-2   6086247

Appro Number	Project Number	Division	Location	Department
	PROPOSAL I			

A. RAW MATERIAL – (INCREASE) OR DECREASE $ _____

B. PACKAGING MATERIAL – (INCREASE) OR DECREASE $ (12,510) _____

$ .50/pallet x 25,020 pallets/year = $12,510

C. MAINTENANCE AND REPAIRS – (INCREASE) OR DECREASE $ (18,800) _____

.04 x $470,000   = $18,800

D. INSURANCE AND TAXES – (INCREASE) OR DECREASE $ (9,020) _____

19.99/M x $470,000   = $9,020

E. OTHER – (INCREASE) OR DECREASE $ _____

FIGURE 16—2a

that an inflation-adjusted study would find the investments even more desirable, since labor costs are expected to increase steadily over that period.

An investment tax credit was not considered, even though such a provision was in effect at the time of the study. Inclusion of the credit would have made the investments more desirable, as was indicated in Chapter 8.

In conclusion, from this case the reader should not only derive the flavor of the context of a real study but also realize that real-world practice is far from uniformly perfect or ideal in its application of readily acknowledged principles.

SUPPLEMENTAL MATERIAL AND OVERHEAD INFORMATION

FORM C103-2 6086247

Appro Number	Project Number	Division	Location	Department
	PROPOSAL II			

A RAW MATERIAL – (INCREASE) OR DECREASE $ _____

B PACKAGING MATERIAL – (INCREASE) OR DECREASE $ (12,510)

   $ .50/pallet x 25,020 pallets/year = $12,510

C. MAINTENANCE AND REPAIRS – (INCREASE) OR DECREASE $ (19,900)

   (4% of purchase price)

   .04 x $497,500 = $19,900

D INSURANCE AND TAXES – (INCREASE) OR DECREASE $ (9,547)

   $19.19/M x $497,500 = $9,547

E OTHER – (INCREASE) OR DECREASE $ _____

FIGURE 16–2b

The last statement is not intended to denigrate the company or its analysts but merely to indicate the reality that the reader should be prepared to face.

There is yet another interesting aspect of this case. It should make clear that the time required to undertake the typical study is devoted to a very great degree to gauging the technological requirements, generating the design alternatives, and estimating the economic characteristics of each design alternative. Analysis of the resulting data, which was the lesson of this book, takes relatively little time in practice, especially if computers are used.

SCHEDULE OF DISCOUNTED CASH FLOW

FORM C 104

	NET SAVINGS BEFORE TAX OR DEPRECIATION (1)	DEPRECIATION (TAX METHOD) (2)	NET BEFORE TAX COL. 1 - COL. 2 (3)	NET AFTER TAX COL. 3 X ___% (4)	DEPRECIATION (TAX METHOD) COL. 2 (5)	CASH FLOW COL. 4 + COL. 5 (6)	25% RETURN FORMULA (7)	25% RETURN FORMULA COL. 6 X COL. 7 (6 × 7)	40% RETURN FORMULA (8)	40% RETURN FORMULA COL. 6 X COL. 8 (6 × 8)	60% RETURN* FORMULA (9)	60% RETURN* FORMULA COL. 6 X COL. 9 (6 × 9)
0	0	-587500				-587500	1.000	-587500	1.000	-587500	1.000	-587500
1	212135	88125	124010	66965	89125	155090	.885	137255	.824	127794	.7520	116628
2	212135	129250	82885	44758	129250	174008	.689	119891	.553	96226	.4127	71813
3	212135	123375	88760	47930	123375	171305	.537	91991	.370	63393	.2265	38801
4	212135	123375	88760	47930	123375	171305	.418	71606	.248	42484	.1243	21293
5	212135	123375	88760	47930	123375	171305	.326	55846	.166	28437	.0682	11683
6	212135	0	212135	114553	0	114553	.254	29096	.112	12830	.0374	4284
7	212135	0	212135	114553	0	114553	.197	22567	.075	8591	.0205	2348
8	212135	0	212135	114553	0	114553	.154	17641	.050	5728	.0113	1294
9	212135	0	212135	114553	0	114553	.119	13632	.034	3895	.0062	710
10	212135	0	212135	114553	0	114553	.093	10653	.023	2635	.0034	389
11	212135	0	212135	114553	0	114553	.073	8362	.015	1718	.0019	218
12	212135	0	212135	114553	0	114553	.057	6530	.010	1146	.0010	115
13	212135	0	212135	114553	0	114553	.044	5040	.007	802	.0006	69
14	212135	0	212135	114553	0	114553	.034	3895	.005	573	.0003	34
15	212135	0	212135	114553	0	114553	.027	3093	.003	344	.0002	23
TOTALS						1401044		9598		-190915		-317997

* UNIFORMLY OVER THE YEARS CONTINUOUS TABLES

SCHEDULE OF DISCOUNTED CASH FLOW

FORM C104

| APPROPRIATION NUMBER | LOCAL PROJECT NUMBER | DIVISION | LOCATION/COUNTRY | DEPARTMENT |

PROJECT TITLE

	NET SAVINGS BEFORE TAX OR DEPRECIATION (1)	DEPRECIATION (TAX METHOD) (2)	NET BEFORE TAX COL. 1 - COL. 2 (3)	NET AFTER TAX COL. 3 X ___% (4)	DEPRECIATION (TAX METHOD) COL. 2 (5)	CASH FLOW COL. 4+COL. 5 (6)	25% RETURN* FORMULA (7)	25% RETURN* FORMULA COL. 6 X COL. 7	40% RETURN* FORMULA (8)	40% RETURN* FORMULA COL. 6 X COL. 8	60% RETURN* FORMULA (9)	60% RETURN* FORMULA COL. 6 X COL. 9
0	0	-557820				-557820	1.000	-557820	1.000	-557820	1.000	-557820
1	205982	83673	122309	66047	83673	149720	.885	132502	.824	123369	.7520	112589
2	205982	122720	83262	44961	122720	167682	.689	115533	.553	92728	.4127	69222
3	205982	117142	88840	47973	117142	165116	.537	88667	.370	61093	.2265	37399
4	205982	117142	88840	47973	117142	165116	.418	69018	.248	40949	.1243	20524
5	205982	117142	88840	47973	117142	165116	.326	53828	.166	27409	.0682	11261
6	205982	0	205982	111230	0	111230	.254	28252	.112	12458	.0374	4160
7	205982	0	205982	111230	0	111230	.197	21912	.075	8342	.0205	2280
8	205982	0	205982	111230	0	111230	.154	17129	.050	5562	.0113	1257
9	205982	0	205982	111230	0	111230	.119	13236	.034	3782	.0062	690
10	205982	0	205982	111230	0	111230	.093	10349	.023	2558	.0034	378
11	205982	0	205982	111230	0	111230	.073	8120	.015	1668	.0019	211
12	205982	0	205982	111230	0	111230	.057	6340	.010	1112	.0010	111
13	205982	0	205982	111230	0	111230	.044	4894	.007	779	.0006	67
14	205982	0	205982	111230	0	111230	.034	3782	.005	556	.0003	33
15	205982	0	205982	111230	0	111230	.027	3003	.003	334	.0002	22
TOTALS						1367231		187142		-175121		-297635

* UNIFORMLY OVER THE YEARS CONTINUOUS TABLES

FIGURE 16-3

353

TABLE 16—4   After-tax Cash Flow; Incremental Benefits   (in dollars)

End of Year	Before-tax Cash Flow	Depreciation Charges	Other Tax Effects	Taxable Income	Income Taxes	After-tax Cash Flow
(1)	(2)	(3)	(4)	(5)	(6)	(7)
				=(2) − (3)	=(5) × 46%	=(2) − (6)
0	− 29,700					− 29,700
1	6,153	4,455		1,698	781	5,372
2	6,153	6,534	‣	− 381	− 175	6,328
3–5	6,153	6,237		− 84	− 39	6,192
6–15	6,153	0		6,153	2,380	3,323

Student Project

A further illustration of how the tools of this text may be professionally adapted to practical design situations is provided by a project done by a student in 1984. The student, a graduate chemical engineer working for a major American chemical company, was in his first engineering economics course and undertook this project after reading "Insulation Without Economics," an article in the May 3, 1982 issue of *Chemical Engineering*. The article outlined the pertinent technical features of the heat losses that would be experienced in pipes used for transmission of steam in chemical plants. Variables such as the outer diameter of pipe, length of the pipe run, average ambient temperature, temperature of the process steam, and efficiency of the fuel used to produce the steam were used to generate equations to indicate energy losses from initial energy input to final use.

Using these thermodynamic equations, the student wrote a Fortran program to aid engineers in determining the economically optimal thickness of insulation to be applied to steam pipes of this sort. The program evaluates the factors described in the article mentioned above, using the given equations as a basis for an economic analysis. The result is a table of yearly costs and cash flows, on an after-tax basis, for any design configuration. The equivalent uniform annual cost of each alternative is computed (on an after-tax basis) so that the least-cost option may be identified.

A Final Caveat

Although it is the author's firm conviction that wise use of the concepts presented in this book is of great value and importance, there are some dangers that the reader should be made aware of. One is that, as in all sorts of modeling, the recommendations generated are reliable only if the modeling process has fairly reflected reality. Thus, a pitfall exists in oversimplifying a real situation.

This point can be taken a step further. A widespread problem of U.S. industry beginning in the 1970s has been a lack of competitiveness in world markets.

Although many factors have been suggested as contributing to this, such as higher wage rates, unfair foreign competition, and older physical plants, some critics have argued that improper management practices have also contributed to this problem.*

Among management's shortcomings, it has been suggested, is too-strict reliance on evaluative models, such as time value analysis as extensively developed in this text. Since it involves "discounting the future," such a practice, if oversimplified and applied mechanically, can lead to an overemphasis on short-term results at the expense of sound long-run strategic planning. Long-term advantages may then be sacrificed for short-run profits. However, such criticisms may be directed at any scientific procedure: it may be used wisely to advantage, or unwisely. Careful use of the tools of this text in the context of sincere commitment to innovation, leadership, and creation of greater economic value are the only roads to future success.

Selected References

Happel, J. and D. Jordan, *Chemical Process Economics,* 2nd ed. New York: Marcel Dekker, 1975.

---

*See, for example, Hayes and Abernathy, "Managing Our Way to Economic Decline," *Harvard Business Review,* July-Aug., 1980, pp. 67–77.

Appendix A

Discrete Compounding Tables

Discrete Compounding; i = 0.5%

N	(F/P)	(P/F)	(F/A)	(P/A)	(A/F)	(A/P)
1	1.00500	.99500	1.00000	.99505	1.00000	1.00500
2	1.01003	.99007	2.00501	1.98510	.49875	.50375
3	1.01508	.98515	3.01514	2.97036	.33166	.33666
4	1.02015	.98025	4.03004	3.95044	.24814	.25314
5	1.02525	.97537	5.05028	4.92589	.19801	.20301
6	1.03038	.97052	6.07548	5.89636	.16460	.16960
7	1.03553	.96569	7.10602	6.86220	.14073	.14573
8	1.04071	.96089	8.14133	7.82289	.12283	.12783
9	1.04591	.95611	9.18198	8.77894	.10891	.11391
10	1.05114	.95135	10.22797	9.73036	.09777	.10277
11	1.05640	.94662	11.27911	10.67697	.08866	.09366
12	1.06168	.94191	12.33540	11.61879	.08107	.08607
13	1.06699	.93722	13.39722	12.55613	.07464	.07964
14	1.07232	.93256	14.46419	13.48868	.06914	.07414
15	1.07768	.92792	15.53669	14.41675	.06436	.06936
16	1.08307	.92330	16.61415	15.33986	.06019	.06519
17	1.08849	.91871	17.69715	16.25850	.05651	.06151
18	1.09393	.91414	18.78567	17.17267	.05323	.05823
19	1.09940	.90959	19.87973	18.08236	.05030	.05530
20	1.10489	.90506	20.97893	18.98727	.04767	.05267
21	1.11042	.90056	22.08405	19.88801	.04528	.05028
22	1.11597	.89608	23.19432	20.78397	.04311	.04811
23	1.12155	.89162	24.31050	21.67576	.04113	.04613
24	1.12716	.88719	25.43183	22.56277	.03932	.04432
25	1.13279	.88277	26.55888	23.44546	.03765	.04265
26	1.13846	.87838	27.69185	24.32397	.03611	.04111
27	1.14415	.87401	28.83015	25.19786	.03469	.03969
28	1.14987	.86966	29.97418	26.06743	.03336	.03836
29	1.15562	.86534	31.12431	26.93296	.03213	.03713
30	1.16140	.86103	32.27997	27.79402	.03098	.03598
35	1.19073	.83982	38.14526	32.03529	.02622	.03122
40	1.22079	.81914	44.15856	36.17204	.02265	.02765
45	1.25162	.79896	50.32388	40.20702	.01987	.02487
50	1.28322	.77929	56.64483	44.14259	.01765	.02265
55	1.31563	.76009	63.12562	47.98136	.01584	.02084
60	1.34885	.74137	69.76949	51.72527	.01433	.01933
65	1.38291	.72311	76.58159	55.37722	.01306	.01806
70	1.41783	.70530	83.56573	58.93923	.01197	.01697
75	1.45363	.68793	90.72592	62.41337	.01102	.01602
80	1.49034	.67099	98.06711	65.80204	.01020	.01520
85	1.52797	.65446	105.59389	69.10734	.00947	.01447
90	1.56655	.63834	113.31046	72.33111	.00883	.01383
95	1.60611	.62262	121.22214	75.47559	.00825	.01325
100	1.64666	.60729	129.33276	78.54231	.00773	.01273
120	1.81939	.54963	163.87848	90.07320	.00610	.01110
150	2.11304	.47325	222.60804	105.34966	.00449	.00949
180	2.45408	.40748	290.81656	118.50317	.00344	.00844
200	2.71150	.36880	342.30068	126.24020	.00292	.00792
240	3.31019	.30210	462.03756	139.58049	.00216	.00716
250	3.47947	.28740	495.89396	142.52000	.00202	.00702
300	4.46494	.22397	692.98759	155.20657	.00144	.00644
360	6.02252	.16604	1004.50499	166.79137	.00100	.00600

Discrete Compounding; i = 1.0%

N	(F/P)	(P/F)	(F/A)	(P/A)	(A/F)	(A/P)
1	1.01000	.99010	1.00000	.99012	1.00000	1.01000
2	1.02010	.98030	2.01006	1.97045	.49750	.50750
3	1.03030	.97059	3.03020	2.94108	.33001	.34001
4	1.04061	.96098	4.06055	3.90211	.24627	.25627
5	1.05101	.95146	5.10120	4.85361	.19603	.20603
6	1.06152	.94204	6.15225	5.79568	.16254	.17254
7	1.07214	.93272	7.21378	6.72841	.13862	.14862
8	1.08286	.92348	8.28600	7.65196	.12069	.13069
9	1.09369	.91434	9.36889	8.56632	.10674	.11674
10	1.10463	.90528	10.46257	9.47160	.09558	.10558
11	1.11567	.89632	11.56730	10.36801	.08645	.09645
12	1.12683	.88745	12.68301	11.25547	.07885	.08885
13	1.13810	.87866	13.80987	12.13416	.07241	.08241
14	1.14948	.86996	14.94798	13.00413	.06690	.07690
15	1.16098	.86134	16.09754	13.86553	.06212	.07212
16	1.17259	.85282	17.25855	14.71837	.05794	.06794
17	1.18431	.84437	18.43118	15.56278	.05426	.06426
18	1.19616	.83601	19.61555	16.39883	.05098	.06098
19	1.20812	.82773	20.81175	17.22659	.04805	.05805
20	1.22020	.81954	22.01986	18.04613	.04541	.05541
21	1.23240	.81142	23.24018	18.85763	.04303	.05303
22	1.24473	.80339	24.47261	19.66104	.04086	.05086
23	1.25717	.79544	25.71735	20.45648	.03888	.04888
24	1.26975	.78756	26.97458	21.24408	.03707	.04707
25	1.28244	.77976	28.24440	22.02388	.03541	.04541
26	1.29527	.77204	29.52680	22.79590	.03387	.04387
27	1.30822	.76440	30.82218	23.56036	.03244	.04244
28	1.32130	.75683	32.13043	24.31720	.03112	.04112
29	1.33452	.74933	33.45184	25.06660	.02989	.03989
30	1.34786	.74192	34.78631	25.80849	.02875	.03875
35	1.41662	.70591	41.66211	29.40949	.02400	.03400
40	1.48889	.67164	48.88858	32.83568	.02045	.03045
45	1.56484	.63904	56.48364	36.09555	.01770	.02770
50	1.64466	.60803	64.46628	39.19726	.01551	.02551
55	1.72856	.57852	72.85594	42.14835	.01373	.02373
60	1.81674	.55044	81.67361	44.95623	.01224	.02224
65	1.90941	.52372	90.94122	47.62786	.01100	.02100
70	2.00681	.49830	100.68148	50.16979	.00993	.01993
75	2.10919	.47412	110.91869	52.58836	.00902	.01902
80	2.21678	.45110	121.67805	54.88953	.00822	.01822
85	2.32986	.42921	132.98633	57.07902	.00752	.01752
90	2.44871	.40838	144.87131	59.16222	.00690	.01690
95	2.57363	.38856	157.36282	61.14434	.00635	.01635
100	2.70491	.36970	170.49138	63.03024	.00587	.01587
120	3.30053	.30298	230.05329	69.70185	.00435	.01435
150	4.44867	.22479	344.86698	77.52136	.00290	.01290
180	5.99620	.16677	499.62007	83.32276	.00200	.01200
200	7.31656	.13668	631.65607	86.33235	.00158	.01158
240	10.89353	.09180	989.35245	90.82022	.00101	.01101
250	12.03327	.08310	1103.32643	91.68969	.00091	.01091
300	19.79066	.05053	1879.06616	94.94710	.00053	.01053
360	35.95444	.02781	3495.44315	97.21869	.00029	.01029

Discrete Compounding; i = 2.0%

N	(F/P)	(P/F)	(F/A)	(P/A)	(A/F)	(A/P)
1	1.02000	.98039	1.00000	.98041	1.00000	1.02000
2	1.04040	.96117	2.02003	1.94158	.49504	.51504
3	1.06121	.94232	3.06044	2.88392	.32675	.34675
4	1.08243	.92384	4.12164	3.80775	.24262	.26262
5	1.10408	.90573	5.20411	4.71351	.19216	.21216
6	1.12616	.88797	6.30817	5.60147	.15852	.17852
7	1.14869	.87056	7.43437	6.47206	.13451	.15451
8	1.17166	.85349	8.58302	7.32552	.11651	.13651
9	1.19509	.83675	9.75471	8.16229	.10251	.12251
10	1.21900	.82035	10.94980	8.98264	.09133	.11133
11	1.24338	.80426	12.16879	9.78690	.08218	.10218
12	1.26824	.78849	13.41219	10.57540	.07456	.09456
13	1.29361	.77303	14.68048	11.34846	.06812	.08812
14	1.31948	.75787	15.97405	12.10631	.06260	.08260
15	1.34587	.74301	17.29360	12.84936	.05782	.07782
16	1.37279	.72844	18.63942	13.57778	.05365	.07365
17	1.40024	.71416	20.01224	14.29196	.04997	.06997
18	1.42825	.70016	21.41247	14.99211	.04670	.06670
19	1.45681	.68643	22.84074	15.67855	.04378	.06378
20	1.48595	.67297	24.29757	16.35152	.04116	.06116
21	1.51567	.65977	25.78354	17.01131	.03878	.05878
22	1.54598	.64684	27.29921	17.65814	.03663	.05663
23	1.57690	.63415	28.84522	18.29231	.03467	.05467
24	1.60844	.62172	30.42212	18.91402	.03287	.05287
25	1.64061	.60953	32.03058	19.52356	.03122	.05122
26	1.67342	.59758	33.67119	20.12114	.02970	.04970
27	1.70689	.58586	35.34460	20.70699	.02829	.04829
28	1.74103	.57437	37.05154	21.28138	.02699	.04699
29	1.77585	.56311	38.79261	21.84450	.02578	.04578
30	1.81137	.55207	40.56840	22.39655	.02465	.04465
35	1.99990	.50003	49.99495	24.99874	.02000	.04000
40	2.20805	.45289	60.40254	27.35559	.01656	.03656
45	2.43787	.41019	71.89350	29.49029	.01391	.03391
50	2.69160	.37153	84.58023	31.42372	.01182	.03182
55	2.97175	.33650	98.58761	33.17491	.01014	.03014
60	3.28106	.30478	114.05282	34.76101	.00877	.02877
65	3.62255	.27605	131.12774	36.19759	.00763	.02763
70	3.99959	.25003	149.97974	37.49873	.00667	.02667
75	4.41588	.22646	170.79382	38.67722	.00586	.02586
80	4.87549	.20511	193.77442	39.74462	.00516	.02516
85	5.38294	.18577	219.14692	40.71139	.00456	.02456
90	5.94320	.16826	247.16011	41.58703	.00405	.02405
95	6.56178	.15240	278.08910	42.38012	.00360	.02360
100	7.24474	.13803	312.23698	43.09844	.00320	.02320
120	10.76533	.09289	488.26643	45.35546	.00205	.02205
150	19.49998	.05128	924.99926	47.43590	.00108	.02108
180	35.32166	.02831	1716.08280	48.58444	.00058	.02058
200	52.48622	.01905	2574.31112	49.04737	.00039	.02039
240	115.89226	.00863	5744.61301	49.56857	.00017	.02017
250	141.27220	.00708	7013.61023	49.64607	.00014	.02014
300	380.24902	.00263	18962.45160	49.86851	.00005	.02005
360	1247.61816	.00080	62330.90960	49.95992	.00002	.02002

Discrete Compounding; i = 3.0%

N	(F/P)	(P/F)	(F/A)	(P/A)	(A/F)	(A/P)
1	1.03000	.97087	1.00000	.97087	1.00000	1.03000
2	1.06090	.94260	2.02999	1.91346	.49261	.52261
3	1.09273	.91514	3.09089	2.82860	.32353	.35353
4	1.12551	.88849	4.18361	3.71709	.23903	.26903
5	1.15927	.86261	5.30910	4.57968	.18836	.21836
6	1.19405	.83748	6.46839	5.41718	.15460	.18460
7	1.22987	.81309	7.66242	6.23026	.13051	.16051
8	1.26677	.78941	8.89231	7.01968	.11246	.14246
9	1.30477	.76642	10.15908	7.78609	.09843	.12843
10	1.34391	.74409	11.46383	8.53018	.08723	.11723
11	1.38423	.72242	12.80775	9.25260	.07808	.10808
12	1.42576	.70138	14.19198	9.95398	.07046	.10046
13	1.46853	.68095	15.61772	10.63492	.06403	.09403
14	1.51259	.66112	17.08628	11.29605	.05853	.08853
15	1.55797	.64186	18.59884	11.93791	.05377	.08377
16	1.60470	.62317	20.15683	12.56108	.04961	.07961
17	1.65285	.60502	21.76151	13.16609	.04595	.07595
18	1.70243	.58740	23.41436	13.75349	.04271	.07271
19	1.75350	.57029	25.11679	14.32377	.03981	.06981
20	1.80611	.55368	26.87028	14.87745	.03722	.06722
21	1.86029	.53755	28.67638	15.41499	.03487	.06487
22	1.91610	.52189	30.53668	15.93689	.03275	.06275
23	1.97358	.50669	32.45274	16.44357	.03081	.06081
24	2.03279	.49193	34.42637	16.93552	.02905	.05905
25	2.09377	.47761	36.45916	17.41312	.02743	.05743
26	2.15659	.46370	38.55289	17.87681	.02594	.05594
27	2.22128	.45019	40.70949	18.32700	.02456	.05456
28	2.28792	.43708	42.93076	18.76408	.02329	.05329
29	2.35656	.42435	45.21866	19.18842	.02211	.05211
30	2.42726	.41199	47.57528	19.60042	.02102	.05102
35	2.81386	.35538	60.46187	21.48719	.01654	.04654
40	3.26203	.30656	75.40099	23.11475	.01326	.04326
45	3.78158	.26444	92.71946	24.51868	.01079	.04079
50	4.38389	.22811	112.79640	25.72974	.00887	.03887
55	5.08213	.19677	136.07095	26.77440	.00735	.03735
60	5.89158	.16973	163.05265	27.67554	.00613	.03613
65	6.82996	.14641	194.33186	28.45287	.00515	.03515
70	7.91779	.12630	230.59297	29.12340	.00434	.03434
75	9.17888	.10895	272.62944	29.70181	.00367	.03367
80	10.64084	.09398	321.36137	30.20075	.00311	.03311
85	12.33564	.08107	377.85478	30.63114	.00265	.03265
90	14.30038	.06993	443.34613	31.00239	.00226	.03226
95	16.57806	.06032	519.26878	31.32264	.00193	.03193
100	19.21852	.05203	607.28401	31.59889	.00165	.03165
120	34.71074	.02881	1123.69126	32.37301	.00089	.03089
150	84.25194	.01187	2775.06449	32.93769	.00036	.03036
180	204.50119	.00489	6783.37274	33.17033	.00015	.03015
200	369.35156	.00271	12278.38493	33.24308	.00008	.03008
240	1204.83618	.00083	40127.87113	33.30567	.00002	.03002
250	1619.19678	.00062	53939.89043	33.31275	.00002	.03002
300	.70984E+04	.00014	.23658E+06	33.32864	.00000	.03000
360	.41821E+05	.00002	.13940E+07	33.33253	.00000	.03000

Discrete Compounding; i = 4.0%

N	(F/P)	(P/F)	(F/A)	(P/A)	(A/F)	(A/P)
1	1.04000	.96154	1.00000	.96154	1.00000	1.04000
2	1.08160	.92456	2.04000	1.88610	.49019	.53019
3	1.12486	.88900	3.12161	2.77510	.32035	.36035
4	1.16986	.85480	4.24647	3.62990	.23549	.27549
5	1.21665	.82193	5.41632	4.45182	.18463	.22463
6	1.26532	.79031	6.63300	5.24215	.15076	.19076
7	1.31593	.75992	7.89833	6.00208	.12661	.16661
8	1.36857	.73069	9.21426	6.73276	.10853	.14853
9	1.42331	.70259	10.58283	7.43535	.09449	.13449
10	1.48025	.67556	12.00614	8.11091	.08329	.12329
11	1.53946	.64958	13.48641	8.76050	.07415	.11415
12	1.60103	.62460	15.02585	9.38509	.06655	.10655
13	1.66508	.60057	16.62688	9.98566	.06014	.10014
14	1.73168	.57747	18.29197	10.56314	.05467	.09467
15	1.80095	.55526	20.02368	11.11842	.04994	.08994
16	1.87298	.53391	21.82462	11.65232	.04582	.08582
17	1.94790	.51337	23.69762	12.16570	.04220	.08220
18	2.02582	.49363	25.64552	12.65932	.03899	.07899
19	2.10685	.47464	27.67136	13.13397	.03614	.07614
20	2.19113	.45639	29.77820	13.59035	.03358	.07358
21	2.27877	.43883	31.96931	14.02918	.03128	.07128
22	2.36992	.42195	34.24811	14.45114	.02920	.06920
23	2.46472	.40573	36.61807	14.85687	.02731	.06731
24	2.56331	.39012	39.08279	15.24699	.02559	.06559
25	2.66584	.37512	41.64610	15.62211	.02401	.06401
26	2.77248	.36069	44.31193	15.98279	.02257	.06257
27	2.88338	.34682	47.08448	16.32962	.02124	.06124
28	2.99871	.33348	49.96781	16.66309	.02001	.06001
29	3.11866	.32065	52.96652	16.98374	.01888	.05888
30	3.24341	.30832	56.08521	17.29206	.01783	.05783
35	3.94611	.25341	73.65265	18.66464	.01358	.05358
40	4.80104	.20829	95.02609	19.79280	.01052	.05052
45	5.84120	.17120	121.03009	20.72006	.00826	.04826
50	7.10672	.14071	152.66812	21.48221	.00655	.04655
55	8.64642	.11565	191.16061	22.10863	.00523	.04523
60	10.51970	.09506	237.99248	22.62351	.00420	.04420
65	12.79883	.07813	294.97076	23.04670	.00339	.04339
70	15.57174	.06422	364.29363	23.39453	.00275	.04275
75	18.94542	.05278	448.63549	23.68042	.00223	.04223
80	23.05002	.04338	551.25047	23.91540	.00181	.04181
85	28.04387	.03566	676.09674	24.10854	.00148	.04148
90	34.11967	.02931	827.99189	24.26729	.00121	.04121
95	41.51186	.02409	1012.79642	24.39776	.00099	.04099
100	50.50554	.01980	1237.63850	24.50501	.00081	.04081
120	110.66412	.00904	2741.60315	24.77409	.00036	.04036
150	358.92896	.00279	8948.22408	24.93035	.00011	.04011
180	1164.15308	.00086	29078.82755	24.97853	.00003	.04003
200	2550.80981	.00039	63745.24679	24.99020	.00002	.04002
240	.12247E+05	.00008	.30614E+06	24.99796	.00000	.04000
250	.18128E+05	.00006	.45317E+06	24.99862	.00000	.04000
300	.12883E+06	.00001	.32207E+07	24.99981	.00000	.04000
360	.13553E+07	.00000	.33881E+08	24.99998	.00000	.04000

Discrete Compounding; i = 5.0%

N	(F/P)	(P/F)	(F/A)	(P/A)	(A/F)	(A/P)
1	1.05000	.95238	1.00000	.95238	1.00000	1.05000
2	1.10250	.90703	2.05000	1.85941	.48781	.53781
3	1.15763	.86384	3.15250	2.72325	.31721	.36721
4	1.21551	.82270	4.31013	3.54595	.23201	.28201
5	1.27628	.78353	5.52565	4.32949	.18097	.23097
6	1.34010	.74622	6.80191	5.07569	.14702	.19702
7	1.40710	.71068	8.14201	5.78638	.12282	.17282
8	1.47746	.67684	9.54912	6.46322	.10472	.15472
9	1.55133	.64461	11.02657	7.10783	.09069	.14069
10	1.62889	.61391	12.57790	7.72174	.07950	.12950
11	1.71034	.58468	14.20681	8.30642	.07039	.12039
12	1.79586	.55684	15.91715	8.86326	.06283	.11283
13	1.88565	.53032	17.71301	9.39358	.05646	.10646
14	1.97993	.50507	19.59864	9.89864	.05102	.10102
15	2.07893	.48102	21.57860	10.37967	.04634	.09634
16	2.18288	.45811	23.65753	10.83778	.04227	.09227
17	2.29202	.43630	25.84042	11.27408	.03870	.08870
18	2.40662	.41552	28.13242	11.68959	.03555	.08555
19	2.52695	.39573	30.53905	12.08533	.03274	.08274
20	2.65330	.37689	33.06602	12.46222	.03024	.08024
21	2.78597	.35894	35.71934	12.82116	.02800	.07800
22	2.92526	.34185	38.50527	13.16301	.02597	.07597
23	3.07153	.32557	41.43055	13.48858	.02414	.07414
24	3.22510	.31007	44.50209	13.79865	.02247	.07247
25	3.38636	.29530	47.72718	14.09395	.02095	.07095
26	3.55568	.28124	51.11353	14.37519	.01956	.06956
27	3.73346	.26785	54.66925	14.64304	.01829	.06829
28	3.92014	.25509	58.40271	14.89814	.01712	.06712
29	4.11614	.24295	62.32284	15.14108	.01605	.06605
30	4.32195	.23138	66.43894	15.37246	.01505	.06505
35	5.51603	.18129	90.32051	16.37420	.01107	.06107
40	7.04000	.14205	120.80007	17.15909	.00828	.05828
45	8.98503	.11130	159.70060	17.77408	.00626	.05626
50	11.46743	.08720	209.34856	18.25593	.00478	.05478
55	14.63567	.06833	272.71349	18.63348	.00367	.05367
60	18.67924	.05354	353.58489	18.92929	.00283	.05283
65	23.83998	.04195	456.79961	19.16107	.00219	.05219
70	30.42653	.03287	588.53057	19.34268	.00170	.05170
75	38.83284	.02575	756.65679	19.48497	.00132	.05132
80	49.56165	.02018	971.23290	19.59646	.00103	.05103
85	63.25465	.01581	1245.09306	19.68382	.00080	.05080
90	80.73071	.01239	1594.61423	19.75226	.00063	.05063
95	103.03520	.00971	2040.70401	19.80589	.00049	.05049
100	131.50197	.00760	2610.03933	19.84791	.00038	.05038
120	348.91431	.00287	6958.28603	19.94268	.00014	.05014
150	1507.98901	.00066	30139.77982	19.98674	.00003	.05003
180	.65175E+04	.00015	.13033E+06	19.99693	.00001	.05001
200	.17293E+05	.00006	.34584E+06	19.99884	.00000	.05000
240	.12174E+06	.00001	.24348E+07	19.99984	.00000	.05000
250	.19830E+06	.00001	.39661E+07	19.99990	.00000	.05000
300	.22740E+07	.00000	.45481E+08	19.99999	.00000	.05000
360	.42477E+08	.00000	.84954E+09	20.00000	.00000	.05000

Discrete Compounding; $i = 6.0\%$

N	(F/P)	(P/F)	(F/A)	(P/A)	(A/F)	(A/P)
1	1.06000	.94340	1.00000	.94340	1.00000	1.06000
2	1.12360	.89000	2.06002	1.83341	.48543	.54543
3	1.19102	.83962	3.18362	2.67303	.31411	.37411
4	1.26248	.79209	4.37465	3.46513	.22859	.28859
5	1.33823	.74726	5.63714	4.21239	.17740	.23740
6	1.41852	.70496	6.97538	4.91736	.14336	.20336
7	1.50363	.66506	8.39389	5.58241	.11913	.17913
8	1.59385	.62741	9.89755	6.20983	.10104	.16104
9	1.68948	.59190	11.49141	6.80173	.08702	.14702
10	1.79085	.55839	13.18091	7.36012	.07587	.13587
11	1.89831	.52679	14.97177	7.88691	.06679	.12679
12	2.01221	.49697	16.87010	8.38388	.05928	.11928
13	2.13294	.46884	18.88232	8.85272	.05296	.11296
14	2.26092	.44230	21.01528	9.29503	.04758	.10758
15	2.39657	.41726	23.27619	9.71229	.04296	.10296
16	2.54037	.39364	25.67279	10.10594	.03895	.09895
17	2.69279	.37136	28.21317	10.47730	.03544	.09544
18	2.85436	.35034	30.90598	10.82764	.03236	.09236
19	3.02562	.33051	33.76034	11.15815	.02962	.08962
20	3.20716	.31180	36.78600	11.46996	.02718	.08718
21	3.39959	.29415	39.99319	11.76412	.02500	.08500
22	3.60357	.27750	43.39282	12.04162	.02305	.08305
23	3.81978	.26180	46.99636	12.30342	.02128	.08128
24	4.04897	.24698	50.81620	12.55040	.01968	.07968
25	4.29191	.23300	54.86519	12.78339	.01823	.07823
26	4.54943	.21981	59.15713	13.00320	.01690	.07690
27	4.82239	.20737	63.70656	13.21057	.01570	.07570
28	5.11174	.19563	68.52904	13.40620	.01459	.07459
29	5.41845	.18455	73.64083	13.59076	.01358	.07358
30	5.74356	.17411	79.05932	13.76487	.01265	.07265
35	7.68619	.13010	111.43646	14.49828	.00897	.06897
40	10.28588	.09722	154.76459	15.04632	.00646	.06646
45	13.76485	.07265	212.74752	15.45585	.00470	.06470
50	18.42050	.05429	290.34170	15.76188	.00344	.06344
55	24.65083	.04057	394.18056	15.99056	.00254	.06254
60	32.98845	.03031	533.14083	16.16144	.00188	.06188
65	44.14607	.02265	719.10122	16.28913	.00139	.06139
70	59.07753	.01693	967.95885	16.38455	.00103	.06103
75	79.05916	.01265	1300.98600	16.45585	.00077	.06077
80	105.79922	.00945	1746.65379	16.50914	.00057	.06057
85	141.58353	.00706	2343.05883	16.54895	.00043	.06043
90	189.47101	.00528	3141.18354	16.57870	.00032	.06032
95	253.55537	.00394	4209.25633	16.60094	.00024	.06024
100	339.31494	.00295	5638.58248	16.61755	.00018	.06018
120	1088.23755	.00092	18120.62622	16.65135	.00006	.06006
150	.62504E+04	.00016	.10416E+06	16.66400	.00001	.06001
180	.35899E+05	.00003	.59830E+06	16.66620	.00000	.06000
200	.11513E+06	.00001	.19189E+07	16.66652	.00000	.06000
240	.11843E+07	.00000	.19738E+08	16.66665	.00000	.06000
250	.21208E+07	.00000	.35347E+08	16.66666	.00000	.06000
300	.39067E+08	.00000	.65112E+09	16.66667	.00000	.06000
360	.12888E+10	.00000	.21479E+11	16.66667	.00000	.06000

Discrete Compounding; i = 8.0%

N	(F/P)	(P/F)	(F/A)	(P/A)	(A/F)	(A/P)
1	1.08000	.92593	1.00000	.92593	1.00000	1.08000
2	1.16640	.85734	2.08000	1.78326	.48077	.56077
3	1.25971	.79383	3.24640	2.57710	.30803	.38803
4	1.36049	.73503	4.50611	3.31213	.22192	.30192
5	1.46933	.68058	5.86660	3.99271	.17046	.25046
6	1.58687	.63017	7.33593	4.62288	.13632	.21632
7	1.71382	.58349	8.92281	5.20637	.11207	.19207
8	1.85093	.54027	10.63663	5.74664	.09401	.17401
9	1.99900	.50025	12.48756	6.24689	.08008	.16008
10	2.15893	.46319	14.48657	6.71008	.06903	.14903
11	2.33164	.42888	16.64549	7.13897	.06008	.14008
12	2.51817	.39711	18.97713	7.53608	.05269	.13269
13	2.71962	.36770	21.49530	7.90378	.04652	.12652
14	2.93719	.34046	24.21492	8.24424	.04130	.12130
15	3.17217	.31524	27.15211	8.55948	.03683	.11683
16	3.42594	.29189	30.32429	8.85137	.03298	.11298
17	3.70002	.27027	33.75023	9.12164	.02963	.10963
18	3.99602	.25025	37.45025	9.37189	.02670	.10670
19	4.31570	.23171	41.44628	9.60360	.02413	.10413
20	4.66096	.21455	45.76196	9.81815	.02185	.10185
21	5.03383	.19866	50.42292	10.01680	.01983	.09983
22	5.43654	.18394	55.45675	10.20075	.01803	.09803
23	5.87146	.17032	60.89330	10.37106	.01642	.09642
24	6.34118	.15770	66.76477	10.52876	.01498	.09498
25	6.84847	.14602	73.10593	10.67478	.01368	.09368
26	7.39635	.13520	79.95443	10.80998	.01251	.09251
27	7.98806	.12519	87.35079	10.93517	.01145	.09145
28	8.62711	.11591	95.33885	11.05108	.01049	.09049
29	9.31727	.10733	103.96595	11.15841	.00962	.08962
30	10.06265	.09938	113.28319	11.25779	.00883	.08883
35	14.78535	.06763	172.31685	11.65457	.00580	.08580
40	21.72452	.04603	259.05653	11.92462	.00386	.08386
45	31.92044	.03133	386.50559	12.10840	.00259	.08259
50	46.90161	.02132	573.77026	12.23349	.00174	.08174
55	68.91383	.01451	848.92310	12.31862	.00118	.08118
60	101.25705	.00988	1253.21338	12.37655	.00080	.08080
65	148.77980	.00672	1847.24789	12.41599	.00054	.08054
70	218.60631	.00457	2720.07942	12.44282	.00037	.08037
75	321.20459	.00311	4002.55821	12.46109	.00025	.08025
80	471.95459	.00212	5886.93360	12.47352	.00017	.08017
85	693.45605	.00144	8655.70249	12.48198	.00012	.08012
90	1018.91479	.00098	12723.93759	12.48773	.00008	.08008
95	1497.11987	.00067	18701.50231	12.49165	.00005	.08005
100	2199.76025	.00045	27484.50891	12.49432	.00004	.08004
120	.10253E+05	.00010	.12815E+06	12.49878	.00001	.08001
150	.10317E+06	.00001	.12896E+07	12.49988	.00000	.08000
180	.10382E+07	.00000	.12977E+08	12.49999	.00000	.08000
200	.48389E+07	.00000	.60487E+08	12.50000	.00000	.08000
240	.10512E+09	.00000	.13140E+10	12.50000	.00000	.08000
250	.22695E+09	.00000	.28369E+10	12.50000	.00000	.08000
300	.10645E+11	.00000	.13306E+12	12.50000	.00000	.08000
360	.10778E+13	.00000	.13473E+14	12.50000	.00000	.08000

Discrete Compounding; $i = 10.0\%$

N	(F/P)	(P/F)	(F/A)	(P/A)	(A/F)	(A/P)
1	1.10000	.90909	1.00000	.90909	1.00000	1.10000
2	1.21000	.82645	2.10001	1.73554	.47619	.57619
3	1.33100	.75131	3.31001	2.48686	.30211	.40211
4	1.46410	.68301	4.64103	3.16988	.21547	.31547
5	1.61051	.62092	6.10514	3.79080	.16380	.26380
6	1.77157	.56447	7.71565	4.35527	.12961	.22961
7	1.94872	.51316	9.48723	4.86843	.10540	.20540
8	2.14360	.46651	11.43596	5.33494	.08744	.18744
9	2.35796	.42410	13.57957	5.75904	.07364	.17364
10	2.59375	.38554	15.93753	6.14458	.06274	.16274
11	2.85313	.35049	18.53130	6.49508	.05396	.15396
12	3.13845	.31863	21.38445	6.81371	.04676	.14676
13	3.45229	.28966	24.52291	7.10337	.04078	.14078
14	3.79752	.26333	27.97521	7.36670	.03575	.13575
15	4.17728	.23939	31.77275	7.60609	.03147	.13147
16	4.59501	.21763	35.95006	7.82372	.02782	.12782
17	5.05451	.19784	40.54508	8.02157	.02466	.12466
18	5.55996	.17986	45.59962	8.20142	.02193	.12193
19	6.11596	.16351	51.15959	8.36493	.01955	.11955
20	6.72756	.14864	57.27561	8.51358	.01746	.11746
21	7.40032	.13513	64.00320	8.64871	.01562	.11562
22	8.14036	.12284	71.40353	8.77155	.01400	.11400
23	8.95440	.11168	79.54393	8.88323	.01257	.11257
24	9.84984	.10152	88.49839	8.98475	.01130	.11130
25	10.83483	.09229	98.34826	9.07705	.01017	.11017
26	11.91832	.08390	109.18314	9.16095	.00916	.10916
27	13.11015	.07628	121.10150	9.23723	.00826	.10826
28	14.42118	.06934	134.21178	9.30657	.00745	.10745
29	15.86330	.06304	148.63300	9.36961	.00673	.10673
30	17.44963	.05731	164.49627	9.42692	.00608	.10608
35	28.10287	.03558	271.02868	9.64416	.00369	.10369
40	45.26009	.02209	442.60076	9.77905	.00226	.10226
45	72.89200	.01372	718.91981	9.86281	.00139	.10139
50	117.39357	.00852	1163.93542	9.91481	.00086	.10086
55	189.06390	.00529	1880.63859	9.94711	.00053	.10053
60	304.48999	.00328	3034.89918	9.96716	.00033	.10033
65	490.38525	.00204	4893.85137	9.97961	.00020	.10020
70	789.77222	.00127	7887.72029	9.98734	.00013	.10013
75	1271.93896	.00079	12709.38662	9.99214	.00008	.10008
80	2048.47559	.00049	20474.75098	9.99512	.00005	.10005
85	3299.09814	.00030	32980.97358	9.99697	.00003	.10003
90	5313.24219	.00019	53122.40921	9.99812	.00002	.10002
95	8557.04687	.00012	85560.44835	9.99883	.00001	.10001
100	.13781E+05	.00007	.13780E+06	9.99927	.00001	.10001
120	.92714E+05	.00001	.92713E+06	9.99989	.00000	.10000
150	.16178E+07	.00000	.16178E+08	9.99999	.00000	.10000
180	.28231E+08	.00000	.28231E+09	10.00000	.00000	.10000
200	.18992E+09	.00000	.18992E+10	10.00000	.00000	.10000
240	.85959E+10	.00000	.85959E+11	10.00000	.00000	.10000
250	.22296E+11	.00000	.22296E+12	10.00000	.00000	.10000
300	.26174E+13	.00000	.26174E+14	10.00000	.00000	.10000
360	.79696E+15	.00000	.79696E+16	10.00000	.00000	.10000

Discrete Compounding; i = 12.0%

N	(F/P)	(P/F)	(F/A)	(P/A)	(A/F)	(A/P)
1	1.12000	.89286	1.00000	.89286	1.00000	1.12000
2	1.25440	.79719	2.11999	1.69005	.47170	.59170
3	1.40493	.71178	3.37439	2.40183	.29635	.41635
4	1.57352	.63552	4.77931	3.03734	.20924	.32924
5	1.76234	.56743	6.35283	3.60477	.15741	.27741
6	1.97382	.50663	8.11516	4.11140	.12323	.24323
7	2.21068	.45235	10.08898	4.56375	.09912	.21912
8	2.47596	.40388	12.29965	4.96763	.08130	.20130
9	2.77307	.36061	14.77560	5.32824	.06768	.18768
10	3.10584	.32197	17.54866	5.65022	.05698	.17698
11	3.47854	.28748	20.65450	5.93769	.04842	.16842
12	3.89596	.25668	24.13303	6.19437	.04144	.16144
13	4.36348	.22917	28.02899	6.42354	.03568	.15568
14	4.88709	.20462	32.39245	6.62816	.03087	.15087
15	5.47355	.18270	37.27955	6.81086	.02682	.14682
16	6.13037	.16312	42.75306	6.97398	.02339	.14339
17	6.86601	.14564	48.88342	7.11962	.02046	.14046
18	7.68993	.13004	55.74940	7.24966	.01794	.13794
19	8.61272	.11611	63.43933	7.36577	.01576	.13576
20	9.64624	.10367	72.05201	7.46944	.01388	.13388
21	10.80379	.09256	81.69824	7.56200	.01224	.13224
22	12.10024	.08264	92.50199	7.64464	.01081	.13081
23	13.55227	.07379	104.60224	7.71843	.00956	.12956
24	15.17853	.06588	118.15442	7.78431	.00846	.12846
25	16.99995	.05882	133.33295	7.84314	.00750	.12750
26	19.03993	.05252	150.33276	7.89566	.00665	.12665
27	21.32474	.04689	169.37281	7.94255	.00590	.12590
28	23.88368	.04187	190.69734	7.98442	.00524	.12524
29	26.74973	.03738	214.58104	8.02180	.00466	.12466
30	29.95969	.03338	241.33071	8.05518	.00414	.12414
35	52.79913	.01894	431.65943	8.17550	.00232	.12232
40	93.04997	.01075	767.08307	8.24378	.00130	.12130
45	163.98566	.00610	1358.21375	8.28252	.00074	.12074
50	288.99829	.00346	2399.98566	8.30450	.00042	.12042
55	509.31323	.00196	4235.94344	8.31697	.00024	.12024
60	897.58228	.00111	7471.51866	8.32405	.00013	.12013
65	1581.84521	.00063	13173.70960	8.32806	.00008	.12008
70	2787.74805	.00036	23222.89947	8.33034	.00004	.12004
75	4912.95703	.00020	40932.97363	8.33164	.00002	.12002
80	8658.29687	.00012	72144.13776	8.33237	.00001	.12001
85	.15259E+05	.00007	.12715E+06	8.33279	.00001	.12001
90	.26891E+05	.00004	.22409E+06	8.33302	.00000	.12000
95	.47392E+05	.00002	.39492E+06	8.33316	.00000	.12000
100	.83520E+05	.00001	.69599E+06	8.33323	.00000	.12000
120	.80565E+06	.00000	.67138E+07	8.33332	.00000	.12000
150	.24137E+08	.00000	.20114E+09	8.33333	.00000	.12000
180	.72314E+09	.00000	.60262E+10	8.33333	.00000	.12000
200	.69756E+10	.00000	.58130E+11	8.33333	.00000	.12000
240	.64908E+12	.00000	.54090E+13	8.33333	.00000	.12000
250	.20159E+13	.00000	.16799E+14	8.33333	.00000	.12000
300	.58260E+15	.00000	.48550E+16	8.33333	.00000	.12000
360	.52293E+18	.00000	.43578E+19	8.33333	.00000	.12000

Discrete Compounding; $i = 15.0\%$

N	(F/P)	(P/F)	(F/A)	(P/A)	(A/F)	(A/P)
1	1.15000	.86957	1.00000	.86956	1.00000	1.15000
2	1.32250	.75614	2.15000	1.62571	.46512	.61512
3	1.52087	.65752	3.47249	2.28322	.28798	.43798
4	1.74900	.57175	4.99336	2.85497	.20027	.35027
5	2.01135	.49718	6.74236	3.35215	.14832	.29832
6	2.31306	.43233	8.75371	3.78448	.11424	.26424
7	2.66002	.37594	11.06677	4.16042	.09036	.24036
8	3.05902	.32690	13.72678	4.48732	.07285	.22285
9	3.51787	.28426	16.78578	4.77158	.05957	.20957
10	4.04555	.24719	20.30365	5.01876	.04925	.19925
11	4.65238	.21494	24.34919	5.23371	.04107	.19107
12	5.35023	.18691	29.00156	5.42062	.03448	.18448
13	6.15277	.16253	34.35178	5.58314	.02911	.17911
14	7.07568	.14133	40.50453	5.72447	.02469	.17469
15	8.13703	.12289	47.58021	5.84737	.02102	.17102
16	9.35758	.10687	55.71721	5.95423	.01795	.16795
17	10.76121	.09293	65.07477	6.04716	.01537	.16537
18	12.37539	.08081	75.83596	6.12796	.01319	.16319
19	14.23170	.07027	88.21135	6.19823	.01134	.16134
20	16.36646	.06110	102.44305	6.25933	.00976	.15976
21	18.82141	.05313	118.80943	6.31246	.00842	.15842
22	21.64462	.04620	137.63084	6.35866	.00727	.15727
23	24.89131	.04017	159.27544	6.39884	.00628	.15628
24	28.62498	.03493	184.16659	6.43377	.00543	.15543
25	32.91872	.03038	212.79148	6.46415	.00470	.15470
26	37.85652	.02642	245.71018	6.49056	.00407	.15407
27	43.53500	.02297	283.56674	6.51353	.00353	.15353
28	50.06523	.01997	327.10159	6.53351	.00306	.15306
29	57.57500	.01737	377.16671	6.55088	.00265	.15265
30	66.21123	.01510	434.74159	6.56598	.00230	.15230
35	133.17427	.00751	881.16194	6.61661	.00113	.15113
40	267.86060	.00373	1779.07092	6.64178	.00056	.15056
45	538.76270	.00186	3585.08521	6.65429	.00028	.15028
50	1083.64258	.00092	7217.61833	6.66052	.00014	.15014
55	2179.59033	.00046	14523.93786	6.66361	.00007	.15007
60	4383.92578	.00023	29219.50985	6.66515	.00003	.15003
65	8817.63281	.00011	58777.56143	6.66591	.00002	.15002
70	.17735E+05	.00006	.11823E+06	6.66629	.00001	.15001
75	.35672E+05	.00003	.23781E+06	6.66648	.00000	.15000
80	.71749E+05	.00001	.47832E+06	6.66657	.00000	.15000
85	.14431E+06	.00001	.96208E+06	6.66662	.00000	.15000
90	.29027E+06	.00000	.19351E+07	6.66664	.00000	.15000
95	.58383E+06	.00000	.38922E+07	6.66666	.00000	.15000
100	.11743E+07	.00000	.78285E+07	6.66666	.00000	.15000
120	.19219E+08	.00000	.12813E+09	6.66667	.00000	.15000
150	.12725E+10	.00000	.84833E+10	6.66667	.00000	.15000
180	.84254E+11	.00000	.56169E+12	6.66667	.00000	.15000
200	.13789E+13	.00000	.91929E+13	6.66667	.00000	.15000
240	.36936E+15	.00000	.24624E+16	6.66667	.00000	.15000
250	.14943E+16	.00000	.99618E+16	6.66667	.00000	.15000
300	.16193E+19	.00000	.10795E+20	6.66667	.00000	.15000
360	.70987E+22	.00000	.47325E+23	6.66667	.00000	.15000

Discrete Compounding; i = 16.0%

N	(F/P)	(P/F)	(F/A)	(P/A)	(A/F)	(A/P)
1	1.16000	.86207	1.00000	.86207	1.00000	1.16000
2	1.34560	.74316	2.15999	1.60523	.46296	.62296
3	1.56089	.64066	3.50559	2.24589	.28526	.44526
4	1.81064	.55229	5.06648	2.79818	.19738	.35738
5	2.10034	.47611	6.87712	3.27429	.14541	.30541
6	2.43639	.41044	8.97745	3.68473	.11139	.27139
7	2.82621	.35383	11.41384	4.03856	.08761	.24761
8	3.27841	.30503	14.24005	4.34359	.07022	.23022
9	3.80295	.26295	17.51845	4.60654	.05708	.21708
10	4.41142	.22668	21.32139	4.83322	.04690	.20690
11	5.11725	.19542	25.73281	5.02864	.03886	.19886
12	5.93601	.16846	30.85005	5.19710	.03241	.19241
13	6.88577	.14523	36.78604	5.34233	.02718	.18718
14	7.98749	.12520	43.67179	5.46753	.02290	.18290
15	9.26549	.10793	51.65928	5.57545	.01936	.17936
16	10.74796	.09304	60.92473	5.66849	.01641	.17641
17	12.46763	.08021	71.67268	5.74870	.01395	.17395
18	14.46245	.06914	84.14027	5.81785	.01188	.17188
19	16.77643	.05961	98.60266	5.87745	.01014	.17014
20	19.46066	.05139	115.37912	5.92884	.00867	.16867
21	22.57436	.04430	134.83970	5.97314	.00742	.16742
22	26.18623	.03819	157.41393	6.01132	.00635	.16635
23	30.37604	.03292	183.60020	6.04424	.00545	.16545
24	35.23619	.02838	213.97616	6.07262	.00467	.16467
25	40.87398	.02447	249.21232	6.09709	.00401	.16401
26	47.41379	.02109	290.08613	6.11818	.00345	.16345
27	55.00000	.01818	337.49994	6.13636	.00296	.16296
28	63.79997	.01567	392.49976	6.15204	.00255	.16255
29	74.00795	.01351	456.29961	6.16555	.00219	.16219
30	85.84918	.01165	530.30730	6.17720	.00189	.16189
35	180.31245	.00555	1120.70266	6.21534	.00089	.16089
40	378.71729	.00264	2360.73265	6.23350	.00042	.16042
45	795.43457	.00126	4965.21525	6.24214	.00020	.16020
50	1670.68188	.00060	10435.51007	6.24626	.00010	.16010
55	3508.99829	.00028	21924.98573	6.24822	.00005	.16005
60	7370.08594	.00014	46056.77956	6.24915	.00002	.16002
65	15479.67969	.00006	96741.73219	6.24960	.00001	.16001
70	.32513E+05	.00003	.20320E+06	6.24981	.00000	.16000
75	.68287E+05	.00001	.42679E+06	6.24991	.00000	.16000
80	.14343E+06	.00001	.89641E+06	6.24996	.00000	.16000
85	.30124E+06	.00000	.18828E+07	6.24998	.00000	.16000
90	.63272E+06	.00000	.39545E+07	6.24999	.00000	.16000
95	.13289E+07	.00000	.83057E+07	6.24999	.00000	.16000
100	.27912E+07	.00000	.17445E+08	6.25000	.00000	.16000
120	.54318E+08	.00000	.33949E+09	6.25000	.00000	.16000
150	.46632E+10	.00000	.29145E+11	6.25000	.00000	.16000
180	.40033E+12	.00000	.25021E+13	6.25000	.00000	.16000
200	.77907E+13	.00000	.48692E+14	6.25000	.00000	.16000
240	.29505E+16	.00000	.18440E+17	6.25000	.00000	.16000
250	.13016E+17	.00000	.81348E+17	6.25000	.00000	.16000
300	.21745E+20	.00000	.13591E+21	6.25000	.00000	.16000
360	.16026E+24	.00000	.10016E+25	6.25000	.00000	.16000

Discrete Compounding; $i = 20.0\%$

N	(F/P)	(P/F)	(F/A)	(P/A)	(A/F)	(A/P)
1	1.20000	.83333	1.00000	.83333	1.00000	1.20000
2	1.44000	.69444	2.20000	1.52778	.45455	.65455
3	1.72800	.57870	3.63999	2.10648	.27473	.47473
4	2.07360	.48225	5.36799	2.58873	.18629	.38629
5	2.48832	.40188	7.44159	2.99061	.13438	.33438
6	2.98598	.33490	9.92991	3.32551	.10071	.30071
7	3.58318	.27908	12.91588	3.60459	.07742	.27742
8	4.29981	.23257	16.49906	3.83716	.06061	.26061
9	5.15977	.19381	20.79887	4.03097	.04808	.24808
10	6.19173	.16151	25.95864	4.19247	.03852	.23852
11	7.43007	.13459	32.15035	4.32706	.03110	.23110
12	8.91609	.11216	39.58043	4.43922	.02527	.22526
13	10.69930	.09346	48.49652	4.53268	.02062	.22062
14	12.83916	.07789	59.19579	4.61057	.01689	.21689
15	15.40699	.06491	72.03494	4.67547	.01388	.21388
16	18.48839	.05409	87.44195	4.72956	.01144	.21144
17	22.18607	.04507	105.93033	4.77463	.00944	.20944
18	26.62328	.03756	128.11639	4.81219	.00781	.20781
19	31.94791	.03130	154.73954	4.84350	.00646	.20646
20	38.33749	.02608	186.68748	4.86958	.00536	.20536
21	46.00499	.02174	225.02496	4.89132	.00444	.20444
22	55.20598	.01811	271.02991	4.90943	.00369	.20369
23	66.24716	.01509	326.23583	4.92453	.00307	.20307
24	79.49660	.01258	392.48301	4.93710	.00255	.20255
25	95.39590	.01048	471.97955	4.94759	.00212	.20212
26	114.47507	.00874	567.37537	4.95632	.00176	.20176
27	137.37001	.00728	681.85009	4.96360	.00147	.20147
28	164.84406	.00607	819.22032	4.96967	.00122	.20122
29	197.81287	.00506	984.06439	4.97472	.00102	.20102
30	237.37534	.00421	1181.87675	4.97894	.00085	.20085
35	590.66553	.00169	2948.32781	4.99154	.00034	.20034
40	1469.76416	.00068	7343.82124	4.99660	.00014	.20014
45	3657.24170	.00027	18281.20959	4.99863	.00005	.20005
50	9100.38281	.00011	45496.91677	4.99945	.00002	.20002
55	.22645E+05	.00004	.11322E+06	4.99978	.00001	.20001
60	.56347E+05	.00002	.28173E+06	4.99991	.00000	.20000
65	.14021E+06	.00001	.70104E+06	4.99996	.00000	.20000
70	.34889E+06	.00000	.17444E+07	4.99999	.00000	.20000
75	.86814E+06	.00000	.43407E+07	4.99999	.00000	.20000
80	.21602E+07	.00000	.10801E+08	5.00000	.00000	.20000
85	.53753E+07	.00000	.26876E+08	5.00000	.00000	.20000
90	.13375E+08	.00000	.66877E+08	5.00000	.00000	.20000
95	.33282E+08	.00000	.16641E+09	5.00000	.00000	.20000
100	.82817E+08	.00000	.41408E+09	5.00000	.00000	.20000
120	.31750E+10	.00000	.15875E+11	5.00000	.00000	.20000
150	.75367E+12	.00000	.37683E+13	5.00000	.00000	.20000
180	.17890E+15	.00000	.89451E+15	5.00000	.00000	.20000
200	.68587E+16	.00000	.34293E+17	5.00000	.00000	.20000
240	.10081E+20	.00000	.50403E+20	5.00000	.00000	.20000
250	.62416E+20	.00000	.31208E+21	5.00000	.00000	.20000
300	.56801E+24	.00000	.28401E+25	5.00000	.00000	.20000
360	.32006E+29	.00000	.16003E+30	5.00000	.00000	.20000

Discrete Compounding; i = 25.0%

N	(F/P)	(P/F)	(F/A)	(P/A)	(A/F)	(A/P)
1	1.25000	.80000	1.00000	.80000	1.00000	1.25000
2	1.56250	.64000	2.25000	1.44000	.44444	.69444
3	1.95313	.51200	3.81250	1.95200	.26230	.51230
4	2.44141	.40960	5.76563	2.36160	.17344	.42344
5	3.05176	.32768	8.20703	2.68928	.12185	.37185
6	3.81470	.26214	11.25879	2.95142	.08882	.33882
7	4.76837	.20972	15.07349	3.16114	.06634	.31634
8	5.96046	.16777	19.84186	3.32891	.05040	.30040
9	7.45058	.13422	25.80232	3.46313	.03876	.28876
10	9.31323	.10737	33.25290	3.57050	.03007	.28007
11	11.64153	.08590	42.56613	3.65640	.02349	.27349
12	14.55192	.06872	54.20766	3.72512	.01845	.26845
13	18.18990	.05498	68.75958	3.78010	.01454	.26454
14	22.73737	.04398	86.94946	3.82408	.01150	.26150
15	28.42171	.03518	109.68683	3.85926	.00912	.25912
16	35.52713	.02815	138.10852	3.88741	.00724	.25724
17	44.40891	.02252	173.63562	3.90993	.00576	.25576
18	55.51114	.01801	218.04456	3.92794	.00459	.25459
19	69.38893	.01441	273.55573	3.94235	.00366	.25366
20	86.73616	.01153	342.94464	3.95388	.00292	.25292
21	108.42020	.00922	429.68079	3.96311	.00233	.25233
22	135.52524	.00738	538.10095	3.97049	.00186	.25186
23	169.40656	.00590	673.62622	3.97639	.00148	.25148
24	211.75819	.00472	843.03278	3.98111	.00119	.25119
25	264.69775	.00378	1054.79102	3.98489	.00095	.25095
26	330.87207	.00302	1319.48828	3.98791	.00076	.25076
27	413.59033	.00242	1650.36133	3.99033	.00061	.25061
28	516.98779	.00193	2063.95117	3.99226	.00048	.25048
29	646.23486	.00155	2580.93945	3.99381	.00039	.25039
30	807.79346	.00124	3227.17383	3.99505	.00031	.25031
35	2465.18945	.00041	9856.75781	3.99838	.00010	.25010
40	7523.16016	.00013	30088.64062	3.99947	.00003	.25003
45	22958.86719	.00004	91831.46875	3.99983	.00001	.25001
50	.70065E+05	.00001	.28026E+06	3.99994	.00000	.25000
55	.21382E+06	.00000	.85528E+06	3.99998	.00000	.25000
60	.65253E+06	.00000	.26101E+07	3.99999	.00000	.25000
65	.19914E+07	.00000	.79655E+07	4.00000	.00000	.25000
70	.60772E+07	.00000	.24309E+08	4.00000	.00000	.25000
75	.18546E+08	.00000	.74184E+08	4.00000	.00000	.25000
80	.56598E+08	.00000	.22639E+09	4.00000	.00000	.25000
85	.17272E+09	.00000	.69089E+09	4.00000	.00000	.25000
90	.52711E+09	.00000	.21084E+10	4.00000	.00000	.25000
95	.16086E+10	.00000	.64344E+10	4.00000	.00000	.25000
100	.49091E+10	.00000	.19636E+11	4.00000	.00000	.25000
120	.42580E+12	.00000	.17032E+13	4.00000	.00000	.25000
150	.34395E+15	.00000	.13758E+16	4.00000	.00000	.25000
180	.27784E+18	.00000	.11114E+19	4.00000	.00000	.25000
200	.24099E+20	.00000	.96397E+20	4.00000	.00000	.25000
240	.18130E+24	.00000	.72521E+24	4.00000	.00000	.25000
250	.16885E+25	.00000	.67540E+25	4.00000	.00000	.25000
300	.11830E+30	.00000	.47322E+30	4.00000	.00000	.25000
360	.77197E+35	.00000	.30879E+36	4.00000	.00000	.25000

Discrete Compounding; i = 30.0%

N	(F/P)	(P/F)	(F/A)	(P/A)	(A/F)	(A/P)
1	1.30000	.76923	1.00000	.76923	1.00000	1.30000
2	1.69000	.59172	2.30000	1.36095	.43478	.73478
3	2.19700	.45517	3.99000	1.81611	.25063	.55063
4	2.85610	.35013	6.18701	2.16624	.16163	.46163
5	3.71293	.26933	9.04311	2.43557	.11058	.41058
6	4.82681	.20718	12.75605	2.64275	.07839	.37839
7	6.27486	.15937	17.58286	2.80211	.05687	.35687
8	8.15732	.12259	23.85773	2.92470	.04192	.34192
9	10.60452	.09430	32.01505	3.01900	.03124	.33124
10	13.78587	.07254	42.61957	3.09154	.02346	.32346
11	17.92163	.05580	56.40543	3.14734	.01773	.31773
12	23.29813	.04292	74.32708	3.19026	.01345	.31345
13	30.28758	.03302	97.62527	3.22328	.01024	.31024
14	39.37386	.02540	127.91285	3.24867	.00782	.30782
15	51.18600	.01954	167.28668	3.26821	.00598	.30598
16	66.54184	.01503	218.47279	3.28324	.00458	.30458
17	86.50441	.01156	285.01469	3.29480	.00351	.30351
18	112.45575	.00889	371.51915	3.30369	.00269	.30269
19	146.19246	.00684	483.97484	3.31053	.00207	.30207
20	190.05028	.00526	630.16757	3.31579	.00159	.30159
21	247.06543	.00405	820.21807	3.31984	.00122	.30122
22	321.18506	.00311	1067.28349	3.32295	.00094	.30094
23	417.54053	.00239	1388.46837	3.32535	.00072	.30072
24	542.80298	.00184	1806.00986	3.32719	.00055	.30055
25	705.64404	.00142	2348.81338	3.32861	.00043	.30043
26	917.33716	.00109	3054.45707	3.32970	.00033	.30033
27	1192.53833	.00084	3971.79428	3.33054	.00025	.30025
28	1550.30029	.00065	5164.33410	3.33118	.00019	.30019
29	2015.39136	.00050	6714.63759	3.33168	.00015	.30015
30	2620.00879	.00038	8730.02895	3.33206	.00011	.30011
35	9727.91406	.00010	32423.04559	3.33299	.00003	.30003
40	.36119E+05	.00003	.12039E+06	3.33324	.00001	.30001
45	.13411E+06	.00001	.44702E+06	3.33331	.00000	.30000
50	.49793E+06	.00000	.16598E+07	3.33333	.00000	.30000
55	.18488E+07	.00000	.61626E+07	3.33333	.00000	.30000
60	.68644E+07	.00000	.22881E+08	3.33333	.00000	.30000
65	.25487E+08	.00000	.84957E+08	3.33333	.00000	.30000
70	.94632E+08	.00000	.31544E+09	3.33333	.00000	.30000
75	.35136E+09	.00000	.11712E+10	3.33333	.00000	.30000
80	.13046E+10	.00000	.43486E+10	3.33333	.00000	.30000
85	.48439E+10	.00000	.16146E+11	3.33333	.00000	.30000
90	.17985E+11	.00000	.59950E+11	3.33333	.00000	.30000
95	.66777E+11	.00000	.22259E+12	3.33333	.00000	.30000
100	.24794E+12	.00000	.82646E+12	3.33333	.00000	.30000
120	.47121E+14	.00000	.15707E+15	3.33333	.00000	.30000
150	.12346E+18	.00000	.41152E+18	3.33333	.00000	.30000
180	.32346E+21	.00000	.10782E+22	3.33333	.00000	.30000
200	.61473E+23	.00000	.20491E+24	3.33333	.00000	.30000
240	.22204E+28	.00000	.74012E+28	3.33333	.00000	.30000
250	.30609E+29	.00000	.10203E+30	3.33333	.00000	.30000
300	.15241E+35	.00000	.50805E+35	3.33333	.00000	.30000
360	.10462E+42	.00000	.34875E+42	3.33333	.00000	.30000

Discrete Compounding; i = 40.0%

N	(F/P)	(P/F)	(F/A)	(P/A)	(A/F)	(A/P)
1	1.40000	.71429	1.00000	.71429	1.00000	1.40000
2	1.96000	.51020	2.40000	1.22449	.41667	.81667
3	2.74400	.36443	4.35999	1.58892	.22936	.62936
4	3.84160	.26031	7.10399	1.84923	.14077	.54077
5	5.37823	.18593	10.94559	2.03516	.09136	.49136
6	7.52953	.13281	16.32382	2.16797	.06126	.46126
7	10.54133	.09486	23.85333	2.26284	.04192	.44192
8	14.75786	.06776	34.39466	2.33060	.02907	.42907
9	20.66101	.04840	49.15253	2.37900	.02034	.42034
10	28.92540	.03457	69.81350	2.41357	.01432	.41432
11	40.49554	.02469	98.73887	2.43826	.01013	.41013
12	56.69376	.01764	139.23440	2.45590	.00718	.40718
13	79.37125	.01260	195.92813	2.46850	.00510	.40510
14	111.11972	.00900	275.29932	2.47750	.00363	.40363
15	155.56754	.00643	386.41886	2.48393	.00259	.40259
16	217.79456	.00459	541.98642	2.48852	.00185	.40185
17	304.91235	.00328	759.78093	2.49180	.00132	.40132
18	426.87720	.00234	1064.69306	2.49414	.00094	.40094
19	597.62769	.00167	1491.56930	2.49582	.00067	.40067
20	836.67871	.00120	2089.19690	2.49701	.00048	.40048
21	1171.35010	.00085	2925.87542	2.49787	.00034	.40034
22	1639.88965	.00061	4097.22437	2.49848	.00024	.40024
23	2295.84448	.00044	5737.11155	2.49891	.00017	.40017
24	3214.18237	.00031	8032.95641	2.49922	.00012	.40012
25	4499.85547	.00022	11247.13934	2.49944	.00009	.40009
26	6299.79297	.00016	15746.98336	2.49960	.00006	.40006
27	8819.71094	.00011	22046.77866	2.49972	.00005	.40005
28	12347.58984	.00008	30866.47645	2.49980	.00003	.40003
29	17286.62500	.00006	43214.06508	2.49986	.00002	.40002
30	24201.26953	.00004	60500.67743	2.49990	.00002	.40002
35	.13016E+06	.00001	.32540E+06	2.49998	.00000	.40000
40	.70003E+06	.00000	.17501E+07	2.50000	.00000	.40000
45	.37649E+07	.00000	.94123E+07	2.50000	.00000	.40000
50	.20249E+08	.00000	.50622E+08	2.50000	.00000	.40000
55	.10890E+09	.00000	.27226E+09	2.50000	.00000	.40000
60	.58570E+09	.00000	.14643E+10	2.50000	.00000	.40000
65	.31500E+10	.00000	.78751E+10	2.50000	.00000	.40000
70	.16942E+11	.00000	.42354E+11	2.50000	.00000	.40000
75	.91116E+11	.00000	.22779E+12	2.50000	.00000	.40000
80	.49004E+12	.00000	.12251E+13	2.50000	.00000	.40000
85	.26356E+13	.00000	.65889E+13	2.50000	.00000	.40000
90	.14175E+14	.00000	.35437E+14	2.50000	.00000	.40000
95	.76235E+14	.00000	.19059E+15	2.50000	.00000	.40000
100	.41001E+15	.00000	.10250E+16	2.50000	.00000	.40000
120	.34305E+18	.00000	.85762E+18	2.50000	.00000	.40000
150	.83022E+22	.00000	.20755E+23	2.50000	.00000	.40000
180	.20092E+27	.00000	.50231E+27	2.50000	.00000	.40000
200	.16811E+30	.00000	.42027E+30	2.50000	.00000	.40000
240	.11768E+36	.00000	.29420E+36	2.50000	.00000	.40000
250	.34040E+37	.00000	.85099E+37	2.50000	.00000	.40000
300	.68926E+44	.00000	.17231E+45	2.50000	.00000	.40000
360	.40370E+53	.00000	.10092E+54	2.50000	.00000	.40000

Discrete Compounding; $i = 50.0\%$

N	(F/P)	(P/F)	(F/A)	(P/A)	(A/F)	(A/P)
1	1.50000	.66667	1.00000	.66667	1.00000	1.50000
2	2.25000	.44444	2.50000	1.11111	.40000	.90000
3	3.37500	.29630	4.75000	1.40741	.21053	.71053
4	5.06250	.19753	8.12500	1.60494	.12308	.62308
5	7.59375	.13169	13.18750	1.73663	.07583	.57583
6	11.39063	.08779	20.78125	1.82442	.04812	.54812
7	17.08594	.05853	32.17188	1.88294	.03108	.53108
8	25.62891	.03902	49.25781	1.92196	.02030	.52030
9	38.44336	.02601	74.88672	1.94798	.01335	.51335
10	57.66504	.01734	113.33008	1.96532	.00882	.50882
11	86.49756	.01156	170.99512	1.97688	.00585	.50585
12	129.74634	.00771	257.49268	1.98459	.00388	.50388
13	194.61951	.00514	387.23901	1.98972	.00258	.50258
14	291.92920	.00343	581.85840	1.99315	.00172	.50172
15	437.89380	.00228	873.78760	1.99543	.00114	.50114
16	656.84082	.00152	1311.68164	1.99696	.00076	.50076
17	985.26123	.00101	1968.52246	1.99797	.00051	.50051
18	1477.89185	.00068	2953.78369	1.99865	.00034	.50034
19	2216.83789	.00045	4431.67578	1.99910	.00023	.50023
20	3325.25659	.00030	6648.51318	1.99940	.00015	.50015
21	4987.88672	.00020	9973.77344	1.99960	.00010	.50010
22	7481.82813	.00013	14961.65625	1.99973	.00007	.50007
23	11222.74219	.00009	22443.48438	1.99982	.00004	.50004
24	16834.11328	.00006	33666.22656	1.99988	.00003	.50003
25	25251.16797	.00004	50500.33594	1.99992	.00002	.50002
26	37876.75000	.00003	75751.50000	1.99995	.00001	.50001
27	.56815E+05	.00002	.11363E+06	1.99996	.00001	.50001
28	.85223E+05	.00001	.17044E+06	1.99998	.00001	.50001
29	.12783E+06	.00001	.25567E+06	1.99998	.00000	.50000
30	.19175E+06	.00001	.38350E+06	1.99999	.00000	.50000
35	.14561E+07	.00000	.29122E+07	2.00000	.00000	.50000
40	.11057E+08	.00000	.22115E+08	2.00000	.00000	.50000
45	.83967E+08	.00000	.16793E+09	2.00000	.00000	.50000
50	.63762E+09	.00000	.12752E+10	2.00000	.00000	.50000
55	.48419E+10	.00000	.96839E+10	2.00000	.00000	.50000
60	.36768E+11	.00000	.73537E+11	2.00000	.00000	.50000
65	.27921E+12	.00000	.55842E+12	2.00000	.00000	.50000
70	.21203E+13	.00000	.42405E+13	2.00000	.00000	.50000
75	.16101E+14	.00000	.32201E+14	2.00000	.00000	.50000
80	.12226E+15	.00000	.24453E+15	2.00000	.00000	.50000
85	.92845E+15	.00000	.18569E+16	2.00000	.00000	.50000
90	.70504E+16	.00000	.14101E+17	2.00000	.00000	.50000
95	.53539E+17	.00000	.10708E+18	2.00000	.00000	.50000
100	.40656E+18	.00000	.81312E+18	2.00000	.00000	.50000
120	.13519E+22	.00000	.27038E+22	2.00000	.00000	.50000
150	.25923E+27	.00000	.51846E+27	2.00000	.00000	.50000
180	.49708E+32	.00000	.99416E+32	2.00000	.00000	.50000
200	.16529E+36	.00000	.33058E+36	2.00000	.00000	.50000
240	.18277E+43	.00000	.36554E+43	2.00000	.00000	.50000
250	.10539E+45	.00000	.21079E+45	2.00000	.00000	.50000
300	.67201E+53	.00000	.13440E+54	2.00000	.00000	.50000
360	.24709E+64	.00000	.49418E+64	2.00000	.00000	.50000

Appendix B

Continuous Compounding Tables

Continuous Compounding (Discrete Flow); $i = 0.5\%$

N	(F/P)	(P/F)	(F/A)	(P/A)	(A/F)	(A/P)
1	1.00501	.99501	1.00000	.99501	1.00000	1.00501
2	1.01005	.99005	2.00495	1.98500	.49877	.50378
3	1.01511	.98511	3.01503	2.97014	.33167	.33668
4	1.02020	.98020	4.03006	3.95026	.24814	.25315
5	1.02531	.97531	5.05023	4.92554	.19801	.20302
6	1.03045	.97045	6.07553	5.89598	.16459	.16961
7	1.03562	.96561	7.10597	6.86157	.14073	.14574
8	1.04081	.96079	8.14155	7.82233	.12283	.12784
9	1.04603	.95600	9.18227	8.77824	.10891	.11392
10	1.05127	.95123	10.22831	9.72949	.09777	.10278
11	1.05654	.94649	11.27968	10.67606	.08865	.09367
12	1.06183	.94177	12.33600	11.61763	.08106	.08608
13	1.06716	.93707	13.39783	12.55470	.07464	.07965
14	1.07251	.93240	14.46499	13.48709	.06913	.07414
15	1.07788	.92775	15.53748	14.41483	.06436	.06937
16	1.08328	.92312	16.61530	15.33789	.06019	.06520
17	1.08871	.91851	17.69862	16.25644	.05650	.06151
18	1.09417	.91393	18.78729	17.17033	.05323	.05824
19	1.09966	.90938	19.88147	18.07971	.05030	.05531
20	1.10517	.90484	20.98097	18.98444	.04766	.05267
21	1.11071	.90033	22.08618	19.88480	.04528	.05029
22	1.11627	.89584	23.19673	20.78050	.04311	.04812
23	1.12187	.89137	24.31317	21.67200	.04113	.04614
24	1.12749	.88692	25.43494	22.55884	.03932	.04433
25	1.13314	.88250	26.56221	23.44118	.03765	.04266
26	1.13882	.87810	27.69539	24.31931	.03611	.04112
27	1.14453	.87372	28.83447	25.19322	.03468	.03969
28	1.15027	.86936	29.97868	26.06235	.03336	.03837
29	1.15603	.86503	31.12900	26.92740	.03212	.03714
30	1.16183	.86071	32.28500	27.78809	.03097	.03599
35	1.19124	.83946	38.15259	32.02762	.02621	.03122
40	1.22140	.81874	44.16856	36.16238	.02264	.02765
45	1.25231	.79852	50.33676	40.19501	.01987	.02488
50	1.28402	.77881	56.66133	44.12823	.01765	.02266
55	1.31652	.75958	63.14594	47.96429	.01584	.02085
60	1.34985	.74083	69.79453	51.70555	.01433	.01934
65	1.38402	.72253	76.61148	55.35448	.01305	.01807
70	1.41905	.70470	83.60103	58.91333	.01196	.01697
75	1.45498	.68730	90.76787	62.38449	.01102	.01603
80	1.49181	.67033	98.11549	65.76965	.01019	.01520
85	1.52957	.65378	105.64955	69.07143	.00947	.01448
90	1.56829	.63764	113.37402	72.29155	.00882	.01383
95	1.60799	.62189	121.29431	75.43228	.00824	.01326
100	1.64869	.60654	129.41495	78.49544	.00773	.01274
120	1.82208	.54882	164.00627	90.01030	.00610	.01111
150	2.11695	.47238	222.83200	105.26096	.00449	.00950
180	2.45953	.40658	291.17749	118.38744	.00343	.00845
200	2.71819	.36789	342.78076	126.10614	.00292	.00793
240	3.31999	.30121	462.83936	139.40999	.00216	.00717
250	3.49020	.28652	496.79687	142.34053	.00201	.00703
300	4.48147	.22314	694.55615	154.98395	.00144	.00645
360	6.04930	.16531	1007.33862	166.52158	.00099	.00601

Continuous Compounding (Discrete Flow); $i = 1.0\%$

N	(F/P)	(P/F)	(F/A)	(P/A)	(A/F)	(A/P)
1	1.01005	.99005	1.00000	.99005	1.00000	1.01005
2	1.02020	.98020	2.01005	1.97026	.49750	.50755
3	1.03045	.97045	3.03027	2.94072	.33000	.34005
4	1.04081	.96079	4.06073	3.90151	.24626	.25631
5	1.05127	.95123	5.10154	4.85274	.19602	.20607
6	1.06183	.94177	6.15278	5.79448	.16253	.17258
7	1.07251	.93240	7.21465	6.72691	.13861	.14866
8	1.08328	.92312	8.28715	7.65002	.12067	.13072
9	1.09417	.91393	9.37047	8.56399	.10672	.11677
10	1.10517	.90484	10.46460	9.46879	.09556	.10561
11	1.11627	.89584	11.56974	10.36461	.08643	.09648
12	1.12749	.88692	12.68609	11.25159	.07883	.08888
13	1.13882	.87810	13.81354	12.12965	.07239	.08244
14	1.15027	.86936	14.95237	12.99902	.06688	.07693
15	1.16183	.86071	16.10268	13.85976	.06210	.07215
16	1.17350	.85215	17.26447	14.71189	.05792	.06797
17	1.18530	.84367	18.43794	15.55553	.05424	.06429
18	1.19721	.83528	19.62326	16.39082	.05096	.06101
19	1.20924	.82696	20.82047	17.21777	.04803	.05808
20	1.22140	.81874	22.02980	18.03658	.04539	.05544
21	1.23367	.81059	23.25119	18.84717	.04301	.05306
22	1.24607	.80252	24.48482	19.64966	.04084	.05089
23	1.25859	.79454	25.73088	20.44420	.03886	.04891
24	1.27124	.78663	26.98947	21.23083	.03705	.04710
25	1.28402	.77881	28.26077	22.00967	.03538	.04543
26	1.29692	.77106	29.54469	22.78069	.03385	.04390
27	1.30995	.76339	30.84163	23.54407	.03242	.04247
28	1.32312	.75579	32.15164	24.29990	.03110	.04115
29	1.33642	.74827	33.47476	25.04817	.02987	.03992
30	1.34985	.74083	34.81116	25.78899	.02873	.03878
35	1.41905	.70470	41.69739	29.38399	.02398	.03403
40	1.49181	.67033	48.93671	32.80368	.02043	.03048
45	1.56829	.63764	56.54716	36.05659	.01768	.02773
50	1.64869	.60654	64.54784	39.15088	.01549	.02554
55	1.73322	.57696	72.95871	42.09427	.01371	.02376
60	1.82208	.54882	81.80083	44.89412	.01222	.02227
65	1.91550	.52206	91.09624	47.55740	.01098	.02103
70	2.01371	.49660	100.86810	50.09077	.00991	.01996
75	2.11695	.47238	111.14111	52.50063	.00900	.01905
80	2.22548	.44934	121.94080	54.79295	.00820	.01825
85	2.33958	.42743	133.29427	56.97351	.00750	.01755
90	2.45953	.40658	145.22954	59.04767	.00689	.01694
95	2.58563	.38675	157.77699	61.02071	.00634	.01639
100	2.71819	.36789	170.96754	62.89749	.00585	.01590
120	3.31999	.30121	230.84868	69.53297	.00433	.01438
150	4.48147	.22314	346.42114	77.30078	.00289	.01294
180	6.04930	.16531	502.42651	83.05537	.00199	.01204
200	7.38858	.13534	635.69043	86.03696	.00157	.01162
240	11.02231	.09073	997.26294	90.47675	.00100	.01105
250	12.18150	.08209	1112.60718	91.33580	.00090	.01095
300	20.08359	.04979	1898.89893	94.54979	.00053	.01058
360	36.59393	.02733	3541.74731	96.78511	.00028	.01033

Continuous Compounding (Discrete Flow); i = 2.0%

N	(F/P)	(P/F)	(F/A)	(P/A)	(A/F)	(A/P)
1	1.02020	.98020	1.00000	.98020	1.00000	1.02020
2	1.04081	.96079	2.02021	1.94099	.49500	.51520
3	1.06183	.94177	3.06100	2.88274	.32669	.34689
4	1.08328	.92312	4.12284	3.80587	.24255	.26275
5	1.10517	.90484	5.20612	4.71071	.19208	.21228
6	1.12749	.88692	6.31130	5.59764	.15845	.17865
7	1.15027	.86936	7.43877	6.46699	.13443	.15463
8	1.17350	.85215	8.58904	7.31913	.11643	.13663
9	1.19721	.83528	9.76253	8.15440	.10243	.12263
10	1.22140	.81874	10.95978	8.97316	.09124	.11144
11	1.24607	.80252	12.18115	9.77567	.08209	.10229
12	1.27124	.78663	13.42720	10.56229	.07448	.09468
13	1.29692	.77106	14.69842	11.33334	.06803	.08824
14	1.32312	.75579	15.99538	12.08915	.06252	.08272
15	1.34985	.74083	17.31848	12.82996	.05774	.07794
16	1.37711	.72616	18.66830	13.55611	.05357	.07377
17	1.40493	.71178	20.04541	14.26789	.04989	.07009
18	1.43331	.69768	21.45033	14.96557	.04662	.06682
19	1.46227	.68387	22.88362	15.64942	.04370	.06390
20	1.49181	.67033	24.34590	16.31976	.04107	.06128
21	1.52194	.65706	25.83769	16.97681	.03870	.05890
22	1.55269	.64405	27.35970	17.62088	.03655	.05675
23	1.58405	.63129	28.91232	18.25215	.03459	.05479
24	1.61605	.61879	30.49637	18.87094	.03279	.05299
25	1.64869	.60654	32.11241	19.47748	.03114	.05134
26	1.68200	.59453	33.76117	20.07204	.02962	.04982
27	1.71598	.58276	35.44312	20.65477	.02821	.04841
28	1.75064	.57122	37.15910	21.22600	.02691	.04711
29	1.78600	.55991	38.90968	21.78589	.02570	.04590
30	1.82208	.54882	40.69574	22.33473	.02457	.04477
35	2.01371	.49660	50.18166	24.92004	.01993	.04013
40	2.22548	.44934	60.66528	27.25938	.01648	.03668
45	2.45953	.40658	72.25139	29.37608	.01384	.03404
50	2.71819	.36789	85.05600	31.29138	.01176	.03196
55	3.00406	.33288	99.20724	33.02440	.01008	.03028
60	3.31999	.30121	114.84676	34.59251	.00871	.02891
65	3.66914	.27254	132.13098	36.01141	.00757	.02777
70	4.05502	.24661	151.23294	37.29527	.00661	.02681
75	4.48147	.22314	172.34380	38.45697	.00580	.02600
80	4.95277	.20191	195.67477	39.50812	.00511	.02531
85	5.47364	.18269	221.45947	40.45926	.00452	.02472
90	6.04930	.16531	249.95619	41.31989	.00400	.02420
95	6.68548	.14958	281.44922	42.09859	.00355	.02375
100	7.38858	.13534	316.25464	42.80321	.00316	.02336
120	11.02231	.09073	496.13623	45.01198	.00202	.02222
150	20.08359	.04979	944.69800	47.03831	.00106	.02126
180	36.59393	.02733	1762.01245	48.15041	.00057	.02077
200	54.59105	.01832	2652.92578	48.59634	.00038	.02058
240	121.49146	.00823	5964.70703	49.09570	.00017	.02037
250	148.38908	.00674	7296.22656	49.16956	.00014	.02034
300	403.35010	.00248	19917.60547	49.38043	.00005	.02025
360	1339.11694	.00075	66241.00000	49.46619	.00002	.02022

Continuous Compounding (Discrete Flow); *i* = 3.0%

N	(F/P)	(P/F)	(F/A)	(P/A)	(A/F)	(A/P)
1	1.03045	.97045	1.00000	.97045	1.00000	1.03045
2	1.06184	.94176	2.03044	1.91219	.49250	.52296
3	1.09417	.91393	3.09228	2.82613	.32339	.35384
4	1.12750	.88692	4.18642	3.71302	.23887	.26932
5	1.16183	.86071	5.31390	4.57372	.18819	.21864
6	1.19722	.83527	6.47573	5.40899	.15442	.18488
7	1.23368	.81059	7.67295	6.21958	.13033	.16078
8	1.27125	.78663	8.90659	7.00618	.11228	.14273
9	1.30996	.76338	10.17784	7.76956	.09825	.12871
10	1.34986	.74082	11.48779	8.51038	.08705	.11750
11	1.39097	.71893	12.83763	9.22930	.07790	.10835
12	1.43333	.69768	14.22857	9.92696	.07028	.10074
13	1.47698	.67706	15.66187	10.60401	.06385	.09430
14	1.52196	.65705	17.13885	11.26106	.05835	.08880
15	1.56831	.63763	18.66080	11.89868	.05359	.08404
16	1.61607	.61879	20.22906	12.51745	.04943	.07989
17	1.66529	.60050	21.84512	13.11794	.04578	.07623
18	1.71600	.58275	23.51039	13.70070	.04253	.07299
19	1.76826	.56553	25.22641	14.26622	.03964	.07010
20	1.82211	.54881	26.99461	14.81503	.03704	.06750
21	1.87760	.53259	28.81668	15.34760	.03470	.06516
22	1.93478	.51685	30.69431	15.86445	.03258	.06303
23	1.99371	.50158	32.62907	16.36604	.03065	.06110
24	2.05442	.48675	34.62273	16.85278	.02888	.05934
25	2.11699	.47237	36.67715	17.32515	.02726	.05772
26	2.18146	.45841	38.79410	17.78355	.02578	.05623
27	2.24790	.44486	40.97557	18.22841	.02440	.05486
28	2.31635	.43171	43.22340	18.66011	.02314	.05359
29	2.38690	.41895	45.53970	19.07906	.02196	.05241
30	2.45959	.40657	47.92662	19.48563	.02087	.05132
35	2.85763	.34994	60.99667	21.34518	.01639	.04685
40	3.32009	.30120	76.18175	22.94571	.01313	.04358
45	3.85739	.25924	93.82437	24.32330	.01066	.04111
50	4.48164	.22313	114.32217	25.50900	.00875	.03920
55	5.20692	.19205	138.13730	26.52956	.00724	.03769
60	6.04957	.16530	165.80621	27.40794	.00603	.03649
65	7.02859	.14228	197.95308	28.16399	.00505	.03551
70	8.16605	.12246	235.30231	28.81471	.00425	.03470
75	9.48758	.10540	278.69604	29.37480	.00359	.03404
80	11.02298	.09072	329.11182	29.85689	.00304	.03349
85	12.80686	.07808	387.68677	30.27180	.00258	.03303
90	14.87944	.06721	455.74146	30.62894	.00219	.03265
95	17.28743	.05785	534.80957	30.93633	.00187	.03232
100	20.08510	.04979	626.67310	31.20090	.00160	.03205
120	36.59727	.02732	1168.86230	31.93851	.00086	.03131
150	90.01422	.01111	2922.84595	32.47093	.00034	.03080
180	221.39774	.00452	7236.91797	32.68741	.00014	.03059
200	403.41113	.00248	13213.46094	32.75432	.00008	.03053
240	1339.36035	.00075	43946.03906	32.81122	.00002	.03048
250	1807.94336	.00055	59332.28125	32.81755	.00002	.03047
300	.81026E+04	.00012	.26602E+06	32.83167	.00000	.03046
360	.49017E+05	.00002	.16095E+07	32.83505	.00000	.03046

Continuous Compounding (Discrete Flow); i = 4.0%

N	(F/P)	(P/F)	(F/A)	(P/A)	(A/F)	(A/P)
1	1.04081	.96079	1.00000	.96079	1.00000	1.04081
2	1.08329	.92312	2.04080	1.88390	.49000	.53081
3	1.12750	.88692	3.12409	2.77082	.32009	.36090
4	1.17351	.85214	4.25158	3.62296	.23521	.27602
5	1.22140	.81873	5.42509	4.44170	.18433	.22514
6	1.27125	.78663	6.64648	5.22832	.15046	.19127
7	1.32313	.75579	7.91772	5.98410	.12630	.16711
8	1.37712	.72615	9.24086	6.71025	.10822	.14903
9	1.43333	.69768	10.61797	7.40793	.09418	.13499
10	1.49182	.67032	12.05129	8.07825	.08298	.12379
11	1.55270	.64404	13.54311	8.72228	.07384	.11465
12	1.61607	.61879	15.09581	9.34107	.06624	.10705
13	1.68202	.59452	16.71190	9.93560	.05984	.10065
14	1.75067	.57121	18.39389	10.50680	.05437	.09518
15	1.82211	.54881	20.14456	11.05561	.04964	.09045
16	1.89647	.52729	21.96667	11.58291	.04552	.08633
17	1.97387	.50662	23.86316	12.08953	.04191	.08272
18	2.05442	.48675	25.83701	12.57628	.03870	.07951
19	2.13827	.46767	27.89143	13.04395	.03585	.07666
20	2.22553	.44933	30.02968	13.49328	.03330	.07411
21	2.31635	.43171	32.25523	13.92500	.03100	.07181
22	2.41088	.41479	34.57155	14.33977	.02893	.06974
23	2.50927	.39852	36.98242	14.73829	.02704	.06785
24	2.61168	.38290	39.49171	15.12119	.02532	.06613
25	2.71826	.36788	42.10338	15.48907	.02375	.06456
26	2.82920	.35346	44.82162	15.84252	.02231	.06312
27	2.94466	.33960	47.65082	16.18213	.02099	.06180
28	3.06483	.32628	50.59547	16.50841	.01976	.06058
29	3.18991	.31349	53.66031	16.82190	.01864	.05945
30	3.32009	.30120	56.85022	17.12309	.01759	.05840
35	4.05516	.24660	74.86198	18.46091	.01336	.05417
40	4.95298	.20190	96.86160	19.55623	.01032	.05113
45	6.04958	.16530	123.73199	20.45300	.00808	.04889
50	7.38896	.13534	156.55153	21.18723	.00639	.04720
55	9.02488	.11080	196.63715	21.78835	.00509	.04590
60	11.02300	.09072	245.59804	22.28052	.00407	.04488
65	13.46351	.07427	305.39893	22.68346	.00327	.04408
70	16.44434	.06081	378.43945	23.01337	.00264	.04345
75	20.08513	.04979	467.65137	23.28346	.00214	.04295
80	24.53201	.04076	576.61548	23.50461	.00173	.04254
85	29.96344	.03337	709.70410	23.68567	.00141	.04222
90	36.59735	.02732	872.25781	23.83391	.00115	.04196
95	44.70004	.02237	1070.80176	23.95528	.00093	.04174
100	54.59669	.01832	1313.30396	24.05464	.00076	.04157
120	121.50656	.00823	2952.82544	24.30177	.00034	.04115
150	403.41260	.00248	9860.49609	24.44270	.00010	.04091
180	1339.36646	.00075	32794.58594	24.48515	.00003	.04084
200	2980.80005	.00034	73015.37500	24.49522	.00001	.04082
240	.14764E+05	.00007	.36174E+06	24.50174	.00000	.04081
250	.22025E+05	.00005	.53966E+06	24.50233	.00000	.04081
300	.16274E+06	.00001	.39877E+07	24.50330	.00000	.04081
360	.17939E+07	.00000	.43957E+08	24.50343	.00000	.04081

Continuous Compounding (Discrete Flow); $i = 5.0\%$

N	(F/P)	(P/F)	(F/A)	(P/A)	(A/F)	(A/P)
1	1.05127	.95123	1.00000	.95123	1.00000	1.05127
2	1.10517	.90484	2.05126	1.85606	.48750	.53877
3	1.16183	.86071	3.15643	2.71677	.31681	.36808
4	1.22140	.81873	4.31826	3.53550	.23157	.28285
5	1.28402	.77880	5.53967	4.31431	.18052	.23179
6	1.34985	.74082	6.82367	5.05512	.14655	.19782
7	1.41906	.70469	8.17353	5.75981	.12235	.17362
8	1.49182	.67032	9.59259	6.43014	.10425	.15552
9	1.56830	.63763	11.08439	7.06777	.09022	.14149
10	1.64871	.60653	12.65270	7.67430	.07903	.13030
11	1.73324	.57695	14.30141	8.25126	.06992	.12119
12	1.82210	.54882	16.03464	8.80007	.06236	.11364
13	1.91552	.52205	17.85675	9.32213	.05600	.10727
14	2.01373	.49659	19.77223	9.81870	.05058	.10185
15	2.11698	.47237	21.78598	10.29108	.04590	.09717
16	2.22552	.44933	23.90295	10.74041	.04184	.09311
17	2.33962	.42742	26.12848	11.16783	.03827	.08954
18	2.45957	.40657	28.46806	11.57440	.03513	.08640
19	2.58568	.38675	30.92766	11.96115	.03233	.08360
20	2.71824	.36788	33.51332	12.32904	.02984	.08111
21	2.85761	.34994	36.23158	12.67898	.02760	.07887
22	3.00412	.33288	39.08914	13.01185	.02558	.07685
23	3.15814	.31664	42.09328	13.32850	.02376	.07503
24	3.32006	.30120	45.25140	13.62969	.02210	.07337
25	3.49028	.28651	48.57146	13.91620	.02059	.07186
26	3.66923	.27254	52.06175	14.18874	.01921	.07048
27	3.85735	.25925	55.73099	14.44798	.01794	.06921
28	4.05512	.24660	59.58832	14.69458	.01678	.06805
29	4.26303	.23457	63.64346	14.92915	.01571	.06698
30	4.48159	.22313	67.90639	15.15228	.01473	.06600
35	5.75446	.17378	92.73294	16.11496	.01078	.06205
40	7.38885	.13534	124.61073	16.86470	.00802	.05930
45	9.48744	.10540	165.54251	17.44859	.00604	.05731
50	12.18207	.08209	218.09959	17.90334	.00459	.05586
55	15.64204	.06393	285.58423	18.25748	.00350	.05477
60	20.08470	.04979	372.23560	18.53329	.00269	.05396
65	25.78918	.03878	483.49805	18.74809	.00207	.05334
70	33.11383	.03020	626.36108	18.91539	.00160	.05287
75	42.51889	.02352	809.80103	19.04567	.00123	.05251
80	54.59514	.01832	1045.34106	19.14714	.00096	.05223
85	70.10133	.01427	1347.77979	19.22617	.00074	.05201
90	90.01157	.01111	1736.11694	19.28770	.00058	.05185
95	115.57669	.00865	2234.74951	19.33565	.00045	.05172
100	148.40291	.00674	2875.00513	19.37297	.00035	.05162
120	403.39551	.00248	7848.48437	19.45605	.00013	.05140
150	1807.85474	.00055	35241.62500	19.49362	.00003	.05130
180	.81021E+04	.00012	.15801E+06	19.50200	.00001	.05128
200	.22023E+05	.00005	.42953E+06	19.50351	.00000	.05127
240	.16273E+06	.00001	.31739E+07	19.50427	.00000	.05127
250	.26829E+06	.00000	.52328E+07	19.50432	.00000	.05127
300	.32683E+07	.00000	.63747E+08	19.50439	.00000	.05127
360	.65644E+08	.00000	.12803E+10	19.50439	.00000	.05127

Continuous Compounding (Discrete Flow); $i = 6.0\%$

N	(F/P)	(P/F)	(F/A)	(P/A)	(A/F)	(A/P)
1	1.06184	.94176	1.00000	.94176	1.00000	1.06184
2	1.12750	.88692	2.06183	1.82868	.48501	.54684
3	1.19722	.83527	3.18933	2.66395	.31355	.37538
4	1.27125	.78663	4.38654	3.45058	.22797	.28981
5	1.34986	.74082	5.65779	4.19140	.17675	.23858
6	1.43333	.69768	7.00763	4.88907	.14270	.20454
7	1.52196	.65705	8.44096	5.54612	.11847	.18031
8	1.61607	.61879	9.96291	6.16490	.10037	.16221
9	1.71600	.58275	11.57898	6.74766	.08636	.14820
10	1.82211	.54881	13.29497	7.29647	.07522	.13705
11	1.93478	.51685	15.11709	7.81332	.06615	.12799
12	2.05442	.48675	17.05185	8.30007	.05864	.12048
13	2.18146	.45841	19.10628	8.75848	.05234	.11418
14	2.31635	.43171	21.28772	9.19019	.04698	.10881
15	2.45959	.40657	23.60408	9.59676	.04237	.10420
16	2.61168	.38290	26.06364	9.97965	.03837	.10020
17	2.77317	.36060	28.67532	10.34025	.03487	.09671
18	2.94466	.33960	31.44849	10.67985	.03180	.09363
19	3.12674	.31982	34.39316	10.99967	.02908	.09091
20	3.32009	.30120	37.51987	11.30086	.02665	.08849
21	3.52539	.28366	40.83998	11.58452	.02449	.08632
22	3.74339	.26714	44.36533	11.85166	.02254	.08438
23	3.97486	.25158	48.10873	12.10324	.02079	.08262
24	4.22065	.23693	52.08356	12.34017	.01920	.08104
25	4.48164	.22313	56.30421	12.56331	.01776	.07960
26	4.75877	.21014	60.78584	12.77344	.01645	.07829
27	5.05303	.19790	65.54463	12.97134	.01526	.07709
28	5.36549	.18638	70.59758	13.15771	.01416	.07600
29	5.69727	.17552	75.96310	13.33324	.01316	.07500
30	6.04957	.16530	81.66032	13.49854	.01225	.07408
35	8.16605	.12246	115.88747	14.19138	.00863	.07047
40	11.02298	.09072	162.08911	14.70465	.00617	.06801
45	14.87944	.06721	224.45486	15.08490	.00446	.06629
50	20.08510	.04979	308.63940	15.36658	.00324	.06508
55	27.11200	.03688	422.27661	15.57527	.00237	.06420
60	36.59727	.02732	575.67017	15.72986	.00174	.06357
65	49.40105	.02024	782.72949	15.84439	.00128	.06311
70	66.68430	.01500	1062.22974	15.92924	.00094	.06278
75	90.01422	.01111	1439.51563	15.99208	.00069	.06253
80	121.50615	.00823	1948.79517	16.03865	.00051	.06235
85	164.01587	.00610	2636.25122	16.07315	.00038	.06222
90	221.39774	.00452	3564.21729	16.09871	.00028	.06212
95	298.85522	.00335	4816.83984	16.11763	.00021	.06204
100	403.41113	.00248	6507.69141	16.13105	.00015	.06199
120	1339.36035	.00075	21643.62109	16.15967	.00005	.06188
150	.81026E+04	.00012	.13102E+06	16.16975	.00001	.06184
180	.49017E+05	.00002	.79267E+06	16.17142	.00000	.06184
200	.16274E+06	.00001	.26318E+07	16.17165	.00000	.06184
240	.17939E+07	.00000	.29010E+08	16.17174	.00000	.06184
250	.32687E+07	.00000	.52860E+08	16.17174	.00000	.06184
300	.65651E+08	.00000	.10617E+10	16.17174	.00000	.06184
360	.24027E+10	.00000	.38855E+11	16.17174	.00000	.06184

Continuous Compounding (Discrete Flow); $i = 8.0\%$

N	(F/P)	(P/F)	(F/A)	(P/A)	(A/F)	(A/P)
1	1.08329	.92312	1.00000	.92312	1.00000	1.08329
2	1.17351	.85214	2.08329	1.77527	.48001	.56330
3	1.27125	.78663	3.25680	2.56190	.30705	.39034
4	1.37712	.72615	4.52805	3.28805	.22085	.30413
5	1.49182	.67032	5.90518	3.95837	.16937	.25263
6	1.61607	.61879	7.39700	4.57716	.13519	.21848
7	1.75067	.57121	9.01307	5.14837	.11095	.19424
8	1.89647	.52729	10.76375	5.67567	.09290	.17619
9	2.05442	.48675	12.66022	6.16242	.07899	.16227
10	2.22553	.44933	14.71465	6.61176	.06796	.15125
11	2.41088	.41479	16.94019	7.02654	.05903	.14232
12	2.61168	.38290	19.35109	7.40944	.05168	.13496
13	2.82920	.35346	21.96277	7.76290	.04553	.12882
14	3.06483	.32628	24.79196	8.08918	.04034	.12362
15	3.32009	.30120	27.85681	8.39038	.03590	.11918
16	3.59661	.27804	31.17091	8.66842	.03208	.11536
17	3.89616	.25666	34.77351	8.92508	.02876	.11204
18	4.22066	.23693	38.66969	9.16201	.02586	.10915
19	4.57218	.21871	42.89035	9.38073	.02332	.10660
20	4.95298	.20190	47.46255	9.58263	.02107	.10436
21	5.36549	.18638	52.41553	9.76900	.01908	.10236
22	5.81237	.17205	57.78104	9.94105	.01731	.10059
23	6.29646	.15882	63.59341	10.09987	.01572	.09901
24	6.82087	.14661	69.88991	10.24648	.01431	.09759
25	7.38896	.13534	76.71082	10.38182	.01304	.09632
26	8.00436	.12493	84.09975	10.50675	.01189	.09518
27	8.67101	.11533	92.10413	10.62207	.01086	.09414
28	9.39319	.10646	100.77522	10.72854	.00992	.09321
29	10.17552	.09828	110.16837	10.82681	.00908	.09236
30	11.02300	.09072	120.34393	10.91753	.00831	.09160
35	16.44434	.06081	185.43675	11.27663	.00539	.08868
40	24.53201	.04076	282.54370	11.51735	.00354	.08683
45	36.59735	.02732	427.40942	11.67870	.00234	.08563
50	54.59669	.01832	643.52368	11.78686	.00155	.08484
55	81.44847	.01228	965.92700	11.85936	.00104	.08432
60	121.50656	.00823	1446.89551	11.90797	.00069	.08398
65	181.26595	.00552	2164.41357	11.94054	.00046	.08375
70	270.41626	.00370	3234.82227	11.96238	.00031	.08360
75	403.41260	.00248	4831.67969	11.97702	.00021	.08349
80	601.81934	.00166	7213.90625	11.98682	.00014	.08342
85	897.80664	.00111	10767.75781	11.99341	.00009	.08338
90	1339.36646	.00075	16069.46875	11.99781	.00006	.08335
95	1998.09497	.00050	23978.67578	12.00077	.00004	.08333
100	2980.80005	.00034	35777.79297	12.00275	.00003	.08331
120	.14764E+05	.00007	.17725E+06	12.00596	.00001	.08329
150	.16274E+06	.00001	.19540E+07	12.00670	.00000	.08329
180	.17939E+07	.00000	.21539E+08	12.00677	.00000	.08329
200	.88852E+07	.00000	.10668E+09	12.00678	.00000	.08329
240	.21797E+09	.00000	.26171E+10	12.00678	.00000	.08329
250	.48510E+09	.00000	.58245E+10	12.00678	.00000	.08329
300	.26485E+11	.00000	.31800E+12	12.00678	.00000	.08329
360	.32181E+13	.00000	.38639E+14	12.00678	.00000	.08329

Continuous Compounding (Discrete Flow); i = 10.0%

N	(F/P)	(P/F)	(F/A)	(P/A)	(A/F)	(A/P)
1	1.10517	.90484	1.00000	.90484	1.00000	1.10517
2	1.22140	.81873	2.10517	1.72357	.47502	.58019
3	1.34986	.74082	3.32657	2.46439	.30061	.40578
4	1.49182	.67032	4.67643	3.13471	.21384	.31901
5	1.64872	.60653	6.16825	3.74124	.16212	.26729
6	1.82211	.54881	7.81697	4.29006	.12793	.23310
7	2.01374	.49659	9.63908	4.78664	.10374	.20891
8	2.22553	.44933	11.65283	5.23598	.08582	.19099
9	2.45959	.40657	13.87836	5.64255	.07205	.17722
10	2.71827	.36788	16.33794	6.01043	.06121	.16638
11	3.00415	.33287	19.05621	6.34330	.05248	.15765
12	3.32009	.30120	22.06036	6.64450	.04533	.15050
13	3.66927	.27253	25.38045	6.91703	.03940	.14457
14	4.05517	.24660	29.04973	7.16363	.03442	.13959
15	4.48165	.22313	33.10489	7.38676	.03021	.13538
16	4.95299	.20190	37.58655	7.58866	.02661	.13178
17	5.47389	.18269	42.53954	7.77135	.02351	.12868
18	6.04958	.16530	48.01343	7.93665	.02083	.12600
19	6.68582	.14957	54.06302	8.08622	.01850	.12367
20	7.38897	.13534	60.74884	8.22155	.01646	.12163
21	8.16607	.12246	68.13777	8.34401	.01468	.11985
22	9.02490	.11080	76.30386	8.45482	.01311	.11828
23	9.97405	.10026	85.32878	8.55508	.01172	.11689
24	11.02302	.09072	95.30286	8.64580	.01049	.11566
25	12.18232	.08209	106.32585	8.72788	.00941	.11458
26	13.46353	.07427	118.50816	8.80216	.00844	.11361
27	14.87950	.06721	131.97169	8.86936	.00758	.11275
28	16.44438	.06081	146.85127	8.93018	.00681	.11198
29	18.17383	.05502	163.29552	8.98520	.00612	.11129
30	20.08519	.04979	181.46948	9.03499	.00551	.11068
35	33.11478	.03020	305.35986	9.22126	.00327	.10845
40	54.59688	.01832	509.62012	9.33424	.00196	.10713
45	90.01474	.01111	846.38696	9.40276	.00118	.10635
50	148.40884	.00674	1401.62085	9.44432	.00071	.10588
55	244.68410	.00409	2317.04370	9.46953	.00043	.10560
60	403.41479	.00248	3826.31616	9.48482	.00026	.10543
65	665.11670	.00150	6314.68359	9.49410	.00016	.10533
70	1096.58838	.00091	10417.28516	9.49972	.00010	.10527
75	1807.96387	.00055	17181.32031	9.50313	.00006	.10523
80	2980.82007	.00034	28333.30078	9.50520	.00004	.10521
85	4914.52734	.00020	46719.75391	9.50646	.00002	.10519
90	8102.66016	.00012	77033.75000	9.50722	.00001	.10518
95	.13359E+05	.00007	.12701E+06	9.50768	.00001	.10518
100	.22026E+05	.00005	.20941E+06	9.50796	.00000	.10518
120	.16274E+06	.00001	.15474E+07	9.50833	.00000	.10517
150	.32687E+07	.00000	.31080E+08	9.50839	.00000	.10517
180	.65653E+08	.00000	.62426E+09	9.50839	.00000	.10517
200	.48511E+09	.00000	.46126E+10	9.50839	.00000	.10517
240	.26485E+11	.00000	.25183E+12	9.50839	.00000	.10517
250	.71994E+11	.00000	.68455E+12	9.50839	.00000	.10517
300	.10685E+14	.00000	.10159E+15	9.50839	.00000	.10517
360	.43103E+16	.00000	.40984E+17	9.50839	.00000	.10517

Continuous Compounding (Discrete Flow); i = 12.0%

N	(F/P)	(P/F)	(F/A)	(P/A)	(A/F)	(A/P)
1	1.12750	.88692	1.00000	.88692	1.00000	1.12750
2	1.27125	.78663	2.12750	1.67355	.47004	.59753
3	1.43333	.69768	3.39874	2.37122	.29423	.42172
4	1.61607	.61878	4.83207	2.99001	.20695	.33445
5	1.82212	.54881	6.44815	3.53882	.15508	.28258
6	2.05443	.48675	8.27027	4.02557	.12092	.24841
7	2.31636	.43171	10.32469	4.45728	.09686	.22435
8	2.61169	.38289	12.64106	4.84018	.07911	.20660
9	2.94468	.33960	15.25276	5.17978	.06556	.19306
10	3.32011	.30119	18.19743	5.48097	.05495	.18245
11	3.74342	.26714	21.51755	5.74810	.04647	.17397
12	4.22069	.23693	25.26096	5.98503	.03959	.16708
13	4.75881	.21014	29.48166	6.19517	.03392	.16142
14	5.36555	.18637	34.24048	6.38154	.02921	.15670
15	6.04964	.16530	39.60600	6.54684	.02525	.15275
16	6.82095	.14661	45.65567	6.69345	.02190	.14940
17	7.69060	.13003	52.47661	6.82348	.01906	.14655
18	8.67112	.11533	60.16721	6.93880	.01662	.14412
19	9.77666	.10228	68.83830	7.04109	.01453	.14202
20	11.02315	.09072	78.61499	7.13181	.01272	.14022
21	12.42857	.08046	89.63817	7.21227	.01116	.13865
22	14.01317	.07136	102.06671	7.28363	.00980	.13729
23	15.79980	.06329	116.07985	7.34692	.00861	.13611
24	17.81422	.05613	131.87964	7.40305	.00758	.13508
25	20.08548	.04979	149.69391	7.45284	.00668	.13418
26	22.64632	.04416	169.77937	7.49700	.00589	.13339
27	25.53365	.03916	192.42575	7.53617	.00520	.13269
28	28.78911	.03474	217.95934	7.57089	.00459	.13208
29	32.45963	.03081	246.74849	7.60171	.00405	.13155
30	36.59813	.02732	279.20825	7.62903	.00358	.13108
35	66.68608	.01500	515.19824	7.72572	.00194	.12944
40	121.50993	.00823	945.20020	7.77879	.00106	.12855
45	221.40543	.00452	1728.71460	7.80791	.00058	.12808
50	403.42676	.00248	3156.36987	7.82390	.00032	.12781
55	735.09082	.00136	5757.72266	7.83267	.00017	.12767
60	1339.42285	.00075	10497.70313	7.83748	.00010	.12759
65	2440.58643	.00041	19134.50000	7.84013	.00005	.12755
70	4447.03516	.00022	34871.76172	7.84158	.00003	.12753
75	8103.01953	.00012	63546.88672	7.84237	.00002	.12751
80	.14765E+05	.00007	.11580E+06	7.84281	.00001	.12751
85	.26903E+05	.00004	.21100E+06	7.84305	.00000	.12750
90	.49020E+05	.00002	.38448E+06	7.84318	.00000	.12750
95	.89321E+05	.00001	.70057E+06	7.84325	.00000	.12750
100	.16275E+06	.00001	.12765E+07	7.84329	.00000	.12750
120	.17941E+07	.00000	.14071E+08	7.84334	.00000	.12750
150	.65659E+08	.00000	.51499E+09	7.84334	.00000	.12750
180	.24030E+10	.00000	.18848E+11	7.84334	.00000	.12750
200	.26489E+11	.00000	.20776E+12	7.84334	.00000	.12750
240	.32186E+13	.00000	.25245E+14	7.84334	.00000	.12750
250	.10686E+14	.00000	.83816E+14	7.84334	.00000	.12750
300	.43111E+16	.00000	.33813E+17	7.84334	.00000	.12750
360	.57744E+19	.00000	.45291E+20	7.84334	.00000	.12750

Continuous Compounding (Discrete Flow); *i* = 15.0%

N	(F/P)	(P/F)	(F/A)	(P/A)	(A/F)	(A/P)
1	1.16183	.86071	1.00000	.86071	1.00000	1.16183
2	1.34986	.74082	2.16183	1.60153	.46257	.62440
3	1.56831	.63763	3.51169	2.23915	.28476	.44660
4	1.82211	.54881	5.08000	2.78797	.19685	.35868
5	2.11699	.47237	6.90211	3.26033	.14488	.30672
6	2.45959	.40657	9.01910	3.66690	.11088	.27271
7	2.85764	.34994	11.47870	4.01684	.08712	.24895
8	3.32010	.30120	14.33633	4.31804	.06975	.23159
9	3.85741	.25924	17.65643	4.57728	.05664	.21847
10	4.48166	.22313	21.51384	4.80041	.04648	.20832
11	5.20695	.19205	25.99550	4.99246	.03847	.20030
12	6.04961	.16530	31.20244	5.15776	.03205	.19388
13	7.02864	.14228	37.25206	5.30004	.02684	.18868
14	8.16611	.12246	44.28067	5.42249	.02258	.18442
15	9.48766	.10540	52.44678	5.52789	.01907	.18090
16	11.02308	.09072	61.93443	5.61861	.01615	.17798
17	12.80699	.07808	72.95750	5.69669	.01371	.17554
18	14.87959	.06721	85.76448	5.76390	.01166	.17349
19	17.28761	.05784	100.64410	5.82175	.00994	.17177
20	20.08533	.04979	117.93166	5.87153	.00848	.17031
21	23.33580	.04285	138.01692	5.91439	.00725	.16908
22	27.11232	.03688	161.35274	5.95127	.00620	.16803
23	31.50000	.03175	188.46503	5.98302	.00531	.16714
24	36.59776	.02732	219.96506	6.01034	.00455	.16638
25	42.52052	.02352	256.56274	6.03386	.00390	.16573
26	49.40175	.02024	299.08325	6.05410	.00334	.16518
27	57.39662	.01742	348.48486	6.07152	.00287	.16470
28	66.68533	.01500	405.88159	6.08652	.00246	.16430
29	77.47728	.01291	472.56689	6.09943	.00212	.16395
30	90.01567	.01111	550.04395	6.11054	.00182	.16365
35	190.56270	.00525	1171.34204	6.14675	.00085	.16269
40	403.42017	.00248	2486.62720	6.16387	.00040	.16224
45	854.03833	.00117	5271.07813	6.17194	.00019	.16202
50	1807.99390	.00055	11165.74219	6.17576	.00009	.16192
55	3827.51245	.00026	23644.71094	6.17757	.00004	.16188
60	8102.82422	.00012	50062.63281	6.17842	.00002	.16185
65	.17154E+05	.00006	.10599E+06	6.17882	.00001	.16184
70	.36314E+05	.00003	.22439E+06	6.17901	.00000	.16184
75	.76877E+05	.00001	.47503E+06	6.17910	.00000	.16184
80	.16275E+06	.00001	.10056E+07	6.17914	.00000	.16183
85	.34454E+06	.00000	.21289E+07	6.17916	.00000	.16183
90	.72938E+06	.00000	.45070E+07	6.17917	.00000	.16183
95	.15441E+07	.00000	.95412E+07	6.17918	.00000	.16183
100	.32688E+07	.00000	.20199E+08	6.17918	.00000	.16183
120	.65656E+08	.00000	.40570E+09	6.17918	.00000	.16183
150	.59100E+10	.00000	.36519E+11	6.17918	.00000	.16183
180	.53200E+12	.00000	.32873E+13	6.17918	.00000	.16183
200	.10685E+14	.00000	.66027E+14	6.17918	.00000	.16183
240	.43107E+16	.00000	.26636E+17	6.17918	.00000	.16183
250	.19319E+17	.00000	.11938E+18	6.17918	.00000	.16183
300	.34929E+20	.00000	.21583E+21	6.17918	.00000	.16183
360	.28302E+24	.00000	.17488E+25	6.17918	.00000	.16183

Continuous Compounding (Discrete Flow); i = 16.0%

N	(F/P)	(P/F)	(F/A)	(P/A)	(A/F)	(A/P)
1	1.17351	.85214	1.00000	.85214	1.00000	1.17351
2	1.37713	.72615	2.17351	1.57829	.46009	.63360
3	1.61607	.61878	3.55064	2.19708	.28164	.45515
4	1.89648	.52729	5.16671	2.72437	.19355	.36706
5	2.22554	.44933	7.06318	3.17370	.14158	.31509
6	2.61169	.38289	9.28872	3.55659	.10766	.28117
7	3.06485	.32628	11.90041	3.88287	.08403	.25754
8	3.59663	.27804	14.96524	4.16091	.06682	.24033
9	4.22068	.23693	18.56187	4.39784	.05387	.22738
10	4.95301	.20190	22.78255	4.59974	.04389	.21740
11	5.81241	.17205	27.73555	4.77178	.03605	.20957
12	6.82092	.14661	33.54794	4.91839	.02981	.20332
13	8.00443	.12493	40.36887	5.04332	.02477	.19828
14	9.39328	.10646	48.37329	5.14978	.02067	.19418
15	11.02311	.09072	57.76657	5.24050	.01731	.19082
16	12.93573	.07731	68.78964	5.31780	.01454	.18805
17	15.18021	.06588	81.72536	5.38368	.01224	.18575
18	17.81413	.05614	96.90552	5.43981	.01032	.18383
19	20.90508	.04784	114.71968	5.48765	.00872	.18223
20	24.53232	.04076	135.62469	5.52841	.00737	.18088
21	28.78893	.03474	160.15700	5.56315	.00624	.17975
22	33.78412	.02960	188.94591	5.59274	.00529	.17880
23	39.64601	.02522	222.73012	5.61797	.00449	.17800
24	46.52499	.02149	262.37598	5.63946	.00381	.17732
25	54.59756	.01832	308.90088	5.65778	.00324	.17675
26	64.07080	.01561	363.49829	5.67339	.00275	.17626
27	75.18776	.01330	427.56909	5.68668	.00234	.17585
28	88.23360	.01133	502.75684	5.69802	.00199	.17550
29	103.54306	.00966	590.99023	5.70768	.00169	.17520
30	121.50887	.00823	694.53345	5.71591	.00144	.17495
35	270.42236	.00370	1552.77197	5.74203	.00064	.17415
40	601.83447	.00166	3462.81201	5.75376	.00029	.17380
45	1339.40479	.00075	7713.67969	5.75904	.00013	.17364
50	2980.89307	.00034	17174.13281	5.76140	.00006	.17357
55	6634.08594	.00015	38228.71484	5.76247	.00003	.17354
60	14764.39453	.00007	85086.43750	5.76295	.00001	.17352
65	.32859E+05	.00003	.18937E+06	5.76316	.00001	.17352
70	.73128E+05	.00001	.42146E+06	5.76326	.00000	.17351
75	.16275E+06	.00001	.93797E+06	5.76330	.00000	.17351
80	.36220E+06	.00000	.20875E+07	5.76332	.00000	.17351
85	.80610E+06	.00000	.46458E+07	5.76333	.00000	.17351
90	.17940E+07	.00000	.10339E+08	5.76334	.00000	.17351
95	.39926E+07	.00000	.23011E+08	5.76334	.00000	.17351
100	.88857E+07	.00000	.51211E+08	5.76334	.00000	.17351
120	.21799E+09	.00000	.12563E+10	5.76334	.00000	.17351
150	.26487E+11	.00000	.15266E+12	5.76334	.00000	.17351
180	.32185E+13	.00000	.18549E+14	5.76334	.00000	.17351
200	.78956E+14	.00000	.45505E+15	5.76334	.00000	.17351
240	.47518E+17	.00000	.27387E+18	5.76334	.00000	.17351
250	.23536E+18	.00000	.13565E+19	5.76334	.00000	.17351
300	.70158E+21	.00000	.40435E+22	5.76334	.00000	.17351
360	.10358E+26	.00000	.59699E+26	5.76334	.00000	.17351

Continuous Compounding (Discrete Flow); i = 20.0%

N	(F/P)	(P/F)	(F/A)	(P/A)	(A/F)	(A/P)
1	1.22140	.81873	1.00000	.81873	1.00000	1.22140
2	1.49183	.67032	2.22140	1.48905	.45017	.67157
3	1.82212	.54881	3.71323	2.03786	.26931	.49071
4	2.22554	.44933	5.53534	2.48719	.18066	.40206
5	2.71828	.36788	7.76088	2.85507	.12885	.35025
6	3.32012	.30119	10.47917	3.15626	.09543	.31683
7	4.05520	.24660	13.79929	3.40286	.07247	.29387
8	4.95304	.20190	17.85448	3.60475	.05601	.27741
9	6.04965	.16530	22.80751	3.77005	.04385	.26525
10	7.38907	.13534	28.85716	3.90539	.03465	.25606
11	9.02503	.11080	36.24622	4.01619	.02759	.24899
12	11.02319	.09072	45.27122	4.10691	.02209	.24349
13	13.46376	.07427	56.29442	4.18118	.01776	.23917
14	16.44467	.06081	69.75813	4.24199	.01434	.23574
15	20.08557	.04979	86.20280	4.29178	.01160	.23300
16	24.53258	.04076	106.28838	4.33254	.00941	.23081
17	29.96417	.03337	130.82097	4.36591	.00764	.22905
18	36.59831	.02732	160.78505	4.39324	.00622	.22762
19	44.70129	.02237	197.38336	4.41561	.00507	.22647
20	54.59828	.01832	242.08455	4.43392	.00413	.22553
21	66.68651	.01500	296.68286	4.44892	.00337	.22477
22	81.45110	.01228	363.36938	4.46120	.00275	.22416
23	99.48462	.01005	444.82031	4.47125	.00225	.22365
24	121.51077	.00823	544.30469	4.47948	.00184	.22324
25	148.41364	.00674	665.81543	4.48622	.00150	.22291
26	181.27283	.00552	814.22900	4.49173	.00123	.22263
27	221.40715	.00452	995.50171	4.49625	.00100	.22241
28	270.42725	.00370	1216.90796	4.49995	.00082	.22222
29	330.30078	.00303	1487.33569	4.50297	.00067	.22208
30	403.43018	.00248	1817.63550	4.50545	.00055	.22195
35	1096.63794	.00091	4948.60937	4.51253	.00020	.22161
40	2980.97266	.00034	13459.48828	4.51513	.00007	.22148
45	8103.12891	.00012	36594.47266	4.51609	.00003	.22143
50	22026.59766	.00005	99481.87500	4.51644	.00001	.22141
55	.59875E+05	.00002	.27043E+06	4.51657	.00000	.22141
60	.16276E+06	.00001	.73511E+06	4.51662	.00000	.22140
65	.44242E+06	.00000	.19982E+07	4.51664	.00000	.22140
70	.12026E+07	.00000	.54318E+07	4.51664	.00000	.22140
75	.32690E+07	.00000	.14765E+08	4.51665	.00000	.22140
80	.88862E+07	.00000	.40136E+08	4.51665	.00000	.22140
85	.24155E+08	.00000	.10910E+09	4.51665	.00000	.22140
90	.65661E+08	.00000	.29657E+09	4.51665	.00000	.22140
95	.17848E+09	.00000	.80615E+09	4.51665	.00000	.22140
100	.48517E+09	.00000	.21913E+10	4.51665	.00000	.22140
120	.26490E+11	.00000	.11964E+12	4.51665	.00000	.22140
150	.10687E+14	.00000	.48268E+14	4.51665	.00000	.22140
180	.43113E+16	.00000	.19473E+17	4.51665	.00000	.22140
200	.23539E+18	.00000	.10632E+19	4.51665	.00000	.22140
240	.70169E+21	.00000	.31693E+22	4.51665	.00000	.22140
250	.51849E+22	.00000	.23418E+23	4.51665	.00000	.22140
300	.11420E+27	.00000	.51582E+27	4.51665	.00000	.22140
360	.18588E+32	.00000	.83953E+32	4.51665	.00000	.22140

Continuous Compounding (Discrete Flow); i = 25.0%

N	(F/P)	(P/F)	(F/A)	(P/A)	(A/F)	(A/P)
1	1.28403	.77880	1.00000	.77880	1.00000	1.28403
2	1.64872	.60653	2.28403	1.38533	.43782	.72185
3	2.11700	.47237	3.93275	1.85770	.25427	.53830
4	2.71829	.36788	6.04976	2.22557	.16530	.44932
5	3.49035	.28650	8.76805	2.51208	.11405	.39808
6	4.48171	.22313	12.25840	2.73521	.08158	.36560
7	5.75463	.17377	16.74011	2.90898	.05974	.34376
8	7.38910	.13533	22.49475	3.04432	.04445	.32848
9	9.48779	.10540	29.88385	3.14972	.03346	.31749
10	12.18258	.08208	39.37166	3.23180	.02540	.30943
11	15.64275	.06393	51.55424	3.29573	.01940	.30342
12	20.08571	.04979	67.19701	3.34551	.01488	.29891
13	25.79057	.03877	87.28271	3.38429	.01146	.29548
14	33.11577	.03020	113.07330	3.41448	.00884	.29287
15	42.52151	.02352	146.18907	3.43800	.00684	.29087
16	54.59875	.01832	188.71066	3.45632	.00530	.28933
17	70.10623	.01426	243.30939	3.47058	.00411	.28814
18	90.01825	.01111	313.41577	3.48169	.00319	.28722
19	115.58580	.00865	403.43408	3.49034	.00248	.28650
20	148.41522	.00674	519.01978	3.49708	.00193	.28595
21	190.56902	.00525	667.43506	3.50233	.00150	.28552
22	244.69565	.00409	858.00439	3.50641	.00117	.28519
23	314.19556	.00318	1102.69971	3.50960	.00091	.28493
24	403.43555	.00248	1416.89600	3.51208	.00071	.28473
25	518.02173	.00193	1820.33179	3.51401	.00055	.28458
26	665.15356	.00150	2338.35425	3.51551	.00043	.28445
27	854.07471	.00117	3003.50732	3.51668	.00033	.28436
28	1096.65479	.00091	3857.58301	3.51759	.00026	.28429
29	1408.13330	.00071	4954.23828	3.51830	.00020	.28423
30	1808.07959	.00055	6362.37109	3.51886	.00016	.28418
35	6310.83984	.00016	22215.70312	3.52024	.00005	.28407
40	22027.07422	.00005	77549.43750	3.52064	.00001	.28404
45	.76882E+05	.00001	.27068E+06	3.52076	.00000	.28403
50	.26835E+06	.00000	.94479E+06	3.52079	.00000	.28403
55	.93662E+06	.00000	.32977E+07	3.52080	.00000	.28403
60	.32692E+07	.00000	.11510E+08	3.52080	.00000	.28403
65	.11411E+08	.00000	.40174E+08	3.52080	.00000	.28403
70	.39827E+08	.00000	.14022E+09	3.52080	.00000	.28403
75	.13901E+09	.00000	.48942E+09	3.52080	.00000	.28403
80	.48519E+09	.00000	.17083E+10	3.52080	.00000	.28403
85	.16935E+10	.00000	.59625E+10	3.52080	.00000	.28403
90	.59109E+10	.00000	.20811E+11	3.52080	.00000	.28403
95	.20631E+11	.00000	.72638E+11	3.52080	.00000	.28403
100	.72010E+11	.00000	.25353E+12	3.52080	.00000	.28403
120	.10687E+14	.00000	.37628E+14	3.52080	.00000	.28403
150	.19324E+17	.00000	.68035E+17	3.52080	.00000	.28403
180	.34939E+20	.00000	.12301E+21	3.52080	.00000	.28403
200	.51854E+22	.00000	.18257E+23	3.52080	.00000	.28403
240	.11422E+27	.00000	.40214E+27	3.52080	.00000	.28403
250	.13915E+28	.00000	.48992E+28	3.52080	.00000	.28403
300	.37340E+33	.00000	.13147E+34	3.52080	.00000	.28403
360	.12207E+40	.00000	.42979E+40	3.52080	.00000	.28403

Continuous Compounding (Discrete Flow); i = 30.0%

N	(F/P)	(P/F)	(F/A)	(P/A)	(A/F)	(A/P)
1	1.34986	.74082	1.00000	.74082	1.00000	1.34986
2	1.82212	.54881	2.34986	1.28963	.42556	.77542
3	2.45961	.40657	4.17198	1.69620	.23969	.58955
4	3.32012	.30119	6.63158	1.99739	.15079	.50065
5	4.48169	.22313	9.95170	2.22052	.10049	.45034
6	6.04966	.16530	14.43339	2.38582	.06928	.41914
7	8.16618	.12246	20.48305	2.50828	.04882	.39868
8	11.02320	.09072	28.64923	2.59899	.03490	.38476
9	14.87976	.06721	39.67242	2.66620	.02521	.37507
10	20.08559	.04979	54.55219	2.71599	.01833	.36819
11	27.11272	.03688	74.63779	2.75287	.01340	.36326
12	36.59833	.02732	101.75044	2.78019	.00983	.35969
13	49.40259	.02024	138.34875	2.80044	.00723	.35709
14	66.68654	.01500	187.75137	2.81543	.00533	.35519
15	90.01746	.01111	254.43782	2.82654	.00393	.35379
16	121.51085	.00823	344.45532	2.83477	.00290	.35276
17	164.02254	.00610	465.96606	2.84087	.00215	.35201
18	221.40732	.00452	629.98853	2.84538	.00159	.35145
19	298.86865	.00335	851.39575	2.84873	.00117	.35103
20	403.43066	.00248	1150.26465	2.85121	.00087	.35073
21	544.57446	.00184	1553.69482	2.85304	.00064	.35050
22	735.09888	.00136	2098.26880	2.85440	.00048	.35034
23	992.28003	.00101	2833.36841	2.85541	.00035	.35021
24	1339.43799	.00075	3825.64819	2.85616	.00026	.35012
25	1808.05273	.00055	5165.08594	2.85671	.00019	.35005
26	2440.61670	.00041	6973.13672	2.85712	.00014	.35000
27	3294.48926	.00030	9413.75781	2.85742	.00011	.34997
28	4447.09375	.00022	12708.23438	2.85765	.00008	.34994
29	6002.94922	.00017	17155.32422	2.85782	.00006	.34992
30	8103.13672	.00012	23158.27734	2.85794	.00004	.34990
35	.36316E+05	.00003	.10380E+06	2.85821	.00001	.34987
40	.16276E+06	.00001	.46520E+06	2.85827	.00000	.34986
45	.72942E+06	.00000	.20849E+07	2.85829	.00000	.34986
50	.32691E+07	.00000	.93439E+07	2.85829	.00000	.34986
55	.14651E+08	.00000	.41877E+08	2.85829	.00000	.34986
60	.65661E+08	.00000	.18768E+09	2.85829	.00000	.34986
65	.29427E+09	.00000	.84111E+09	2.85829	.00000	.34986
70	.13188E+10	.00000	.37696E+10	2.85829	.00000	.34986
75	.59106E+10	.00000	.16894E+11	2.85829	.00000	.34986
80	.26490E+11	.00000	.75715E+11	2.85829	.00000	.34986
85	.11872E+12	.00000	.33933E+12	2.85829	.00000	.34986
90	.53206E+12	.00000	.15208E+13	2.85829	.00000	.34986
95	.23845E+13	.00000	.68157E+13	2.85829	.00000	.34986
100	.10687E+14	.00000	.30546E+14	2.85829	.00000	.34986
120	.43113E+16	.00000	.12323E+17	2.85829	.00000	.34986
150	.34935E+20	.00000	.99856E+20	2.85829	.00000	.34986
180	.28309E+24	.00000	.80914E+24	2.85829	.00000	.34986
200	.11421E+27	.00000	.32643E+27	2.85829	.00000	.34986
240	.18588E+32	.00000	.53129E+32	2.85829	.00000	.34986
250	.37335E+33	.00000	.10671E+34	2.85829	.00000	.34986
300	.12205E+40	.00000	.34885E+40	2.85829	.00000	.34986
360	.80138E+47	.00000	.22906E+48	2.85829	.00000	.34986

Continuous Compounding (Discrete Flow); $i = 40.0\%$

N	(F/P)	(P/F)	(F/A)	(P/A)	(A/F)	(A/P)
1	1.49182	.67032	1.00000	.67032	1.00000	1.49182
2	2.22554	.44933	2.49182	1.11965	.40131	.89314
3	3.32011	.30119	4.71736	1.42084	.21198	.70381
4	4.95302	.20190	8.03747	1.62274	.12442	.61624
5	7.38904	.13534	12.99050	1.75808	.07698	.56880
6	11.02315	.09072	20.37955	1.84879	.04907	.54089
7	16.44461	.06081	31.40271	1.90961	.03184	.52367
8	24.53246	.04076	47.84731	1.95037	.02090	.51272
9	36.59811	.02732	72.37976	1.97769	.01382	.50564
10	54.59795	.01832	108.97786	1.99601	.00918	.50100
11	81.45053	.01228	163.57582	2.00828	.00611	.49794
12	121.50986	.00823	245.02637	2.01651	.00408	.49591
13	181.27135	.00552	366.53613	2.02203	.00273	.49455
14	270.42505	.00370	547.80762	2.02573	.00183	.49365
15	403.42676	.00248	818.23291	2.02821	.00122	.49305
16	601.84131	.00166	1221.65869	2.02987	.00082	.49264
17	897.84131	.00111	1823.49976	2.03098	.00055	.49237
18	1339.42139	.00075	2721.34106	2.03173	.00037	.49219
19	1998.18115	.00050	4060.76270	2.03223	.00025	.49207
20	2980.93506	.00034	6058.94531	2.03257	.00017	.49199
21	4447.03125	.00022	9039.87891	2.03279	.00011	.49193
22	6634.18750	.00015	13486.90625	2.03294	.00007	.49190
23	9897.04687	.00010	20121.10547	2.03304	.00005	.49187
24	14764.64453	.00007	30018.14063	2.03311	.00003	.49186
25	22026.25781	.00005	44782.80078	2.03316	.00002	.49185
26	32859.30078	.00003	66809.06250	2.03319	.00001	.49184
27	49020.29297	.00002	99668.31250	2.03321	.00001	.49183
28	.73130E+05	.00001	.14869E+06	2.03322	.00001	.49183
29	.10910E+06	.00001	.22182E+06	2.03323	.00000	.49183
30	.16275E+06	.00001	.33091E+06	2.03323	.00000	.49183
35	.12026E+07	.00000	.24452E+07	2.03325	.00000	.49182
40	.88860E+07	.00000	.18067E+08	2.03325	.00000	.49182
45	.65659E+08	.00000	.13350E+09	2.03325	.00000	.49182
50	.48516E+09	.00000	.98644E+09	2.03325	.00000	.49182
55	.35848E+10	.00000	.72889E+10	2.03325	.00000	.49182
60	.26488E+11	.00000	.53858E+11	2.03325	.00000	.49182
65	.19572E+12	.00000	.39796E+12	2.03325	.00000	.49182
70	.14462E+13	.00000	.29405E+13	2.03325	.00000	.49182
75	.10686E+14	.00000	.21728E+14	2.03325	.00000	.49182
80	.78961E+14	.00000	.16055E+15	2.03325	.00000	.49182
85	.58344E+15	.00000	.11863E+16	2.03325	.00000	.49182
90	.43111E+16	.00000	.87655E+16	2.03325	.00000	.49182
95	.31855E+17	.00000	.64769E+17	2.03325	.00000	.49182
100	.23538E+18	.00000	.47858E+18	2.03325	.00000	.49182
120	.70164E+21	.00000	.14266E+22	2.03325	.00000	.49182
150	.11419E+27	.00000	.23218E+27	2.03325	.00000	.49182
180	.18585E+32	.00000	.37789E+32	2.03325	.00000	.49182
200	.55402E+35	.00000	.11265E+36	2.03325	.00000	.49182
240	.49230E+42	.00000	.10010E+43	2.03325	.00000	.49182
250	.26879E+44	.00000	.54651E+44	2.03325	.00000	.49182
300	.13040E+53	.00000	.26514E+53	2.03325	.00000	.49182
360	.34542E+63	.00000	.70232E+63	2.03325	.00000	.49182

Continuous Compounding (Discrete Flow); i = 50.0%

N	(F/P)	(P/F)	(F/A)	(P/A)	(A/F)	(A/P)
1	1.64872	.60653	1.00000	.60653	1.00000	1.64872
2	2.71828	.36788	2.64872	.97441	.37754	1.02626
3	4.48169	.22313	5.36700	1.19754	.18632	.83505
4	7.38906	.13534	9.84869	1.33287	.10154	.75026
5	12.18250	.08208	17.23775	1.41496	.05801	.70673
6	20.08556	.04979	29.42026	1.46475	.03399	.68271
7	33.11549	.03020	49.50581	1.49494	.02020	.66892
8	54.59822	.01832	82.62128	1.51326	.01210	.66083
9	90.01727	.01111	137.21954	1.52437	.00729	.65601
10	148.41339	.00674	227.23676	1.53111	.00440	.65312
11	244.69237	.00409	375.65015	1.53519	.00266	.65138
12	403.42969	.00248	620.34253	1.53767	.00161	.65033
13	665.14307	.00150	1023.77197	1.53918	.00098	.64970
14	1096.63574	.00091	1688.91455	1.54009	.00059	.64931
15	1808.04712	.00055	2785.55005	1.54064	.00036	.64908
16	2980.96582	.00034	4593.59766	1.54098	.00022	.64894
17	4914.78125	.00020	7574.55859	1.54118	.00013	.64885
18	8103.10938	.00012	12489.34375	1.54130	.00008	.64880
19	13359.76953	.00007	20592.44922	1.54138	.00005	.64877
20	22026.53906	.00005	33952.21094	1.54142	.00003	.64875
21	36315.62891	.00003	55978.76563	1.54145	.00002	.64874
22	59874.35547	.00002	92294.37500	1.54147	.00001	.64873
23	.98716E+05	.00001	.15217E+06	1.54148	.00001	.64873
24	.16276E+06	.00001	.25088E+06	1.54148	.00000	.64873
25	.26834E+06	.00000	.41364E+06	1.54149	.00000	.64872
26	.44242E+06	.00000	.68198E+06	1.54149	.00000	.64872
27	.72942E+06	.00000	.11244E+07	1.54149	.00000	.64872
28	.12026E+07	.00000	.18538E+07	1.54149	.00000	.64872
29	.19828E+07	.00000	.30564E+07	1.54149	.00000	.64872
30	.32690E+07	.00000	.50392E+07	1.54149	.00000	.64872
35	.39825E+08	.00000	.61390E+08	1.54149	.00000	.64872
40	.48517E+09	.00000	.74788E+09	1.54149	.00000	.64872
45	.59106E+10	.00000	.91111E+10	1.54149	.00000	.64872
50	.72006E+11	.00000	.11100E+12	1.54149	.00000	.64872
55	.87721E+12	.00000	.13522E+13	1.54149	.00000	.64872
60	.10687E+14	.00000	.16473E+14	1.54149	.00000	.64872
65	.13019E+15	.00000	.20069E+15	1.54149	.00000	.64872
70	.15860E+16	.00000	.24449E+16	1.54149	.00000	.64872
75	.19322E+17	.00000	.29784E+17	1.54149	.00000	.64872
80	.23539E+18	.00000	.36285E+18	1.54149	.00000	.64872
85	.28676E+19	.00000	.44204E+19	1.54149	.00000	.64872
90	.34935E+20	.00000	.53852E+20	1.54149	.00000	.64872
95	.42559E+21	.00000	.65605E+21	1.54149	.00000	.64872
100	.51848E+22	.00000	.79923E+22	1.54149	.00000	.64872
120	.11420E+27	.00000	.17604E+27	1.54149	.00000	.64872
150	.37333E+33	.00000	.57549E+33	1.54149	.00000	.64872
180	.12204E+40	.00000	.18813E+40	1.54149	.00000	.64872
200	.26882E+44	.00000	.41439E+44	1.54149	.00000	.64872
240	.13042E+53	.00000	.20105E+53	1.54149	.00000	.64872
250	.19357E+55	.00000	.29838E+55	1.54149	.00000	.64872
300	.13938E+66	.00000	.21485E+66	1.54149	.00000	.64872

Appendix C

Program Code: General-Purpose Time Value Problem-Solver (Pre-Tax)

This program was described in Chapter 5. It is written in Applesoft BASIC for the Apple II series. The code can easily be modified to run on other machines.

```
]PR#0
]LIST

10   DIM F(15,7),F#(15),A(15,51),S(51),Y(10)
20   DEF  FN A(N) = (1 + D3) ^ N
30   REM        = (F/P)
40   DEF  FN B(N) = 1 /  FN A(N)
50   REM      = (P/F)
60   DEF  FN C(N) = ( FN A(N) - 1) / D3
70   REM      = (F/A)
80   DEF  FN D(N) = 1 /  FN C(N)
90   REM      = (A/F)
100  DEF  FN E(N) =  FN D(N) *  FN A(N)
110  REM      = (A/P)
120  DEF  FN F(N) = 1 /  FN E(N)
130  REM      = (P/A)
135 SWITCH = O
136 Y2 = O
140  HOME : PRINT "GENERAL PURPOSE TIME VALUE": PRINT "  PROBLEM SOLVER   (PRE-TA
X)"
145  PRINT : PRINT "     COPYRIGHT 1985, IRA H. KLEINFELD"
150  PRINT : PRINT " ENTER DEPENDENT VARIABLE:"
160  PRINT "  1 - INTERNAL RATE OF RETURN"
170  PRINT "  2 - PRESENT OR FUTURE VALUE"
180  PRINT "  3 - PERIODIC WORTH"
190  PRINT "  4 - EXTERNAL RATE OF RETURN"
```

393

```
200   INPUT D
210   IF D = 1 GOTO 440
220   IF D = 2 GOTO 270
230   IF D = 3 GOTO 340
240   IF D = 4 GOTO 420
250   PRINT "INVALID DEP VAR CODE"
260   GOTO 140
270   REM  NET PRESENT VALUE REQUESTED
280   PRINT "ENTER PERIOD FOR WHICH VALUE WILL BE": PRINT "COMPUTED"
290   INPUT D1
300   D1 = D1 + 1
310   PRINT "ENTER DISCOUNT RATE: (E.G., 10.34% IS": PRINT "ENTERED AS .1034)"
320   INPUT D3
330   GOTO 440
340   PRINT "ENTER PERIOD WHEN PAYMENTS WILL START"
350   INPUT D1
360 D1 = D1 + 1
370   PRINT "ENTER NUMBER OF PERIODS"
380   INPUT D2
390   PRINT "ENTER DISCOUNT RATE   (E.G. 10.34% IS ENTERED AS 0.1034)"
400   INPUT D3
410   GOTO 440
420   PRINT "ENTER DISCOUNT RATE FOR EXTERNAL RATE": PRINT "OF RETURN CALCULATION
."
430   INPUT D3
440   PRINT : PRINT "DOES DATA FOR THE FACTORS ALREADY EXIST?": PRINT "  (IS THIS
A SUBSEQUENT ITERATION?  ANS Y OR N)"
442   INPUT B$: IF B$ = "Y" OR B$ = "YES" GOTO 630
448   PRINT "NOW ENTER DATA FOR A MAXIMUM OF FIFTEEN FACTORS"
450   FOR J = 1 TO 15
460   GOSUB 2100
470   IF J > 14 GOTO 550
480   PRINT : PRINT "ARE THERE ANY MORE FACTORS?    ANSWER YES OR NO."
490   INPUT A$
500   IF A$ = "YES" GOTO 540
510   IF A$ = "Y" GOTO 540
520 FC = J
530   GOTO 570
540   NEXT J
570   FOR J2 = 1 TO FC
580   GOSUB 2860
590   REM  *** GO AND EVALUATE ANNUAL COST DUE TO FACTOR "J2".
600   NEXT J2
610   GOSUB 3150
612   REM  COMPUTE FLOWS/PD, # OF PDS
615   GOSUB 2720
617   REM  *:ECHO INPUT DATA
620   PRINT : PRINT "DO YOU WISH TO ALTER OR ADD A FACTOR?"
622   INPUT A$: IF A$ = "Y" OR A$ = "YES" GOTO 940
624   GOSUB 3305
630   Y2 = Y2 + 1
640   IF D <  > 1 GOTO 680
650   GOSUB 1370
660   REM  ** D = 1 ==> COMPUTE IRR
670   GOTO 860
680   IF D = 4 GOTO 790
690   GOSUB 1210
700   REM  ** COMPUTE NPV OF COST FLOWS   (N1 = NPV)
710   IF D = 2 GOTO 820
720   REM  ** D=3 ==> COMPUTE PERIOD COST
730 Y1 = (N1  *  FN B(1))  *  FN E(D2)
740   REM  ** CONVERT NPV TO PERIOD COST   (Y1 NOW EQUALS PERIOD COST)
750 U1 = D1 - 1
760   PRINT : PRINT "PERIOD WORTH STARTING IN PERIOD ";U1;" FOR ";D2;" PERIODS =
";Y1
770   REM
780   GOTO 860
790   REM  NEXT LINE CALLS ERR SUBR
800   GOSUB 3380
```

```
810    GOTO 860
820    Y1 = N1
830    T1 = D1 - 1
840    PRINT : PRINT "VALUE IN PERIOD ";T1;" = ";Y1
850    REM
860    Y(Y2) = Y1
865    IF Y2 = 10 GOTO 1150
870    PRINT : PRINT "DO YOU WANT TO ALTER OR ADD A FACTOR?"
880    INPUT A#
890    IF A# = "YES" GOTO 940
900    IF A# = "Y" GOTO 940
902    IF SWITCH = 1 GOTO 1130
904    PRINT : PRINT "DO YOU WISH TO COMPUTE A DIFFERENT TIME VALUE MEASURE FOR TH
E SAME DATA?": PRINT "  (IF SO, PRESS Y OR YES)"
906    INPUT B#
908    IF B# = "Y" OR B# = "YES" GOTO 150
910    IF Y2 = 1 GOTO 3600
920    GOTO 1150
930    REM    *** GO AND LIST THE SENSITIVITY VALUES
940    PRINT : PRINT "ENTER NO. OF FACTOR TO BE ADDED OR": PRINT "  ALTERED.": PRI
NT "(MUST BE 1 TO 15.": PRINT "IF TO BE ADDED, ENTER ";FC + 1;")":SWITCH = 1
950    INPUT J
960    IF J > 15 GOTO 1000
970    IF J > FC THEN 990
980    GOTO 1010
990 FC = J: GOTO 1010
1000   PRINT "INVALID FACTOR #": GOTO 940
1010   FOR K3 = 1 TO 7
1020   F(J,K3) = 0
1030   REM  THIS LOOP REMOVES OLD DATA ON FACTOR "J"
1040   NEXT K3
1050   GOSUB 2100
1060   GOSUB 2720
1070   J2 = J
1080   FOR K1 = 1 TO P
1090   S(K1) = 0
1100   A(J,K1) = 0
1110   NEXT K1
1120   GOSUB 2860
1125   GOTO 870
1130   GOSUB 3150
1135   GOSUB 3305
1140   GOTO 630
1150   REM   START LISTING OUTPUT HERE
1160   FOR K4 = 1 TO 10
1170   PRINT "ITERATION: ";K4;" VALUE OF DEP VAR: ";Y(K4)
1180   REM
1190   NEXT K4
1200   GOTO 3600
1210   REM    ****************************
1220   REM    THIS SUBR COMPUTES THE NPV FOR THE FLOWS IN THE
1230   REM    DESIRED START PERIOD.
1240 N1 = 0
1250   FOR J8 = 1 TO P
1260   T1 = J8 - 1
1270   Z9 =  FN B(T1)
1280   Z9 = S(J8) * Z9
1290   N1 = N1 + Z9
1300   NEXT J8
1310   T1 = D1 - 1
1320   N1 = N1 *  FN A(T1)
1330   REM  ** N1 NOW CONTAINS NPV IN DESIRED PERIOD
1340   RETURN
1350   REM    ****************************
1360   REM   THIS SUBR CALLS THE IRR SUBR, AND THEN ROUNDS THE IRR.
1370   GOSUB 1520
1380   IF D3 < O GOTO 1410
1390   D3 = D3 + .00005
1400   GOTO 1420
```

```
1410 D3 = D3 - .00005
1420 D3 = D3 * 10000
1430 D3 =   INT (D3)
1440 D3 = D3 / 10000
1450 Y1 = 100 * D3
1460  PRINT : PRINT "INTERNAL RATE OF RETURN= ";Y1;"%"
1470  REM
1480  RETURN
1490  REM  *****************************
1500  REM  ** THIS SUBR COMPUTES THE INTERNAL RATE OF RETURN OF THE INPUT
1510  REM  ** FLOWS.
1520 D3 = 0.0001
1530  REM  ** D3 = TEMP VALUE OF IRR
1540 D1 = 1
1550  REM  ** D1 = 1 ==> IRR COMPUTED FOR THE FIRST PERIOD
1560  GOSUB 1210
1570  REM  ** EVALUATE NPV
1580 V1 = N1
1590  IF V1 = 0 GOTO 1870
1600 D3 = 0.0002
1610  GOSUB 1210
1620 V2 = N1
1630  GOSUB 1920
1640  IF M1 = 1 GOTO 1690
1650  REM  ** IF M1 =0, IRR IS MOVING IN WRONG DIRECTION
1660  REM  ** IF M1 = 1, IRR IS MOVING IN RIGHT DIRECTION
1670 I1 =  - .1
1680  GOTO 1700
1690 I1 = .1
1700 D3 = 0
1710  PRINT : PRINT "I'M WORKING ON IT!"
1720  FOR J9 = 1 TO 30
1730 V1 = V2
1740 D3 = D3 + I1
1750  GOSUB 1210
1760  REM  ** GO AND COMPUTE NPV
1770 V2 = N1
1780  GOSUB 1920
1790  REM  ** PRINT ITERATIONS OF NPV, IRR?
1800  REM
1810  IF M1 = 1 GOTO 1860
1820  REM  ** M1=1 ==> IRR IS MOVING IN CORRECT DIRECTION
1830 I1 = ( - 1) * (I1) / (10)
1840  IF  ABS (I1) < .00001 GOTO 1870
1850  REM  ** M1=2 ==> IRR HAS CAUSED NPV TO CROSS THE ZERO VALUE
1860  NEXT J9
1870  RETURN
1880  REM  *****************************
1890  REM  THIS IS A SUBR TO SEE IF THE "IRR" IS GOING IN THE
1900  REM  RIGHT DIRECTION, THE WRONG DIRECTION, OR HAS MADE THE NPV
1910  REM  CROSS THE ZERO VALUE
1920  IF V1 > 0 GOTO 1950
1930  IF V2 >  = 0 GOTO 2050
1940  GOTO 1960
1950  IF V2 <  = 0 GOTO 2050
1960 V3 =  ABS (V1)
1970 V4 =  ABS (V2)
1980  IF V4 < V3 GOTO 2020
1990 M1 = 0
2000  REM  ** M1 = 0 ==> "IRR" GOING IN WRONG DIRECTION
2010  GOTO 2070
2020 M1 = 1
2030  REM  ** M1 = 1 ==> "IRR" GOING IN CORRECT DIRECTION
2040  GOTO 2070
2050 M1 = 2
2060  REM  ** M1 = 2 ==> "NPV" HAS CROSSED THE ZERO VALUE
2070  RETURN
2080  REM  *****************************
2090  REM  THIS SUBR GATHERS DATA FOR AN INPUT FACTOR
```

```
2100    PRINT : PRINT "ENTER NAME OF FACTOR ";J
2110    INPUT F#(J)
2120    PRINT : PRINT "ENTER FACTOR TYPE FOR FACTOR ";J
2130    PRINT "   1 - SINGLE PERIOD AMOUNT"
2140    PRINT "   2 - UNIFORM SERIES"
2150    PRINT "   3 - UNIFORM GRADIENT SERIES"
2160    INPUT F(J,3)
2170    IF F(J,3) = 1 GOTO 2220
2180    IF F(J,3) = 2 GOTO 2250
2190    IF F(J,3) = 3 GOTO 2300
2200    PRINT "*** INVALID FACTOR TYPE ***"
2210    GOTO 2120
2220    F(J,2) = 1: GOSUB 2370
2230    GOSUB 2420
2240    GOTO 2690
2250    GOSUB 2370
2260    GOSUB 2420
2270    GOSUB 2490
2280    GOSUB 2530
2290    GOTO 2690
2300    GOSUB 2370
2310    GOSUB 2420
2320    GOSUB 2490
2330    GOSUB 2530
2340    GOSUB 2560
2350    GOTO 2690
2360    REM ****** THIS ROUTINE OBTAINS STARTING CASH FLOW VALUE
2370    PRINT "ENTER VALUE IF CASH FLOW OCCURS ONCE, OR"
2380    PRINT "STARTING VALUE IF CASH FLOW IS A SERIES"
2390    INPUT F(J,5)
2400    RETURN
2410    REM  ****** THIS ROUTINE OBTAINS THE 1ST PERIOD FOR CASH FLOW
2420    PRINT : PRINT "ENTER PERIOD IF CASH FLOW OCCURS ONCE,OR"
2430    PRINT "STARTING PERIOD IF CASH FLOW IS A SERIES"
2440    INPUT F(J,1)
2450    F(J,1) = F(J,1) + 1
2460    PRINT " "
2470    RETURN
2480    REM  ****** THIS ROUTINE OBTAINS THE PERIOD INCREMENT FOR A SERIES
2490    PRINT "ENTER PERIOD INCREMENT FOR THE CASH FLOW SERIES"
2500    INPUT F(J,4)
2510    RETURN
2520    REM  ****** THIS ROUTINE OBTAINS THE NO. OF OCCURRENCES IN A SERIES
2530    PRINT : PRINT "ENTER NUMBER OF OCCURRENCES OF CASH FLOW IN SERIES"
2540    INPUT F(J,2)
2550    RETURN
2560    PRINT "ENTER TYPE OF GRADIENT:"
2570    PRINT "    1 - PERCENTAGE INCREASE/DECREASE"
2580    PRINT "    2 - ABSOLUTE INCREASE/DECREASE"
2590    INPUT F(J,7)
2600    IF F(J,7) = 1 GOTO 2620
2610    IF F(J,7) < > 2 GOTO 2670
2620    PRINT "TO ENTER VALUE INCREMENT FOR GRADIENT": PRINT "SERIES: "
2630    PRINT "   FOR PERCENTAGE CHANGE - ENTER -.102": PRINT "   FOR A -10.2 PER
CENT DECREASE IN": PRINT "   SUCCESSIVE AMOUNTS, OR"
2640    PRINT "   FOR AN ABSOLUTE INC/DEC, ENTER THAT": PRINT "   AMOUNT"
2650    INPUT F(J,6)
2660    GOTO 2690
2670    PRINT "***** ERROR *****    INVALID GRADIENT TYPE ENTERED"
2680    GOTO 2560
2690    RETURN
2700    REM  *****************************
2710    REM  THIS SUBR ECHOES THE INPUT DATA
2720    PRINT " "
2730    PRINT "FACTOR/PERIOD /=====VALUE=====/  #  DISC"
2740    PRINT "# NAME/1ST INC/ 1ST       INC / OCC RATE"
2750    FOR T = 1 TO FC
2760    U1 = 0
2770    IF F(T,1) = 0 GOTO 2790
```

```
2780  U1 = F(T,1) - 1
2790  IF F(T,7) < > 1 GOTO 2800
2800  PRINT T; TAB( 4);F$(T); TAB( 9);U1; TAB( 13);F(T,4); TAB( 15);F(T,5); TAB(
      24);F(T,6); TAB( 34);F(T,2); TAB( 37);D3
2810  REM
2820  NEXT T
2830  RETURN
2840  REM  ******************************
2850  REM    THIS SUBR CONVERTS INPUT DATA TO PERIOD FLOWS PER FACTOR
2860  IF F(J2,1) = O GOTO 3120
2870  IF F(J2,3) < > 1 GOTO 2910
2880  REM    FACTOR IS SINGLE PERIOD AMOUNT
2890  A(J2,F(J2,1)) = A(J2,F(J2,1)) + F(J2,5)
2900  GOTO 3120
2910  T1 = F(J2,1)
2920  T2 = F(J2,1) + (F(J2,2) - 1) * F(J2,4)
2930  IF T2 < 52 GOTO 2950
2940  T2 = 51
2950  T3 = F(J2,4)
2960  IF F(J2,3) = 3 GOTO 3010
2970  FOR J3 = T1 TO T2 STEP T3
2980  A(J2,J3) = A(J2,J3) + F(J2,5)
2990  NEXT J3
3000  GOTO 3120
3010  T4 = F(J2,5)
3020  IF F(J2,7) = 1 GOTO 3080
3030  FOR J4 = T1 TO T2 STEP T3
3040  A(J2,J4) = A(J2,J4) + T4
3050  T4 = T4 + F(J2,6)
3060  NEXT J4
3070  GOTO 3120
3080  FOR J5 = T1 TO T2 STEP T3
3090  A(J2,J5) = A(J2,J5) + T4
3100  T4 = T4 * (1 + F(J2,6))
3110  NEXT J5
3120  RETURN
3130  REM  ******************************
3140  REM  THIS SUBR COMPUTES TOTAL NET CASH FLOW PER PERIOD
3150  FOR J6 = 1 TO 51
3160  FOR J7 = 1 TO FC
3170  REM  FI = NUMBER OF FACTORS INPUT
3180  S(J6) = S(J6) + A(J7,J6)
3190  NEXT J7
3200  NEXT J6
3210  REM  HOME
3230  REM  FIND NO. OF PERIODS IN ANALYSIS
3240  FOR K6 = 1 TO 51
3250  T1 = 52 - K6
3260  IF S(T1) = O GOTO 3290
3270  P = T1
3280  GOTO 3300
3290  NEXT K6
3300  RETURN
3305  PRINT : PRINT "PERIOD, NET AMOUNT"
3310  FOR J8 = 1 TO P
3320  U1 = J8 - 1
3330  PRINT  TAB( 3);U1; TAB( 8);S(J8)
3340  REM
3350  NEXT J8
3355  SWITCH = 0
3360  RETURN
3370  REM  ******************************
3380  REM    THIS SUBR COMPUTES E.R.R.
3390  E1 = O
3400  REM    E1 = SUM OF NEG  FLOWS IN TIME O.
3410  E2 = O
3420  REM    E2 = SUM OF POS FLOWS IN TIME P.
3430  FOR K7 = 1 TO P
3440  IF S(K7) < O GOTO 3490
```

```
3450   IF S(K7) = O GOTO 3510
3460   N = P - K7
3470   E2 = E2 + (S(K7) *   FN A(N))
3480   GOTO 3510
3490   N = K7 - 1
3500   E1 = E1 + (S(K7) *   FN B(N))
3510   NEXT K7
3520   PRINT : PRINT "SUM OF - FLOWS = ";E1
3530   PRINT "SUM OF + FLOWS = ";E2
3540   PRINT "NO. OF PERIODS = ";P - 1
3550   T1 = 1 / (P - 1)
3560   E2 = E2 * ( - 1)
3570   Y1 = ((E2 / E1) ^ T1 - 1) * 100
3580   PRINT "E.R.R. = ";Y1;"%"
3590   RETURN
3600   END
```

Appendix D

Computer Program: General-Purpose
Problem-Solver—After-Tax Analysis

This appendix describes a general-purpose problem-solver computer program capable of performing after-tax transformations as described in this book. Because it is an extension of the program presented in Chapter 5 and Appendix C, it has many of the same data-gathering routines and computational features. For instance, it allows the user to enter data on elemental factors and then amalgamate them to obtain the period flows. It treats gradients, both arithmetic and percent. It permits computation of any of the major yardsticks (e.g., annual worth, present or future worth, or rate of return). It facilitates sensitivity analyses and permits graphical display of the output. Since this program uses many of the features and subroutines of the one in Chapter 5, the reader may wish to refer to the program descriptions found there as well as this appendix.

That this program was not described in Chapter 5 is due primarily to pedagogical reasons. The aim in Chapter 5 was to introduce the reader to the basic elements of computer program design useful in engineering economic analysis. At that point, after-tax considerations had not yet been presented. Thus, much of the material in this program would have seemed foreign to the reader.

Program Description

When the program is run, the initial menu appears as follows:

```
GENERAL-PURPOSE TIME VALUE
PROBLEM SOLVER--(AFTER-TAX)
     COPYRIGHT 1985, IRA H. KLEINFELD
  1 - ADD/ALTER FACTORS
  2 - COMPUTE INTERNAL RATE OF RETURN
  3 - COMPUTE PRESENT OR FUTURE VALUE
  4 - COMPUTE PERIODIC WORTH
  5 - COMPUTE EXTERNAL RATE OF RETURN
  6 - ALTER TAX RATE OF 50%
  7 - ALTER GAINS & LOSS RATE OF 25%
  8 - SENSITIVITY ANALYSIS BY RATE
  9 - SENSITIVITY ANALYSIS BY MAGNITUDE
 10 - QUIT
```

The choices involve data entry (or change) as well as computation and parameter changes. (One could use this program for pretax analysis by changing the income-tax rate to 0 percent, for example.) To begin, one must enter data. Thus, menu item 1 would be selected. The resulting series of queries and responses is shown below:

```
CHOOSE ONE ? 1

NO FACTORS HAVE BEEN ENTERED

DO YOU WANT TO ADD/ALTER A FACTOR?
? Y

ENTER NO. OF FACTOR TO BE ADDED/ALTERED
(MUST BE 1 TO 15.
IF TO BE ADDED, ENTER 1)
? 1
```

The request for elemental data includes its classification as to FACTOR TYPE. There are five types which the user designates, as given in the table below:

Type	Designation
Investment	I
Flows other than investment subject to income taxes	NI
Salvage value	Q
Flows not subject to income taxes	NT
Depreciation (nonstandard)	D

Factor designations are necessary in order to differentiate their treatment logically for income-tax purposes. In order to demonstrate the application of these designations, screens applying to Example 8–7 are given below.

```
ENTER NAME AND FACTOR TYPE
? INV, I
```

```
ENTER AMOUNT
?-100
ENTER PERIOD
?0
```

When the user designates I for the factor labeled INV, he is then queried as to the nature of depreciation accounting to be applied to that investment, the depreciation period, and the relevant investment tax credit. (Zero must be entered if there is no applicable credit.) More than one investment factor may be entered, each with its own depreciation schedule.

```
ENTER METHOD OF DEPRECIATION
1 - STRAIGHT LINE
2 - SUM OF THE YEARS DIGITS
3 - ACRS (POST 1985)
4 - DECLINING BALANCE
?3
ENTER THE DEPRECIATION PERIOD
?5
ENTER PERCENTAGE OF ITC
?.1
```

After the data for each factor have been entered, the user has the option to alter it, in case of error, for instance. This is done by responding affirmatively to the question "DO YOU WANT TO ADD/ALTER A FACTOR?" and then entering the number corresponding to the factor number to be changed.

To proceed to the next factor, its number must be keyed in, as shown below.

```
ENTER NO. OF FACTOR TO BE ADDED/ALTERED
(MUST BE 1 TO 15.
IF TO BE ADDED, ENTER 2)
?2
```

The annual savings of $30,000 is designated as NI, since it needs to be transformed to an after-tax amount, but is not an investment or salvage value.

```
ENTER NAME AND FACTOR TYPE
?SAV,NI
ENTER TYPE OF CASH FLOW
1 - SINGLE PERIOD AMOUNT
2 - UNIFORM SERIES
3 - UNIFORM GRADIENT SERIES
?2
ENTER STARTING VALUE
?30
ENTER STARTING PERIOD
?1
ENTER PERIOD INCREMENT
?1
ENTER NUMBER OF OCCURRENCES
?6
```

```
FACTOR/PERIOD /=====VALUE=====/   #   FACTOR
# NAME/1ST INC/ 1ST        INC  / OCC  TYPE
1   INV  0   0 -100       0          1    I
2   SAV  1   1  30        0          6    NI
```

DO YOU WANT TO ADD/ALTER A FACTOR?
? Y

The $15,000 salvage value is designated as Q. This permits the program to compute
the income-tax effect if salvage value is different from the book value.

ENTER NO. OF FACTOR TO BE ADDED/ALTERED
(MUST BE 1 TO 15.
IF TO BE ADDED, ENTER 3)
? 3
ENTER NAME AND FACTOR TYPE
? SV, Q
ENTER AMOUNT
? 15
ENTER PERIOD
? 6
ENTER BOOK VALUE AT PERIOD 6
? 0

```
FACTOR/PERIOD /=====VALUE=====/   #   FACTOR
# NAME/1ST INC/ 1ST        INC  / OCC  TYPE
1   INV  0   0 -100       0          1    I
2   SAV  1   1  30        0          6    NI
3   SV   6   0  15        0          1    Q
```

DO WANT TO ADD/ALTER A FACTOR?
? Y

The loan principal received at time 0 and its repayment at time 6 may be designated
as NT, since they have no effect on income taxes.

ENTER NO. OF FACTOR TO BE ADDED/ALTERED
(MUST BE 1 TO 15.
IF TO BE ADDED, ENTER 4)
? 4
ENTER NAME AND FACTOR TYPE
? LON, NT
ENTER TYPE OF CASH FLOW
1 - SINGLE PERIOD AMOUNT
2 - UNIFORM SERIES
3 - UNIFORM GRADIENT SERIES
? 1
ENTER AMOUNT
? 35

```
ENTER PERIOD
?0


FACTOR/PERIOD /=====VALUE=====/   #   FACTOR
# NAME/1ST INC/ 1ST        INC  / OCC  TYPE
1  INV  0   0 -100      0          1   I
2  SAV  1   1 30        0          6   NI
3  SV   6   0 15        0          1   Q
4  LON  0   0 35        0          1   NT

DO YOU WANT TO ADD/ALTER A FACTOR?
?Y
ENTER NO.  OF FACTOR TO BE ADDED/ALTERED
(MUST BE 1 TO 15.
IF TO BE ADDED, ENTER 5)
?5
ENTER NAME AND FACTOR TYPE
?RLON,NT
ENTER TYPE OF CASH FLOW
1 - SINGLE PERIOD AMOUNT
2 - UNIFORM SERIES
3 - UNIFORM GRADIENT SERIES
?1
ENTER AMOUNT
?-35
ENTER PERIOD
?6


FACTOR/PERIOD /=====VALUE=====/   #   FACTOR
# NAME/1ST INC/ 1ST        INC  / OCC  TYPE
1  INV  0   0 -100      0          1   I
2  SAV  1   1 30        0          6   NI
3  SV   6   0 15        0          1   Q
4  LON  0   0 35        0          1   NT
5  RLON 6   0 -35       0          1   NT

DO YOU WANT TO ADD/ALTER A FACTOR?
?Y
```

The annual $3,150 interest payments can be treated like the savings, since they are transformed into after-tax flows using the same algorithm. They are designated as NI.

```
ENTER NO.  OF FACTOR TO BE ADDED/ALTERED
(MUST BE 1 TO 15.
IF TO BE ADDED, ENTER 6)
?6
ENTER NAME AND FACTOR TYPE
?INT,NI
```

```
ENTER TYPE OF CASH FLOW
1 - SINGLE PERIOD AMOUNT
2 - UNIFORM SERIES
3 - UNIFORM GRADIENT SERIES
?2
ENTER STARTING VALUE
?-3.15
ENTER STARTING PERIOD
?1
ENTER PERIOD INCREMENT
?1
ENTER NUMBER OF OCCURRENCES
?6
```

After the input data for the factors have been displayed to permit editing, and the user has indicated his satisfaction with the data in storage, the amalgamated period flows and the after-tax transformation are displayed.

```
FACTOR/PERIOD /=====VALUE=====/   #   FACTOR
# NAME/1ST INC/ 1ST        INC  / OCC  TYPE
1  INV 0   0 -100      0          1    I
2  SAV 1   1 30        0          6    NI
3  SV  6   0 15        0          1    Q
4  LON 0   0 35        0          1    NT
5  RLON 6  0 -35       0          1    NT
6  INT 1   1 -3.15     0          6    NI
```

```
DO YOU WANT TO ADD/ALTER A FACTOR?
?N
```

PER	BEF TAX	DEP	AFT TAX
0	-65	0	-65
1	26.85	20	33.425
2	26.85	32	29.425
3	26.85	24	25.425
4	26.85	16	21.425
5	26.85	8	17.425
6	6.85	0	-14.075

At that point the menu appears which allows for computation of the time value measures, altering parameters, or sensitivity analysis.

```
1 - ADD/ALTER FACTORS
2 - COMPUTE INTERNAL RATE OF RETURN
3 - COMPUTE PRESENT OR FUTURE VALUE
4 - COMPUTE PERIODIC WORTH
5 - COMPUTE EXTERNAL RATE OF RETURN
```

```
 6 - ALTER TAX RATE OF 50%
 7 - ALTER GAINS & LOSS RATE OF 25%
 8 - SENSITIVITY ANALYSIS BY RATE
 9 - SENSITIVITY ANALYSIS BY MAGNITUDE
10 - QUIT

CHOOSE ONE ?4
PERIODIC WORTH
ENTER PERIOD WHEN PAYMENTS WILL START
?1
ENTER NUMBER OF PERIODS
?6
ENTER DISCOUNT RATE
?.1

PERIOD WORTH STARTING IN PERIOD 1 FOR 6 PERIODS =
6.04208855
```

The factor designation D, nonstandard depreciation, is useful if new depreciation rules are evaluated, or for replacement problems where the asset has already undergone some depreciation and a schedule from the I designation would not be appropriate. This is illustrated with screens for Example 9–2.

⋮

```
ENTER NO. OF FACTOR TO BE ADDED/ALTERED
(MUST BE 1 TO 15.
IF TO BE ADDED, ENTER 1)
?1
ENTER NAME AND FACTOR TYPE
?D1,D
ENTER AMOUNT
?6.4
ENTER PERIOD
?1


FACTOR/PERIOD /=====VALUE=====/  #  FACTOR
# NAME/1ST INC/ 1ST       INC  / OCC  TYPE
1  D1   1   0 6.4      0        1     D

DO YOU WANT TO ADD/ALTER A FACTOR?
?Y
ENTER NO. OF FACTOR TO BE ADDED/ALTERED
(MUST BE 1 TO 15.
IF TO BE ADDED, ENTER 2)
?2
```

```
ENTER NAME AND FACTOR TYPE
?D2,D
ENTER AMOUNT
?4.8
ENTER PERIOD
?2
```

$$\vdots$$

The entry of only two out of the four "nonstandard" depreciation charges was shown in order to save space. The others may be similarly entered. Annual operating cost would be entered with the NI designation. Implicit investment of $5,000 can be handled by entering it as a negative number and designating its type as Q. (Recall that the I designation would result in a depreciation schedule based on a $5,000 first cost, which is definitely not wanted here.)

$$\vdots$$

```
ENTER NO. OF FACTOR TO BE ADDED/ALTERED
 (MUST BE 1 TO 15.
IF TO BE ADDED, ENTER 6)
?6
ENTER NAME AND FACTOR TYPE
?INV,Q
ENTER AMOUNT
?-5
ENTER PERIOD
?0
ENTER BOOK VALUE AT PERIOD 0
?16
```

#	FACTOR NAME	/PERIOD 1ST INC/	/=====VALUE=====/ 1ST	INC	/ # OCC	FACTOR TYPE
1	D1	1 0	6.4	0	1	D
2	D2	2 0	4.8	0	1	D
3	D3	3 0	3.2	0	1	D
4	D4	4 0	1.6	0	1	D
5	COST	1 1	-12	0	5	NI
6	INV	0 0	-5	0	1	Q

```
DO YOU WANT TO ADD/ALTER A FACTOR?
?N
```

PER	BEF TAX	DEP	AFT TAX
0	-5	0	-7.75
1	-12	6.4	-2.8

```
2        -12         4.8        -3.6
3        -12         3.2        -4.4
4        -12         1.6        -5.2
5        -12          0         -6
```

```
 1 - ADD/ALTER FACTORS
 2 - COMPUTE INTERNAL RATE OF RETURN
 3 - COMPUTE PRESENT OR FUTURE VALUE
 4 - COMPUTE PERIODIC WORTH
 5 - COMPUTE EXTERNAL RATE OF RETURN
 6 - ALTER TAX RATE OF 50%
 7 - ALTER GAINS & LOSS RATE OF 25%
 8 - SENSITIVITY ANALYSIS BY RATE
 9 - SENSITIVITY ANALYSIS BY MAGNITUDE
10 - QUIT
```

Sensitivity analyses may be done by varying the discount rate or the magnitude of a factor. These are illustrated by screens applying to Example 3–3 and Figure 8–1. (Note that since Example 3–3 is pre-tax, the factors are entered as NT. This is another way to use the after-tax program for pre-tax analysis.)

Example 3–3

⋮

```
FACTOR/PERIOD /=====VALUE=====/   #   FACTOR
# NAME/1ST INC/ 1ST        INC  / OCC  TYPE
1  F1   0   0 -1000        0        1   NT
2  F2   1   1 380          0        5   NT
```

DO YOU WANT TO ADD/ALTER A FACTOR?
?N

```
PER      BEF TAX      DEP      AFT TAX
 0       -1000        0        -1000
 1        380         0         380
 2        380         0         380
 3        380         0         380
 4        380         0         380
 5        380         0         380
```

```
1 - ADD/ALTER FACTORS
2 - COMPUTE INTERNAL RATE OF RETURN
3 - COMPUTE PRESENT OR FUTURE VALUE
```

```
 4 - COMPUTE PERIODIC WORTH
 5 - COMPUTE EXTERNAL RATE OF RETURN
 6 - ALTER TAX RATE OF 50%
 7 - ALTER GAINS & LOSS RATE OF 25%
 8 - SENSITIVITY ANALYSIS BY RATE
 9 - SENSITIVITY ANALYSIS BY MAGNITUDE
10 - QUIT

CHOOSE ONE ?8
SENSITIVITY ANALYSIS BY RATE
1 - COMPUTE PRESENT OR FUTURE VALUE
2 - COMPUTE PERIODIC WORTH
3 - COMPUTE EXTERNAL RATE OF RETURN
4 - GRAPHICS DISPLAY
5 - RETURN TO MAIN MENU

CHOOSE ONE?1
PRESENT OR FUTURE VALUE
ENTER PERIOD
?0
ENTER STARTING RATE
?.1
ENTER INCREMENT
?.05
ENTER NUMBER OF OCCURRENCES
?5

VALUE IN PERIOD 0 = 440.49897

VALUE IN PERIOD 0 = 273.818936

VALUE IN PERIOD 0 = 136.432613

VALUE IN PERIOD 0 = 21.9264

VALUE IN PERIOD 0 = -74.4834961
```

Having generated these present worth figures, the user can then have them graphed when the subsequent screen appears:

```
1 - COMPUTE PRESENT OR FUTURE VALUE
2 - COMPUTE PERIODIC WORTH
3 - COMPUTE EXTERNAL RATE OF RETURN
4 - GRAPHIC DISPLAY
5 - RETURN TO MAIN MENU
```

Note that when the menu above appears, the user can choose graphic display only if the sensitivity values have already been computed.

Computer screens applicable to Figure 8–1 follow.

Figure 8—1

$\vdots$

```
FACTOR/PERIOD /=====VALUE=====/   #   FACTOR
# NAME/1ST INC/ 1ST        INC   / OCC  TYPE
1  LEAS 0    1 -22        0        6    NI
2  SAV 1     1 30         0        6    NI
```

DO YOU WANT TO ADD/ALTER A FACTOR?
?N

PER	BEF TAX	DEP	AFT TAX
0	-22	0	-11
1	8	0	4
2	8	0	4
3	8	0	4
4	8	0	4
5	8	0	4
6	30	0	15

```
 1 - ADD/ALTER FACTORS
 2 - COMPUTE INTERNAL RATE OF RETURN
 3 - COMPUTE PRESENT OR FUTURE VALUE
 4 - COMPUTE PERIODIC WORTH
 5 - COMPUTE EXTERNAL RATE OF RETURN
 6 - ALTER TAX RATE OF 50%
 7 - ALTER GAINS & LOSS RATE OF 25%
 8 - SENSITIVITY ANALYSIS BY RATE
 9 - SENSITIVITY ANALYSIS BY MAGNITUDE
10 - QUIT
```

CHOOSE ONE ?9
SENSITIVITY ANALYSIS BY MAGNITUDE

```
1 - COMPUTE PRESENT OR FUTURE VALUE
2 - COMPUTE PERIODIC WORTH
3 - COMPUTE EXTERNAL RATE OF RETURN
4 - COMPUTE INTERNAL RATE OF RETURN
5 - GRAPHIC DISPLAY
6 - RETURN TO MAIN MENU
```

CHOOSE ONE?1

```
FACTOR/PERIOD /=====VALUE=====/   #   FACTOR
# NAME/1ST INC/ 1ST         INC / OCC  TYPE
1  LEAS 0   1 -22        0        6    I
2  SAV  1   1 30         0        6    NI


ENTER FACTOR NUMBER
?1
ENTER STARTING VALUE
?-18
ENTER INCREMENT
?-1
ENTER NUMBER OF OCCURRENCES
?6
PRESENT OR FUTURE VALUE
ENTER PERIOD
?0
ENTER DISCOUNT RATE
?.1


VALUE IN PERIOD 0 = 22.2118295

VALUE IN PERIOD 0 = 19.8164361

VALUE IN PERIOD 0 = 17.4210427

VALUE IN PERIOD 0 = 15.0256494

VALUE IN PERIOD 0 = 12.630256

VALUE IN PERIOD 0 = 10.2348626
```

These present worth figures are identical to those depicted in Figure 8–1. Having generated these present worths, the user can then have them graphed when the subsequent screen appears:

```
1 - COMPUTE PRESENT OR FUTURE VALUE
2 - COMPUTE PERIODIC WORTH
3 - COMPUTE EXTERNAL RATE OF RETURN
4 - COMPUTE INTERNAL RATE OF RETURN
5 - GRAPHIC DISPLAY
6 - RETURN TO MAIN MENU
```

The percentage gradient factor is useful for conducting an inflation-adjusted analysis. This is illustrated with screens for Example 10–3.

$$\vdots$$

```
ENTER NO. OF FACTOR TO BE ADDED/ALTERED
(MUST BE 1 TO 15.
IF TO BE ADDED, ENTER 2)
```

```
?2
ENTER NAME AND FACTOR TYPE
?REV,NI
ENTER TYPE OF CASH FLOW
1 - SINGLE PERIOD AMOUNT
2 - UNIFORM SERIES
3 - UNIFORM GRADIENT SERIES
?3
ENTER STARTING VALUE
?44
ENTER STARTING PERIOD
?1
ENTER PERIOD INCREMENT
?1
ENTER NUMBER OF OCCURRENCES
?6
ENTER TYPE OF GRADIENT
1 - PERCENTAGE INCREASE/DECREASE
2 - ABSOLUTE INCREASE/DECREASE
?1
ENTER GRADIENT PERCENTAGE
?.1
```

FACTOR/PERIOD	/=====VALUE=====/		#	FACTOR
# NAME/1ST INC/ 1ST	INC	/ OCC	TYPE	
1 INV 0 0 -100	0	1	I	
2 REV 1 1 44	10	6	NI	

```
DO YOU WANT TO ADD/ALTER A FACTOR?
?Y
```

⋮

FACTOR/PERIOD	/=====VALUE=====/		#	FACTOR
# NAME/1ST INC/ 1ST	INC	/ OCC	TYPE	
1 INV 0 0 -100	0	1	I	
2 REV 1 1 44	10	6	NI	
3 COST 1 1 -10.5	5	6	NI	
4 SV 6 0 18.89	0	1	Q	

```
DO YOU WANT TO ADD/ALTER A FACTOR?
?
```

PER	BEF TAX	DEP	AFT TAX
0	-100	0	-100
1	33.5	14.1666667	23.8333333
2	37.375	14.1666667	25.7708334
3	41.66375	14.1666667	27.9152084
4	46.4089375	14.1666667	30.2878021
5	51.6575844	14.1666667	32.9121255
6	76.3514836	14.1666667	52.7590752

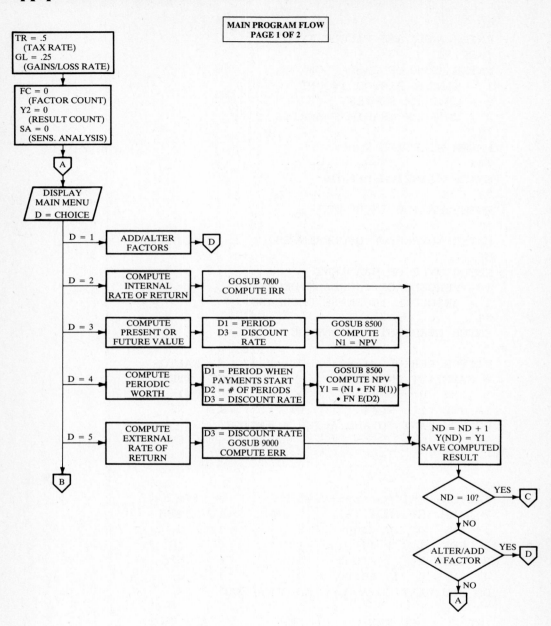

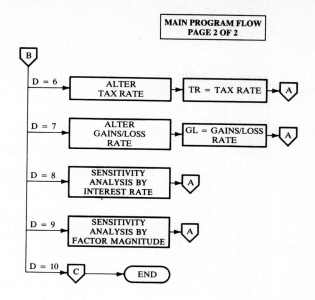

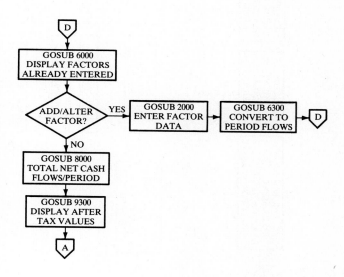

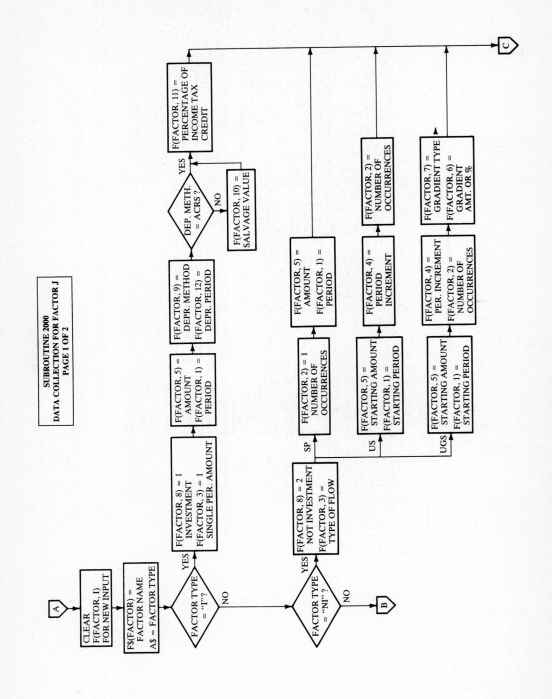

SUBROUTINE 2000
DATA COLLECTION FOR FACTOR J
PAGE 1 OF 2

416

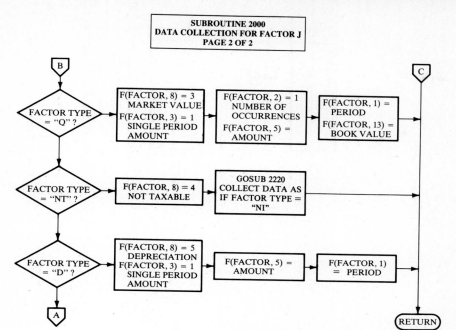

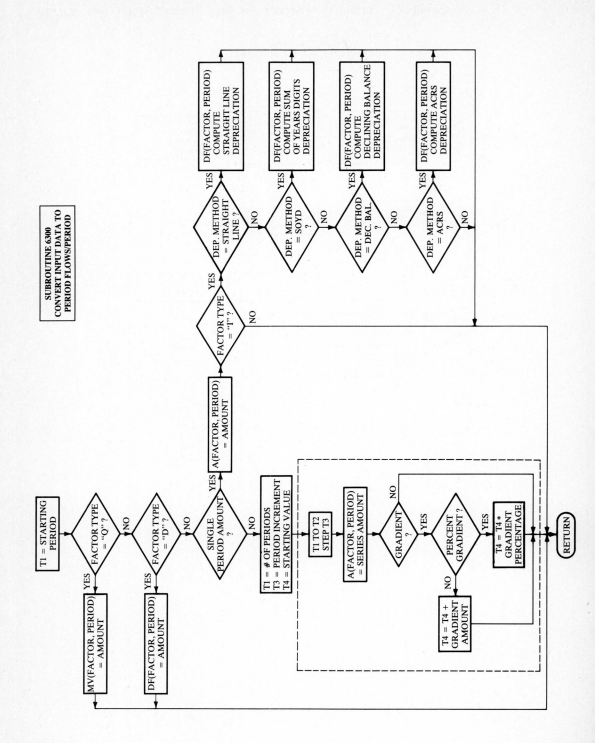

SUBROUTINE 6300
CONVERT INPUT DATA TO
PERIOD FLOWS/PERIOD

418

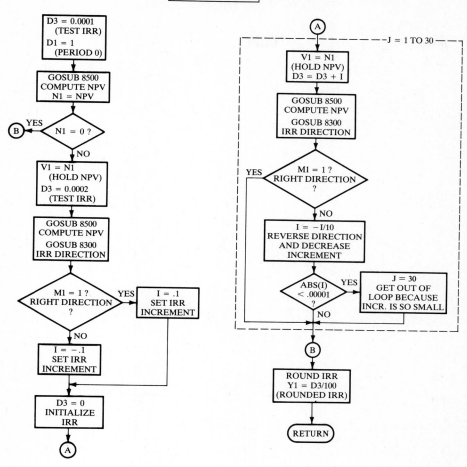

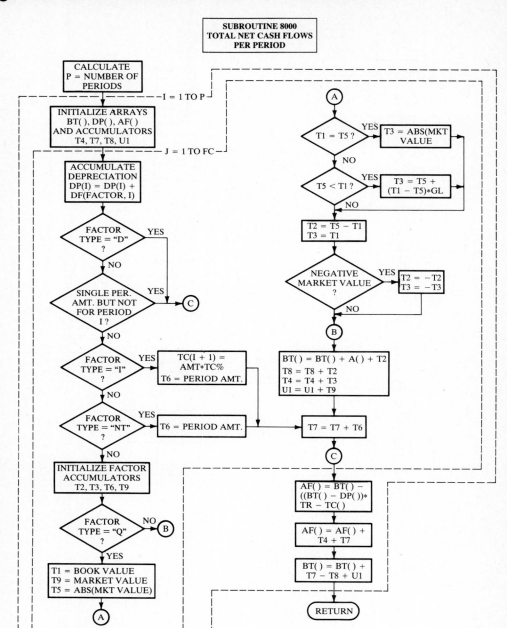

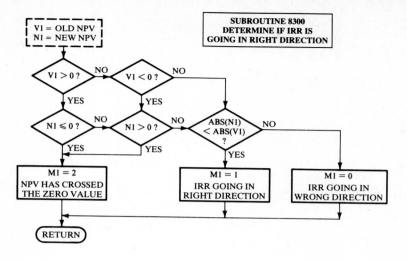

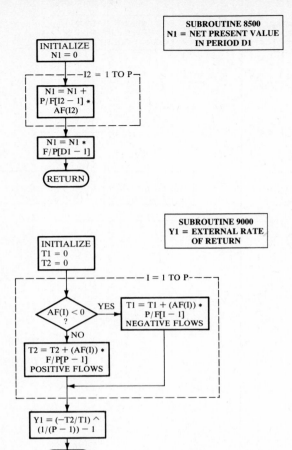

Program Code: General-Purpose Time Value Problem-Solver (After-Tax)

This code in BASIC was prepared for the IBM-PC in order to utilize its graphics capability. With the exception of the graphics portion the code can easily be modified to run on other machines.

```
10    DIM F(15,13)
20    DIM F$(15),A(15,51),Y(10),X(10)
21    DIM AF(51)
22    DIM BT(51)
23    DIM MV(15,51)
24    DIM DP(51)
25    DIM DF(15,51)
26    DIM TC(51)
28    DEF   FN A(N) = (1 + D3) ^ N
30    REM             = (F/P)
40    DEF   FN B(N) = 1 /   FN A(N)
50    REM             = (P/F)
60    DEF   FN C(N) = ( FN A(N) - 1) / D3
70    REM             = (F/A)
80    DEF   FN D(N) = 1 /   FN C(N)
90    REM             = (A/F)
100   DEF   FN E(N) =  FN D(N) *  FN A(N)
110   REM             = (A/P)
120   DEF   FN F(N) = 1 /   FN E(N)
130   REM             = (P/A)
131 TR = .5
132 GL = .25
135 FC = 0
136 ND = 0
137 SA = 0
140   CLS : PRINT "GENERAL PURPOSE TIME VALUE":PRINT"   PROBLEM SOLVER    (AFTER-TAX
)"
145   PRINT : PRINT "       COPYRIGHT 1985, IRA H. KLEINFELD"
150   PRINT
155   PRINT "   1 - ADD/ALTER FACTORS"
160   PRINT "   2 - COMPUTE INTERNAL RATE OF RETURN"
170   PRINT "   3 - COMPUTE PRESENT OR FUTURE VALUE"
180   PRINT "   4 - COMPUTE PERIODIC WORTH"
190   PRINT "   5 - COMPUTE EXTERNAL RATE OF RETURN"
191   PRINT "   6 - ALTER TAX RATE OF ";TR * 100;"%"
192   PRINT "   7 - ALTER GAINS & LOSS RATE OF ";GL * 100;"%"
193   PRINT "   8 - SENSITIVITY ANALYSIS BY RATE"
194   PRINT "   9 - SENSITIVITY ANALYSIS BY MAGNITUDE"
195   PRINT "  10 - QUIT"
196   PRINT : PRINT "CHOOSE ONE ";
200   INPUT D
210   IF D < 1 OR D >10 GOTO 140
220   ON D GOTO 5000,400,600,700,500,300,350,9500,9700,1150
300   REM
301   PRINT "ALTER TAX RATE"
302   REM
310   INPUT TR
320   GOTO 140
350   REM
351   PRINT "ALTER GAINS & LOSS RATE"
352   REM
360   INPUT GL
370   GOTO 140
400   REM
401   PRINT "INTERNAL RATE OF RETURN"
402   REM
410   GOSUB 7000
420   GOTO 1000
```

```
500    REM
501    PRINT "EXTERNAL RATE OF RETURN"
502    REM
510    GOSUB 9400
530    GOSUB 9000
540    GOTO 1000
600    REM
601    PRINT "PRESENT OR FUTURE VALUE"
602    REM
610    PRINT "ENTER PERIOD"
620    INPUT D1
630 D1 = D1 + 1
640    GOSUB 9400
660    GOSUB 8500
665 Y1 = N1
670    PRINT : PRINT "VALUE IN PERIOD ";D1 - 1;" = ";Y1
680    GOTO 1000
700    REM
701    PRINT "PERIODIC WORTH"
702    REM
710    PRINT "ENTER PERIOD WHEN PAYMENTS WILL START"
720    INPUT D1
730 D1 = D1 + 1
740    PRINT "ENTER NUMBER OF PERIODS"
750    INPUT D2
760    GOSUB 9400
805    GOSUB 8500
820    Y1 = (N1 *  FN B(1)) *  FN E(D2)
830    PRINT : PRINT "PERIOD WORTH STARTING IN PERIOD ";D1 - 1;" FOR ";D2;" PERIOD
S = ";Y1
1000 ND = ND + 1:Y(ND) = Y1
1005 IF SA=1 THEN X(ND)=D3
1006 IF SA=2 THEN X(ND)=R2
1010   IF ND = 10 GOTO 1150
1015   ON SA GOTO 9570,9885
1020   PRINT : PRINT "DO YOU WANT TO ADD/ALTER A FACTOR?"
1030   INPUT A$: IF  LEFT$ (A$,1) = "Y" THEN   GOSUB 6000: GOTO 5100
1040   GOTO 150
1150   PRINT
1155   PRINT
1160   FOR I = 1 TO ND
1165 IF SA THEN PRINT I;TAB(5);"IND VAR = ";X(I);TAB(25);"DEP VAR = ";Y(I):GOTO
1190
1170   PRINT "ITERATION#";I;" DEP VAR = ";Y(I)
1190   NEXT I
1200   ON SA GOTO 9500,9700
1210 END
2000   REM
2001   REM   GATHERING OF DATA FOR AN INPUT FACTOR
2002   REM
2005   FOR I = 1 TO 13:F(J,I) = 0: NEXT I
2010   PRINT "ENTER NAME AND FACTOR TYPE"
2020   INPUT F$(J),A$
2030   IF A$ <  > "I" GOTO 2210
2040 F(J,8) = 1
2050 F(J,3) = 1
2060   GOSUB 2310
2070   PRINT "ENTER METHOD OF DEPRECIATION"
2080   PRINT "1 - STRAIGHT LINE"
2090   PRINT "2 - SUM OF THE YEARS DIGITS"
2100   PRINT "3 - ACRS (POST 1985)"
2110   PRINT "4 - DECLINING BALANCE"
2120   INPUT F(J,9): IF F(J,9) < 1 OR F(J,9) > 4 GOTO 2070
2125 IF F(J,9)=3 THEN PRINT "ENTER THE RECOVERY PERIOD":GOTO 2140
2130   PRINT "ENTER THE DEPRECIATION PERIOD"
2140   INPUT F(J,12)
2150   IF F(J,9) = 3 GOTO 2180
2160   PRINT "ENTER SALVAGE VALUE FOR DEPR. PURPOSES"
```

```
2170    INPUT F(J,10)
2180    PRINT "ENTER PERCENTAGE OF ITC"
2190    INPUT F(J,11)
2200    GOTO 3500
2210    IF A$ <  > "NI" GOTO 3200
2215 F(J,8) = 2
2220    PRINT "ENTER TYPE OF CASH FLOW"
2230    PRINT "1 - SINGLE PERIOD AMOUNT"
2240    PRINT "2 - UNIFORM SERIES"
2250    PRINT "3 - UNIFORM GRADIENT SERIES"
2260    INPUT F(J,3):IF F(J,3)<1 OR F(J,3)>3 GOTO 2220
2270 ON F(J,3) GOTO 2310,2350,2400
2300    GOTO 2220
2310 F(J,2) = 1
2315    GOSUB 2600
2320    GOSUB 2700
2340    GOTO 2450
2350    GOSUB 2610
2360    GOSUB 2710
2370    GOSUB 2800
2380    GOSUB 2900
2390    GOTO 2450
2400    GOSUB 2610
2410    GOSUB 2710
2420    GOSUB 2800
2430    GOSUB 2900
2440    GOSUB 3000
2450    RETURN
2600    PRINT "ENTER AMOUNT": GOTO 2620
2610    PRINT "ENTER STARTING VALUE"
2620    INPUT F(J,5)
2630    RETURN
2700    PRINT "ENTER PERIOD": GOTO 2720
2710    PRINT "ENTER STARTING PERIOD"
2720    INPUT F(J,1)
2725 F(J,1) = F(J,1) + 1
2727 IF F(J,1)>51 THEN F(J,1)=51
2730    RETURN
2800    PRINT "ENTER PERIOD INCREMENT"
2810    INPUT F(J,4)
2820    RETURN
2900    PRINT "ENTER NUMBER OF OCCURRENCES"
2910    INPUT F(J,2)
2920    RETURN
3000    PRINT "ENTER TYPE OF GRADIENT"
3010    PRINT "1 - PERCENTAGE INCREASE/DECREASE"
3020    PRINT "2 - ABSOLUTE INCREASE/DECREASE"
3030    INPUT F(J,7)
3040    IF F(J,7) = 1 THEN  PRINT "ENTER GRADIENT PERCENTAGE": GOTO 3090
3045    IF F(J,7) = 2 THEN  PRINT "ENTER GRADIENT AMOUNT": GOTO 3090
3050    PRINT "*** INVALID GRADIENT TYPE ENTERED ***": GOTO 3000
3090    INPUT F(J,6)
3100    RETURN
3200    IF A$ <  > "Q" GOTO 3260
3205 F(J,8) = 3
3210 F(J,3) = 1
3220    GOSUB 2310
3230    PRINT "ENTER BOOK VALUE AT PERIOD ";F(J,1) - 1
3250    INPUT F(J,13)
3255    GOTO 3500
3260    IF A$ <  > "NT" GOTO 3310
3270 F(J,8) = 4
3280 F(J,3) = 1
3290    GOSUB 2220
3300    GOTO 3500
3310    IF A$ <  > "D" GOTO 2000
3320 F(J,8) = 5
3330 F(J,3) = 1
```

```
3340    GOSUB 2310
3500    RETURN
5000    REM
5001    REM   ALTER/ADD FACTORS
5002    REM
5010    GOSUB 6000
5020    PRINT : PRINT "DO YOU WANT TO ADD/ALTER A FACTOR?"
5030    INPUT A$: IF  LEFT$ (A$,1) = "Y" GOTO 5100
5040    GOSUB 8000
5050    GOSUB 9300
5060    GOTO 150
5100    PRINT : PRINT "ENTER NO. OF FACTOR TO BE ADDED/ALTERED": PRINT "(MUST BE 1
        TO 15.": PRINT "IF TO BE ADDED, ENTER ";FC + 1;")"
5110    INPUT J
5120    IF J > 15 OR J > FC + 1 THEN  PRINT "INVALID FACTOR #": GOTO 5100
5130    IF J = FC + 1 THEN FC = J
5170    GOSUB 2000
5240    GOSUB 6300
5250    GOTO 5000
6000    REM
6001    REM   ECHO INPUT DATA
6002    REM
6010    PRINT
6015    IF FC = 0 THEN  PRINT "NO FACTORS HAVE BEEN ENTERED": GOTO 6100
6020    PRINT "FACTOR/PERIOD /=====VALUE=====/  #    FACTOR"
6030    PRINT "# NAME/1ST INC/ 1ST        INC / OCC    TYPE"
6040    FOR J = 1 TO FC
6060    IF F(J,1) = 0 THEN U1=0:GOTO 6080
6070    U1 = F(J,1) - 1
6072    IF F(J,8)=1 THEN A$="I":GOTO 6080
6074    IF F(J,8)=2 THEN A$="NI":GOTO 6080
6076    IF F(J,8)=3 THEN A$="Q":GOTO 6080
6077    IF F(J,8)=4 THEN A$="NT":GOTO 6080
6078    IF F(J,8)=5 THEN A$="D"
6080    PRINT J; TAB( 4);F$(J); TAB( 9);U1; TAB( 13);F(J,4); TAB( 15);F(J,5); TAB
        24);F(J,6); TAB( 34);F(J,2);TAB(40);A$
6090    NEXT J
6100    RETURN
6300    REM
6301    REM   CONVERTS DATA TO PERIOD COSTS PER FACTOR
6302    REM
6303    FOR I = 1 TO 51
6304    A(J,I) = 0
6305    DF(J,I) = 0
6306    MV(J,I) = 0
6307    NEXT I
6310    T1 = F(J,1)
6320    IF F(J,3) = 1 GOTO 6450
6330    T3 = F(J,4)
6340    T4 = F(J,5)
6350    T2 = T1 + (F(J,2) - 1) * T3
6360    IF T2 > 51 THEN T2 = 51
6370    FOR I = T1 TO T2 STEP T3
6380    A(J,I) = T4
6390    IF F(J,3) < > 3 GOTO 6420
6400    IF F(J,7) < > 1 THEN T4 = T4 + F(J,6): GOTO 6420
6410    T4 = T4 * (1 + F(J,6))
6420    NEXT I
6430    GOTO 6900
6450    IF F(J,8) < > 1 THEN 6870
6455    A(J,T1)=F(J,5)
6460    T1 = T1 + 1
6465    T3 = F(J,12)
6480    T2 = T1 + T3 - 1
6485    ON F(J,9) GOTO 6490,6550,6610,6800
6490    REM straight line
6500    T4 = ( ABS (F(J,5)) - F(J,10)) / T3
6510    FOR I = T1 TO T2
6520    DF(J,I) = T4
```

```
6530   NEXT I
6540   GOTO 6900
6550   REM sum of the years digits
6570   FOR I = T1 TO T2
6580 DF(J,I) = ( ABS (F(J,5)) - F(J,10)) * ((2 * (T3 - (I - T1 + 1) + 1)) / (T3
* (T3 + 1)))
6590   NEXT I
6600   GOTO 6900
6610   REM acrs (post 1985)
6620 T5 =   INT (((1 / T3) * 100) + .4) / 100
6630   FOR I = T1 TO T2
6640   IF I = T1 THEN T4 = T5: GOTO 6670
6650 T4 = (1 - T5) * ((2 * (T3 - (I - T1 + 1) + 1)) / (T3 * (T3 - 1)))
6660 T4 =   INT ((T4 * 100) + .4) / 100
6670 DF(J,I) =   ABS (F(J,5)) * T4
6680   NEXT I
6690   GOTO 6900
6800   REM declining balance
6810 T3 =   ABS (F(J,5))
6820 K = 1 - (F(J,10) / T3) ^ (1 / F(J,12))
6830   FOR I = T1 TO T2
6840 DF(J,I) = K * T3
6850 T3 = T3 - DF(J,I)
6860   NEXT I
6865 GOTO 6900
6870 IF F(J,8)=3 THEN MV(J,T1)=F(J,5):GOTO 6900
6875 IF F(J,8)=5 THEN DF(J,T1)=F(J,5):GOTO 6900
6880 A(J,T1)=F(J,5)
6900   RETURN
7000   REM
7001   REM   Y1=ROUNDED INTERNAL RATE OF RETURN
7002   REM
7310 D3 = .0001
7320 D1 = 1
7330   GOSUB 8500
7340   IF N1 = 0 GOTO 7560
7350 V1 = N1
7360 D3 = .0002
7370   GOSUB 8500
7390   GOSUB 8300
7400   IF M1 = 1 THEN I = .1: GOTO 7440
7410 I =   - .1
7440 D3 = 0
7450   PRINT : PRINT "I'M WORKING ON IT!"
7460   FOR J = 1 TO 30
7470 V1 = N1
7480 D3 = D3 + I
7490   GOSUB 8500
7510   GOSUB 8300
7520   IF M1 = 1 GOTO 7550
7530 I = -I / 10
7540   IF   ABS (I) < 9.999999E-06 THEN J = 30
7550   NEXT J
7560   IF D3 < 0 THEN D3 = D3 - .00005: GOTO 7580
7570 D3 = D3 + .00005
7580 D3 =   INT (D3 * 10000) / 10000
7590 Y1 = D3 * 100
7600   PRINT : PRINT "INTERNAL RATE OF RETURN = ";Y1;"%"
7610   RETURN
8000   REM
8001   REM AF()=TOTAL NET CASH FLOW PER PERIOD
8002   REM   P=NUMBER OF PERIODS IN ANALYSIS
8003   REM
8010   FOR I = 1 TO 50
8012 TC(I) = 0
8014   FOR J = 1 TO FC
8016   IF A(J,I) <   > 0 OR DF(J,I) <   > 0 OR MV(J,I) <   > 0 THEN P = I:J = FC
8018   NEXT J
8020   NEXT I
```

```
8022   FOR I = 1 TO P
8024  BT(I) = 0
8030  DP(I) = 0
8031  AF(I) = 0
8035  T4 = 0:T7=0:T8=0:U1=0
8040   FOR J = 1 TO FC
8045  T6=0
8050  DP(I) = DP(I) + DF(J,I): IF F(J,8) = 5 GOTO 8160
8051   IF F(J,3) = 1 AND F(J,1) <  > I GOTO 8160
8055   IF F(J,8) = 1 THEN TC(I + 1) =  ABS (F(J,5)) * F(J,11):T6 = A(J,I): GOTO 8
155
8060   IF F(J,8) = 4 THEN T6 = A(J,I): GOTO 8155
8070  T2 = 0:T3 = 0:T6=0:T9=0
8080   IF F(J,8) <  > 3 GOTO 8140
8100  T1 = F(J,13)
8105  T9=MV(J,I):T5=ABS(T9)
8110   IF T5 = T1 THEN T3 = T5: GOTO 8135
8120   IF T5 < T1 THEN T3 = T5 + (T1 - T5) * GL: GOTO 8135
8130  T2 = T5 - T1:T3 = T1
8135  IF MV(J,I)<0 THEN T2=-T2:T3=-T3
8140  BT(I) = BT(I) + A(J,I) + T2
8146  T8=T8+T2
8150  T4 = T4 + T3
8154  U1=U1+T9
8155  T7=T7+T6
8160   NEXT J
8170  AF(I) = BT(I) - ((BT(I) - DP(I)) * TR - TC(I))
8175  AF(I) = AF(I) + T4 + T7
8180  BT(I) = BT(I) + T7 - T8 + U1
8210   NEXT I
8220   RETURN
8300   REM
8301   REM   M1=0 IRR GOING IN WRONG DIRECTION
8302   REM   M1=1 IRR GOING IN RIGHT DIRECTION
8303   REM   M1=2 NPV HAS CROSSED THE ZERO VALUE
8304   REM
8310   IF (V1 > 0 AND N1 <  = 0) OR (V1 < 0 AND N1 >  = 0) THEN M1 = 2: GOTO 8340
8320   IF  ABS (N1) <  ABS (V1) THEN M1 = 1: GOTO 8340
8330  M1 = 0
8340   RETURN
8500   REM
8501   REM   N1=NPV IN PERIOD D1
8502   REM
8510  N1 = 0
8520   FOR I2 = 1 TO P
8530  N1 = N1 +  FN B(I2 - 1) * AF(I2)
8540   NEXT I2
8550  N1 = N1 *  FN A(D1 - 1)
8560   RETURN
9000   REM
9001   REM   Y1=EXTERNAL RATE OF RETURN
9002   REM
9010  T1 = 0
9020  T2 = 0
9030   FOR I = 1 TO P
9040   IF AF(I) < 0 THEN T1 = T1 + (AF(I) *  FN B(I - 1)): GOTO 9060
9050   IF AF(I) > 0 THEN T2 = T2 + (AF(I) *  FN A(P - I))
9060   NEXT I
9070   PRINT : PRINT "SUM OF - FLOWS = ";T1
9080   PRINT "SUM OF + FLOWS = ";T2
9090   PRINT "NO. OF PERIODS = ";P - 1
9120  Y1 = (( - T2 / T1) ^ (1 / (P - 1)) - 1)*100
9130   PRINT "E.R.R. = ";Y1;"%"
9140   RETURN
9300   REM
9301   REM   DISPLAY PERIOD AMOUNT
9302   REM
9310   PRINT : PRINT "PER   BEF TAX       DEP    AFT TAX"
```

```
9320    FOR I = 1 TO P
9330    PRINT   TAB( 3);I - 1; TAB( 8);BT(I); TAB( 20);DP(I); TAB( 30);AF(I)
9340    NEXT I
9360    RETURN
9400    REM
9401    REM   ENTER DISCOUNT RATE
9402    REM
9410    IF SA = 1 GOTO 9450
9420    PRINT "ENTER DISCOUNT RATE"
9430    INPUT D3
9440    GOTO 9490
9450    PRINT "ENTER STARTING RATE"
9455    INPUT D3
9460    PRINT "ENTER INCREMENT"
9465    INPUT R2
9470    PRINT "ENTER NUMBER OF OCCURRENCES"
9480    INPUT R3:R3 = R3 - 1
9490    RETURN
9500    REM
9501    PRINT "SENSITIVITY ANALYSIS BY RATE"
9502    REM
9505 SA=1
9510    PRINT
9515    PRINT "   1 - COMPUTE PRESENT OR FUTURE VALUE"
9520    PRINT "   2 - COMPUTE PERIODIC WORTH"
9525    PRINT "   3 - COMPUTE EXTERNAL RATE OF RETURN"
9527 PRINT "   4 - GRAPICS DISPLAY"
9530    PRINT "   5 - RETURN TO MAIN MENU"
9535    PRINT : PRINT "CHOOSE ONE";
9536    INPUT D
9540    IF D = 5 GOTO 137
9541 IF D=4 GOTO 11000
9545    IF D < 1 OR D > 3 GOTO 9500
9555    FOR I = 1 TO 10:Y(I) = 0:X(I)=0: NEXT I:ND = 0
9560    ON D GOTO 600,700,500
9570 D3 = D3 + R2:R3 = R3 - 1
9575    IF R3 < 0 GOTO 1150
9580    ON D GOTO 660,805,530
9700    REM
9701    PRINT "SENSITIVITY ANALYSIS BY MAGNITUDE"
9702    REM
9705 SA=2
9710    PRINT
9715    PRINT "   1 - COMPUTE PRESENT OR FUTURE VALUE"
9720    PRINT "   2 - COMPUTE PERIODIC WORTH"
9725    PRINT "   3 - COMPUTE EXTERNAL RATE OF RETURN"
9730    PRINT "   4 - COMPUTE INTERNAL RATE OF RETURN"
9732 PRINT "   5 - GRAPHIC DISPLAY"
9735    PRINT "   6 - RETURN TO MAIN MENU"
9740    PRINT : PRINT "CHOOSE ONE";
9750    INPUT D
9760    IF D = 6 GOTO 137
9765 IF D=5 GOTO 11000
9770    IF D < 1 OR D > 4 GOTO 9700
9790    FOR I = 1 TO 10:Y(I) = 0:X(I)=0: NEXT I:ND = 0
9800    GOSUB 6000
9810    PRINT
9820    PRINT "ENTER FACTOR NUMBER"
9825    INPUT R1
9826    IF R1 < 1 OR R1 > FC GOTO 9820
9830    PRINT "ENTER STARTING VALUE"
9835    INPUT R2
9840    PRINT "ENTER INCREMENT"
9845    INPUT R3
9850    PRINT "ENTER NUMBER OF OCCURRENCES"
9855    INPUT R4:R4 = R4 - 1
9860 J = R1
9865 F(J,5) = R2
```

```
9870   GOSUB 6300
9875   GOSUB 8000
9880   ON D GOTO 600,700,500,400
9885 J = R1
9890 R2 = R2 + R3:R4 = R4 - 1
9895   IF R4 < 0 GOTO 1150
9896 F(J,5) = R2
9897   GOSUB 6300
9898   GOSUB 8000
9900   ON D GOTO 660,805,530,400
11000 REM graphing routine
11005 KEY OFF
11010 SCREEN 1
11100 XO=X(1):YO=Y(1)
11105 X9=XO:Y9=YO
11110 FOR I=2 TO ND
11115 IF X(I)<XO THEN XO=X(I):GOTO 11125
11120 IF X(I)>X9 THEN X9=X(I)
11125 IF Y(I)<YO THEN YO=Y(I):GOTO 11135
11130 IF Y(I)>Y9 THEN Y9=Y(I)
11135 NEXT I
11140 IF (XO<0 AND X9<0) OR (ABS(XO)>ABS(X9)) THEN X9=ABS(XO):GOTO 11150
11145 XO=-X9
11150 IF (YO<0 AND Y9<0) OR (ABS(YO)>ABS(Y9)) THEN Y9=ABS(YO):GOTO 11160
11155 YO=-Y9
11160 TM=ABS(X9)/5
11165 TY=ABS(Y9)/5
11210 XS=250/(X9-XO)
11220 YS=180/(Y9-YO)
11300 C1=YS*Y9:C2=279-XS*X9
11310 LINE (29,C1)-(279,C1)
11320 LINE (C2,0)-(C2,180)
11330 FOR X=C2 TO 279 STEP TM*XS
11340 LINE (X,C1-2)-(X,C1+2)
11350 NEXT X
11360 FOR X=C2 TO 29 STEP -TM*XS
11370 LINE (X,C1-2)-(X,C1+2)
11380 NEXT X
11390 LOCATE 11,1: PRINT USING "+#####.###";XO
11395 LOCATE 11,31:PRINT USING "+#####.###";X9
11400 FOR Y=C1 TO 0 STEP -TY*YS
11410 LINE(C2-2,Y)-(C2+2,Y)
11420 NEXT Y
11430 FOR Y=C1 TO 180 STEP TY*YS
11440 LINE (C2-2,Y)-(C2+2,Y)
11450 NEXT Y
11460 LOCATE 1,21:PRINT USING "+#####.###";Y9
11465 LOCATE 23,21:PRINT USING "+#####.###";YO
11500 CIRCLE (30+(X(1)-XO)*XS,-(Y(1)-YO)*YS+180),2
11510 FOR NP=2 TO ND
11515 T1=30+(X(NP)-XO)*XS
11516 T2=180-(Y(NP)-YO)*YS
11520 LINE -(T1,T2)
11525 CIRCLE (T1,T2),2
11530 NEXT NP
11540 LOCATE 25,8: PRINT "press any key to continue";
11550 A$ = INKEY$
11560 IF LEN(A$)=0 THEN 11550
11600 SCREEN 2
11610 ON SA GOTO 9500,9700
```

Appendix E

Decision Tree Program

Reprinted from *Industrial Engineering*, June 1981, p. 18, 82, with permission. This program was coded in BASIC for the TRS-80.

```
100 REM TREE
110 DIM F(40)
120 DIM T(40)
130 DIM P(40)
140 DIM D(40)
150 DIM DD(40)
160 DIM Y(40)
170 DIM R(40)
180 FF=0 : A=0
190 FOR C=1 TO 12
200 PRINT
210 IF C<>6 THEN 250
220 PRINT TAB(20);"DECISION TREE"
230 PRINT TAB(19);"IEMS DEPARTMENT"
240 PRINT TAB(13);"UNIVERSITY OF CENTRAL FLORIDA"
250 NEXT C
260 FOR C=1 TO 500
270 NEXT C
280 PRINT "TOTAL NUMBER OF BRANCHES ";
290 INPUT T
300 PRINT "INPUT THE DISCOUNT RATE ";
310 INPUT I
320 I=I/100
330 FOR X=1 TO T
340 PRINT "INPUT BRANCH ";X;" (FROM , TO) ";
350 INPUT F(X),T(X)
360 PRINT "WHAT IS THE PROBABILITY OF BRANCH ";X;
```

```
370 INPUT P(X)
380 PRINT "WHAT IS THE DOLLAR VALUE ";
390 INPUT DD(X)
400 PRINT "OVER WHAT # OF YEARS IS THE MONEY INVESTED ";
410 INPUT Y(X)
420 R(X)=I
430 IF R(X)=0 THEN 480
440 IF Y(X)=0 THEN 480
450 Q=(1+R(X))CY(X)
460 D(X)=DD(X)*Q*(Q-1)/(R(X)*G)
470 GOTO 490
480 D(X)=DD(X)
490 NEXT X
500 FOR Z=1 TO T-1
510 J=Z
520 FOR X=Z+1 TO T
530 IF F(X)>F(J) THEN 570
540 IF F(X)<F(J) THEN 560
550 IF T(X)>T(J) THEN 570
560 J=X
570 NEXT X
580 IF J=Z THEN 800
590 K(1)=F(Z)
600 K(2)=T(Z)
610 K(3)=P(Z)
620 K(4)=D(Z)
630 K(5)=Y(Z)
640 K(6)=R(Z)
650 K(7)=DD(Z)
660 F(Z)=F(J)
670 T(Z)=T(J)
680 P(Z)=P(J)
690 D(Z)=D(J)
700 Y(Z)=Y(J)
710 R(Z)=R(J)
720 DD(Z)=DD(J)
730 F(J)=K(1)
740 T(J)=K(2)
750 P(J)=K(3)
760 D(J)=K(4)
770 Y(J)=K(5)
780 R(J)=K(6)
790 DD(J)=K(7)
800 NEXT Z
810 PRINT TAB(12);"ORDERED TREE"
820 GOSUB 1320
830 J=1
840 B=F(J)
850 IF F(J)-B<0 THEN 1750
860 IF F(J)-B>0 THEN 900
870 A=A+P(J)
880 J=J+1
890 IF J<=T THEN 850
900 IF A=1 THEN 920
910 IF A<>0 THEN 1750
920 A=P(J)
930 B=F(J)
940 J=J+1
950 IF J<=T THEN 850
960 X=T+1
970 X=X-1
980 IF X<=0 THEN 1210
990 IF P(X)>0 THEN 1250
```

```
1000 D(X)=D(X)/(1+R(X))↑Y(X)
1010 A=D(X)
1020 B=F(X)
1030 C=X
1040 J=X
1050 J=J-1
1060 IF J=0 THEN 1130
1070 IF (F(X)-F(J))>0 THEN 1130
1080 D(J)=D(J)/(1+R(J))↑Y(J)
1090 IF D(J)-A<=0 THEN 1050
1100 A=D(J)
1110 C=J
1120 GOTO 1050
1130 P(C)=-45.7
1140 X=J
1150 IF X<=0 THEN 1210
1160 J=J+1
1170 J=J-1
1180 IF B-T(J)<>0 THEN 1170
1190 D(J)=A+D(J)
1200 GOTO 960
1210 PRINT TAB(10);"EVALUATED DECISION TREE"
1220 FF=1
1230 GOSUB 1320
1240 GOTO 1510
1250 D(X)=(D(X)/(1+R(X))↑Y(X))*P(X)
1260 J=X
1270 J=J-1
1280 IF J=0 THEN 1300
1290 IF (F(X)-T(J))<>0 THEN 1270
1300 D(J)=D(X)+D(J)
1310 GOTO 970
1320 PRINT
1330 PRINT "FROM";TAB(8);"TO";TAB(21);"DOLLAR";TAB(37);"INTEREST"
1340 PRINT "NODE";TAB(7);"NODE";TAB(14);"PROB";TAB(21);"VALUE";
TAB(30);"YEARS";TAB(37);"RATE"
1350 M=4
1360 FOR X=1 TO T
1370 IF P(X)<=0 THEN 1480
1380 IF FF=1 THEN DD(X)=D(X)/P(X)
1390 PRINT F(X);TAB(7);T(X);TAB(12);P(X);
1400 PRINT TAB(19);DD(X);TAB(33);Y(X);TAB(36);R(X)
1410 M=M+1
1420 IF M<14 THEN 1440
1430 GOSUB 1900
1440 NEXT X
1450 H=1
1460 GOSUB 1900
1470 RETURN
1480 PRINT F(X);TAB(7);T(X);TAB(11);"DECIDE";
1490 IF FF=1 THEN DD(X)=D(X)
1500 GOTO 1400
1510 PRINT " BEST DECISIONS ARE"
1520 PRINT "FROM";TAB(6);"TO";TAB(15);"DOLLAR"
1530 PRINT "NODE";TAB(7);"NODE";TAB(15);"VALUE"
1540 H=1
1550 M=3
1560 FOR X=1 TO T
1570 IF P(X)<>-45.7 THEN 1660
1580 PRINT F(X);TAB(8);T(X);TAB(14);D(X)
1590 M=M+1
1600 IF M<14 THEN 1660
1610 GOSUB 1900
```

```
1620 PRINT " BEST DECISIONS ARE"
1630 PRINT "FROM";TAB(8);"TO";TAB(15)"DOLLAR"
1640 PRINT "NODE"TAB(7);"NODE";TAB(15);"VALUE"
1650 H=1
1660 NEXT V
1670 PRINT
1680 M=M+1
1690 IF M<15 THEN 1670
1700 PRINT "TO RERUN INPUT 0 OR TO STOP INPUT 1 ";
1710 FF=0
1720 INPUT M
1730 IF M=0 THEN 190
1740 END
1750 FOR M=1 TO 13
1760 PRINT
1770 IF M<>6 THEN 1810
1780 PRINT "THE INPUTED DATA IS WRONG"
1790 PRINT "FROM ";F(J-1);" TO ";T(J-1)
1800 PRINT "ALSO CHECK NODE NUMBERS AND PROBABILITY"
1810 NEXT M
1820 FOR M=1 TO 1500
1830 NEXT M
1840 PRINT
1850 PRINT "TO RERUN ENTER 0 OR TO STOP ENTER 1 ";
1860 INPUT Z
1870 FF=0
1880 IF Z=0 THEN 190
1890 GOTO 1740
1900 PRINT
1910 M=M+1
1920 IF M<15 THEN 1900
1930 PRINT "HIT ENTER TO CONTINUE ";
1940 INPUT MM
1950 M=4
1960 IF H=1 THEN 1990
1970 PRINT "FROM";TAB(8);"TO";TAB(21);"DOLLAR";TAB(37);"INTEREST"
1980 PRINT "NODE";TAB(7);"NODE";TAB(14);"PROB";TAB(21);"VALUE";
TAB(30);"YEARS";TAB(37);"RATE"
1990 H=0
2000 RETURN
```

Appendix F

PERT Program

Reprinted from *Industrial Engineering*, January 1981, p. 21, 22, with permission. This program was coded in BASIC for the TRS-80. This program can be used for CPM by making $t_0 = t_{ml} = t_p$.

```
100 REM PERT
110 DIM F(100)
120 DIM T(100)
130 DIM E(100)
140 DIM L(100)
150 DIM A(100)
160 DIM C(100)
170 DIM S(100)
180 DIM V(100)
190 DIM L(100)
200 DIM M(100)
210 DIM ND(100)
220 DIM NM(100)
230 DIM NE(100)
240 DIM Nv(100)
250 DIM NL(100)
260 DIM PR(71)
270 DIM SL(100)
280 FOR X=1 TO 71
290 READ PR(X)
300 NEXT X
310 FOR X=1 TO 12
320 PRINT
330 IF X<>6 THEN 380
340 PRINT TAB(25);"PERT"
350 PRINT
360 PRINT TAB(20);"IEMS DEPARTMENT"
370 PRINT TAB(13);"UNIVERSITY OF CENTRAL FLORIDA"
```

```
380 NEXT X
390 FOR X=1 TO 1000
400 NEXT X
410 PRINT 'TOTAL NUMBER OF ACTIVITIES ';
420 INPUT B
430 IF B<=100 THEN 470
440 PRINT 'THERE IS NOT ENOUGH STORAGE FOR OVER HUNDRED BRANCHES"
450 PRINT '                PLEASE REINPUT'
460 GOTO 410
470 FOR X=1 TO B
480 PRINT 'INPUT ACTIVITY ';X;' (FROM , TO) ';
490 INPUT F(X),T(X)
500 IF F(X)<T(X) THEN 530
510 PRINT 'FROM NODE MUST BE LESS THAN TO NODE'
520 GOTO 480
530 PRINT 'INPUT TIME ESTIMATES FOR ACTIVITY ';X
540 PRINT 'OPT. , M.L. , PESS. ';
550 INPUT O,ML,P
560 IF O<=ML THEN 590
570 PRINT 'TIME ESTIMATES OUT OF SEQUENCE PLEASE REINPUT ACTIVITY ';X
580 GOTO 540
590 IF ML>P THEN 570
600 C(X)=(O+4*ML+P)/6
610 V(X)=((P-O)/6)*((P-O)/6)
620 NEXT X
630 FOR Z=1 TO B-1
640 J=Z
650 FOR X=Z+1 TO B
660 IF F(X)>F(J) THEN 700
670 IF F(X)<F(J) THEN 690
680 IF T(X)>T(J) THEN 700
690 J=X
700 NEXT X
710 IF J=Z THEN 840
720 H(1)=F(Z)
730 H(2)=T(Z)
740 H(3)=C(Z)
750 H(4)=V(Z)
760 F(Z)=F(J)
770 T(Z)=T(J)
780 C(Z)=C(J)
790 V(Z)=V(J)
800 F(J)=H(1)
810 T(J)=H(2)
820 C(J)=H(3)
830 V(J)=H(4)
840 NEXT Z
850 FOR X=1 TO B
860 E(X)=0
870 U(X)=0
880 NEXT X
890 FOR X=1 TO B
900 E(X)=E(X)+C(X)
910 U(X)=U(X)+V(X)
920 FOR D=X+1 TO B
930 IF T(X)<>F(D) THEN 970
940 IF E(D)>E(X) THEN 970
950 E(D)=E(X)
960 U(D)=U(X)
970 NEXT D
980 NEXT X
990 P=0
1000 FOR X=1 TO B
1010 IF P>E(X) THEN 1040
1020 P=E(X)
1030 Q=U(X)
1040 NEXT X
```

```
1050 FOR X=1 TO B
1060 L(X)=P
1070 M(X)=0
1080 NEXT X
1090 FOR Z=1 TO B
1100 X=B-Z+1
1110 FOR D=1 TO B-Z
1120 IF F(X)<>T(D) THEN 1160
1130 IF L(X)-C(X)>L(D) THEN 1160
1140 L(D)=L(X)-C(X)
1150 M(D)=M(X)+V(X)
1160 NEXT D
1170 NEXT Z
1180 FOR X=1 TO B
1190 E(X)=E(X)-C(X)
1200 U(X)=U(X)-V(X)
1210 L(X)=L(X)-C(X)
1220 M(X)=M(X)+V(X)
1230 NEXT X
1240 X=0
1250 NN=0
1260 NN=NN+1
1270 X=X+1
1280 ND(NN)=F(X)
1290 NE(NN)=E(X)
1300 NV(NN)=U(X)
1310 NL(NN)=L(X)
1320 NM(NN)=M(X)
1330 IF X=B THEN 1380
1340 IF F(X)<>F(X+1) THEN 1260
1350 X=X+1
1360 IF NL(NN)<L(X) THEN 1330
1370 GOTO 1290
1380 NN=NN+1
1390 ND(NN)=T(B)
1400 NE(NN)=L(B)+C(B)
1402 FOR JJ=1TOB
1404 IF T(JJ)<>T(B) THEN 1419
1406 IF E(JJ)<>L(JJ) THEN 1419
1410 NV(NN)=U(JJ)+V(JJ)
1419 NEXT JJ
1420 NL(NN)=NE(NN)
1430 NM(NN)=0
1440 PRINT TAB(23);"PERT TABLE"
1450 PRINT
1460 GOSUB 2050 &LFNJ=J+2
1480 FOR X=1 TO NN
1490 IF J<14 THEN 1510
1500 GOSUB 2000
1510 SL(X)=INT((NL(X)-NE(X))*10+.5)/10
1520 ND(X)=INT(ND(X)*10+.5)/10
1530 NE(X)=INT(NE(X)*10+.5)/10
1540 NV(X)=INT(NV(X)*10+.5)/10
1550 NL(X)=INT(NL(X)*10+.5)/10
1560 NM(X)=INT(NM(X)*10+.5)/10
1570 PRINT ND(X);TAB(10);NE(X);TAB(19);NV(X);TAB(31);NL(X);TAB(40);
NM(X);TAB(53);SL(X)
1580 J=J+1
1590 NEXT X
1600 PRINT
1610 J=J+1
1620 IF J<15 THEN 1600
1630 PRINT "HIT ENTER TO CONTINUE ";
1640 INPUT J
1650 FOR XX=1 TO 12
1660 PRINT
```

```
1670 NEXT XX
1680 PRINT "EXPECTED PROJECT DURATION IS ";NE(NN)
1690 PRINT
1700 PRINT "VARIANCE OF PROJECT DURATION ";NV(NN)
1710 PRINT
1720 PRINT "INPUT SCHEDULED COMPLETION TIME ";
1730 INPUT TS
1740 Z=(TS-NE(NN))/SQR(NV(NN))
1750 IF Z<-3.5 THEN 1810
1760 IF Z>3.5 THEN 1830
1770 TT=INT(ABS(Z)/.05)+1
1780 PP=PR(TT)+(((ABS(Z)-(TT-1)*.05)/.05)*(PR(TT+1)-PR(TT)))
1790 IF Z<0 THEN PP=1-PP
1800 GOTO 1840
1810 PP=0
1820 GOTO 1840
1830 PP=1.0
1840 PRINT
1850 PP=INT(PP*1000+.5)/1000
1860 PRINT "PROBABILITY OF MEETING SCHEDULE ";PP
1870 PRINT
1880 PRINT
1890 PRINT "NEXT STEP TO BE PERFORMED"
1900 PRINT
1910 PRINT "1-INPUT NEW DATA"
1920 PRINT "2-CHANGE COMPLETION TIME"
1930 PRINT "3-EXIT PROGRAM"
1940 PRINT
1950 PRINT "INPUT CHOICE ";
1960 INPUT X
1970 IF X=2 THEN 1650
1980 IF X=1 THEN 310
1990 END
2000 PRINT
2010 J=J+1
2020 IF J<15 THEN 2000
2030 PRINT "HIT ENTER TO CONTINUE";
2040 INPUT J
2050 PRINT "EVENT";TAB(11);"EARLIEST TIME";TAB(33);"LATEST TIME"
2060 PRINT "NUMBER";TAB(10);"MEAN";TAB(16);"VARIANCE";TAB(31);"MEAN";
TAB(39);"VARIANCE";TAB(53);"SLACK"
2070 J=2
2080 RETURN
2090 DATA .5,.5199,.5398,.5596,.5793,.5987,.6179,.6368,.6554,.6736
2100 DATA .6915,.7088,.7257,.7422,.758,.7734,.7881,.8023,.8159,.8289
2110 DATA .8413,.8531,.8643,.8749,.8849,.8944,.9032,.9115,.9192,.9265
2120 DATA .9332,.9394,.9452,.9505,.9554,.9599,.9641,.9678,.9713,.9744
2130 DATA .9772,.9798,.9821,.9842,.9861,.9878,.9893,.9906,.9918,.9929
2140 DATA .9938,.9946,.9953*.996,.9965,.997,.9974,.9978,.9981,.9984
2150 DATA .9987,.9989,.999,.9992,.9993,.9994,.9995,.9996,.9997,.9997,
          .9998
```

Appendix G

General References*

Barish, N. and S. Kaplan, *Analysis for Engineering and Managerial Decision Making,* 2nd ed. New York: McGraw-Hill, 1978.

Bussey, L., *The Economic Analysis of Industrial Projects,* Englewood Cliffs, NJ: Prentice-Hall, 1978.

Clifton, D. and L. Fyffe, *Project Feasibility Analysis,* New York: John Wiley & Sons, 1977.

DeGarmo, E., W. Sullivan, and J. Canada, *Engineering Economy,* 7th ed. New York: Macmillan, 1984.

Engineering Economist, The. A quarterly journal published jointly by the Institute of Industrial Engineers and the American Society for Engineering Education.

Fleischer, G., *Engineering Economy: Capital Allocation Theory,* Boston: Wadsworth, 1984.

Grant, E., W. Ireson, and R. Leavenworth, *Principles of Engineering Economy,* 7th ed. New York: John Wiley & Sons, 1982.

Jelen, F. and J. Black, *Cost and Optimization Engineering,* 2nd ed. New York: McGraw-Hill, 1983.

Malik, A., *Engineering Economy with Computer Applications,* Mahomet, IL: Engineering Technology Press, 1979.

*Selected references appear at the end of individual chapters.

Newnan, D., *Engineering Economic Analysis,* rev. ed. San Jose, CA: Engineering Press, 1980.

Riggs, J., *Engineering Economics,* 2nd ed. New York: McGraw-Hill, 1982.

Smith, G., *Engineering Economy: Analysis of Capital Expenditures,* 3rd ed. Ames, IA: Iowa State University Press, 1979.

Tarquin, J. and L. Blank, *Engineering Economy,* 2nd ed. New York: McGraw-Hill, 1983.

Taylor, G., *Managerial and Engineering Economy,* 3rd ed. New York: D. Van Nostrand, 1980.

Theusen, H., W. Fabrycky, and G. Theusen, *Engineering Economy,* 5th ed. Englewood Cliffs, NJ: Prentice-Hall, 1977.

Appendix H

Answers to Odd-numbered Exercises

2–1. $1,028.30
2–3. $5,880.24
2–5. $5,042.29
2–7. $4,144.63
2–9. $8,692.42
2–11. $212.47
2–13. $8,659.11
2–15. $640.42
2–17. $3,164.30; $120,458.10
2–19. $4,317.85
2–21. $5,073.00
2–23. $22,510.16
2–25. $6,251.68; $1,734.28
2–27. $617.17; $663.24
2–29. $2,591.78
2–31. $2,615.10; $53,792.22; $1,365.97
3–1. 15.81%
3–3. 12.55% (per year)
3–5. $97,101.28
3–7. $13,988
3–9. 6.67%
3–11. 10.48%; 11.12%

3–13. 20.82%; Yes
3–15. $$-54,249; 9.09\%; \$-14,335;$$
$$\$-18,611$$
4–1. $AW_A = 4.194$
$AW_B = -.741$
$AW_C = -6.934$
$AW_D = 5.648$
$AW_E = 4.325$
Choose D if $300 thousand is available
Choose A if $200 thousand is available
4–3. $PW_A = -27.384$
$PW_B = 0.746$
$PW_C = 17.848$
$PW_D = -9.957$
Choose C in either case
4–5. $PW_A = -364.929$
$PW_B = -356.208$
$PW_C = -352.488$
$PW_D = -371.891$
Choose C

4–7. AW_A = \$1,275
AW_B = \$21,620
Choose *B*

4–9. PW_A = − \$20,489
PW_B = − \$18,272
Choose *B*

4–11. PW = − \$337; No

4–13. PW_{FULL} = − 5628
PW_{DEF} = − 4383
Choose deferred

4–15. AW_A = − \$239
AW_B = \$49
ERR_A = 9.8%
ERR_B = 12.3%
Choose *B*

6–1. 23,333 units/mo.; \$10,000/mo.

6–3. $Profit_o$ = \$7,000/mo.
$Profit_a$ = \$60,560/mo.
$Profit_b$ = \$35,000/mo.
$Profit_{a+b}$ = \$87,000/mo.

6–5. \$92,118

6–7. 3.3 yrs.; Teflon

6–9. 776 units/year

6–11. \$0.0733/kw-h; Choose *A*

7–1. d_2 = \$3,000
D_2 = \$6,000
BV_2 = \$6,000

7–3. d_2 = \$3,000
D_2 = \$7,500
BV_2 = \$4,500

7–5. d_2 = \$5,400
D_2 = \$9,360
BV_2 = \$2,640

7–7a. d_2 = \$22,500
D_2 = \$47,500
BV_2 = \$202,500
b. d_2 = \$43,345
D_2 = \$99,146
BV_2 = \$150,854
c. d_2 = \$40,000
D_2 = \$90,000
BV_2 = \$160,000

7–9. d_2 = \$13,110
D_2 = \$27,793
BV_2 = \$222,207

7–11.

	SL	SYD	DB	ACRS
d_2	\$2,667	\$3,810	\$3,827	\$5,760
BV_2	\$12,667	\$9,619	\$8,653	\$8,640

7–13. a. \$100,000
b. \$750,000

8–1. 8.74%

8–3. 8.07%

8–5. AW_A = 2.195
AW_B = −.24
AW_C = −3.231
AW_D = 3.151
AW_E = 2.529
Choose *D*; choose *A*

8–7. AW_A = −3.311
AW_B = 2.051
AW_C = 5.233
AW_D = 1.707
Choose *C*

8–9. AW_A = −57.486
AW_B = −55.959
AW_C = −55.216
AW_D = −58.552
Choose *C*

8–11. AW_A = \$2,686
AW_B = \$11,656
Choose *B*

8–13. PW_A = − \$11,823
PW_B = − \$10,301
Choose *B*

8–15. PW = \$662; Yes

8–17. PW_A = − \$300,365
PW_B = − \$299,399
PW_C = − \$289,439
Choose *C*

8–19. AW_{buy} = \$3,117
AW_{lease} = \$6,559

9–1. $EUAC_{min}$ = − \$14,093; 6 years

9–3. a. $EUAC(3)$ = − \$26,451
b. $EUAC(3)$ = − \$13,087

9–5. $EUAC(6)$ = − \$4,836

9–7. a. AW_{def} = \$23,000;
AW_{ch} = \$35,140
b. AW_{def} = \$11,930;
AW_{ch} = \$21,893

9–9. $DC_1 = -\$4,800$; $AW_{ch} = -\$7,389$
   $DC_2 = -\$5,008$
9–11. TIA = \$2,888
9–13. $AW_{def} = -\$4,940$
   $AW_{ch} = -\$10,733$, $\therefore$ overhaul

10–3. $PW = -\$615$
10–5. $AW_A = 1.569$
   $AW_B = -1.448$
   $AW_C = -5.740$
   $AW_D = +0.572$
   $AW_E = -0.477$
   Choose D; choose A
10–7. $AW_A = -5.130$
   $AW_B = -0.231$
   $AW_C = +2.768$
   $AW_D = -2.598$
   Choose C in either case
10–9. $AW_A = -63.00$
   $AW_B = -61.59$
   $AW_C = -61.07$
   $AW_D = -64.98$
   Choose C
10–11. $AW_A = -\$15,699$
   $AW_B = -\$806$
10–13. $AW_{buy} = -\$4,617$
   $AW_{lease} = -\$4,388$
10–15. $AW_A = -\$2,991$
   $AW_B = -\$2,763$
10–17. $AW = \$1,894$
10–19. $AW_A = -\$41,851$
   $AW_B = -\$42,581$
   $AW_C = -\$40,159$
10–21. $PW_{full} = -\$8,527,000$
   $PW_{def} = -\$8,130,000$
10–23. $DC_1 = \$11,828$;
   $DC_2 = -\$8,193$
   $AW_{ch} = \$21,945$
10–25. $DC_1 = -\$5,100$;
   $DC_2 = -\$5,458$
   $AW_{ch} = -\$9,097$

11–3. \$0.538
11–5. a. \$1,600 unfavorable
   b. \$1,585 unfavorable
11–7. \$63,600; \$1,600 favorable;
   \$0.33/hr. unfavorable

11–9. a. base 1,085
   filament 1,096
   bulb 1,097
   b. \$0.08714/unit
11–11. Yes
11–13. Accept
11–15. 32.18 hrs.
11–17. a. $Y_i = 300i^{-.2638}$
   b. 52.3
   c. 71.1
11–19. 26.6 hrs.
   35.7 hrs.

12–1. a. 11.7%
   b. 18.9%
12–3. 13.2%
12–5. a. all
   b. C,E,D
   c. A,B,C,E; \$50,000
12–7. a. IA, IIH
   b. IB, IIH, IIIM
12–9. a. \$91,876.14
   b. \$60,000; \$160,000; \$54,000;
   \$154,000; \$48,000;
   \$148,000 ...

13–1. a. 0.4
   c. 2.0
13–3. a. 0.276
   b. 0.190
   c. 0.034
   d. 0.431
   e. 0.638
   f. 0.353
   g. 0.357
   h. no
   i. 0.076; 0.058
   j. 0.081
   k. 0.133
13–5. a. 0.03; 0.07
   b. 0.04; 0.02
   c. 0.86
   d. 0.05
   e. 0.10
   f. 0.09
   g. 0.022
   h. 0.143; 0; 0.857

13–7. a. 0.221
    b. 7.27 yrs.

13–9. Deferred; $E(PW) = -4,516$

13–11. Choose C. $E(AW_C) = -\$562,532$

13–13. a. 1,600
    b. 1,600
    c. \$9,800

13–15. $EV = 44.13$

13–17. Full development,
    $EV = \$705,000$

13–19. No

14–1. $B/C_{B-A} = 1.52$
    $B/C_{C-B} = 0.76$
    Choose B

14–3. $B/C_{B-A} = 0.53$
    Choose A

14–5. $B/C_{B-A} = 2.67$
    Choose B

14–7. $B/C \gg 1$; No opposition

14–9. Toll $= \$0.492$ for cars if $B = C$
    Practical toll $= \$0.50$

14–11. 77 Car trips

15–3. 1–2–5–8

15–5. A–C–H–M–N

15–7. a. 11.83
    b. 0.998
    c. 0.0475

Index